AH

WITHDRAWN

Medical Physics

Volume I
Physiological Physics, External Probes

612.014
D18m
v.1

Medical Physics

Volume I
Physiological Physics, External Probes

A. C. Damask
Department of Physics
Queens College of the City University of New York
New York, New York

ACADEMIC PRESS New York San Francisco London 1978
A Subsidiary of Harcourt Brace Jovanovich, Publishers

COPYRIGHT © 1978, BY ACADEMIC PRESS, INC.
ALL RIGHTS RESERVED.
NO PART OF THIS PUBLICATION MAY BE REPRODUCED OR TRANSMITTED IN ANY FORM OR BY ANY MEANS, ELECTRONIC OR MECHANICAL, INCLUDING PHOTOCOPY, RECORDING, OR ANY INFORMATION STORAGE AND RETRIEVAL SYSTEM, WITHOUT PERMISSION IN WRITING FROM THE PUBLISHER.

ACADEMIC PRESS, INC.
111 Fifth Avenue, New York, New York 10003

United Kingdom Edition published by
ACADEMIC PRESS, INC. (LONDON) LTD.
24/28 Oval Road, London NW1 7DX

Library of Congress Cataloging in Publication Data

Damask, A C
Medical physics

Includes bibliographies
1. Medical physics. I. Title.
R895.D35 612'.014 78-205
ISBN 0-12-201201-1 (v. 1)

PRINTED IN THE UNITED STATES OF AMERICA

TO MY STUDENTS,
WHO ASKED HOW PHYSICS IS USED IN MEDICAL SCIENCE

CAT May 12 80

4-07-80 MLS 24,37 Pelletier Lib

79-6663

Contents

CHAPTER 11 Cryobiology: Cell Freezing and Cryosurgery

APPENDIX A Chemical Thermodynamics

APPENDIX B The Wave Equation

APPENDIX C Binomial and Poisson Probability Distributions

APPENDIX D Differential Equations

Preface

The interaction of physical with medical science is rapidly increasing. However, there is a retarding factor to the increase, which can be called a language barrier. Students of the physical sciences usually do not know the vocabulary of physiology, while the physiology students are sometimes insufficiently familiar with the mathematical language of the physicist. This book is an attempt to lower the barrier, and is addressed to undergraduates and postgraduates of all scientific disciplines.

The first seven chapters cover physiological functions, while the last four discuss the fundamentals and physiological effects of external probes. After a general introductory background of both the physics and physiology in each chapter, example topics are discussed that illustrate an application in current use and one or more areas of active research in which physics and medicine interact. A notable omission is the physics of perception. The research literature on senses, synapses, and brain is so extensive that perception requires its own volume.

The book was developed from a course given by the author that although originally designed for undergraduate physics students soon came to be dominated by premedical students. Since students in these latter disciplines usually take only one year each of physics and calculus, the level of the book was limited. Very few of the students had physical chemistry so an appendix is included, which covers the limited chemical thermodynamics required. Finally, since the students' knowledge of chemistry was erratic, the course had no chemistry prerequisite and the book reflects this. Although an occasional organic chemical is named for reference or familiarization purposes, the reader should not take alarm. Nor should a reader with but a single year of calculus be overly concerned with an occasional differential equation. The ones used are elementary, and their method of solution is either indicated or the solution is readily verifiable by substitution. If this book is used in a classroom, the instructor may expect to spend a little extra time when a differential equation is used. More extensive derivations than required for understanding of the simple processes are used in Chapters 7 and 8, but these are for completeness and may be skipped without loss of continuity.

There are no problems because the aim of the course and the book is to lower a barrier, not to set tests of ability in a field unfamiliar to all students. The teaching of the course is done in a participatory manner. After each topic the students are asked to give brief reports from original papers, usually further details derived from the references. This method has been very successful, and the students quickly realize that the language barrier has indeed been lowered.

The reader interested in further topics is referred to the encyclopedic "Medical Physics" Vols. I–III, edited by O. Glasser (Year Book Publishers, Chicago, 1944, 1950, 1960). For a greater breadth of the physiological processes "Textbook of Medical Physiology," 4th ed., by Arthur Guyton (Saunders, Philadelphia, 1971), is written in a style very satisfying to a physical scientist.

Acknowledgments

A book of this scope encompasses knowledge beyond that of a single individual. Therefore the various topics have been read and corrected by researchers in these areas. The writer expresses his appreciation to the following individuals: Dr. Herbert Susskind, Brookhaven National Laboratory, Chapters 1 and 2; Dr. Charles Swenberg, National Institute of Mental Health, Chapters 3 and 4; Dr. Richard Wasley, Lawrence Livermore Laboratory, and Professor Larry Hench, University of Florida at Gainesville, Chapter 5; Dr. Robert Hamby, Long Island Jewish Hospital–Hillside Medical Center, and the State University of New York at Stony Brook, Chapters 6 and 7; Dr. David Welch, Brookhaven National Laboratory, Chapters 5–8; Dr. Eugene Church, Picatinny Arsenal, and Dr. Clare Shellabarger, Brookhaven National Laboratory, Chapter 9; Dr. Bryan Westerman, EMI Medical Inc., Chapter 10; Dr. Peter Mazur, Oak Ridge National Laboratory, Chapter 11; and Professor Richard Borg, Lawrence Radiation Laboratory, Appendix A. Dr. Welch also suggested the derivation for radiation from a disk to avoid the use of Bessel functions (in Chapter 8).

In addition, appreciation is expressed to my good friend and colleague, Dr. G. J. Dienes of Brookhaven National Laboratory, for critically reading the entire manuscript, and to my wife, Mary, for editing it.

CHAPTER 1

Newtonian Fluid Flow: Respiration and Micturition

INTRODUCTION

A physiological system is extremely complex and permits very few exact calculations. The vessels and tubes are all elastic so that their volume is pressure dependent, but they are not made of a single Hookean elastic material. Usually they are a complex of fibers with differing elastic constants, and before we can consider real systems we must review the principles that apply to a single idealized system. Fortunately, we can make a reasonable approximation by a simple system which, while it will not yield the details of the complex system, will at least yield understanding of the nature of the complex system's processes.

In addition to the complexities of the fluid-containing organs and vessels, the fluids themselves are complex. Except for air and water, the viscosity of the fluids is not a constant with respect to shearing force. These latter fluids, called non-Newtonian, will be discussed in the next chapter, after the principles for simple, or Newtonian, fluids are established.

PRESSURE

The definition of density ρ of a system is its mass per unit volume, $\rho = m/V$, and should be distinguished from weight per unit volume, which involves the force of gravity. Thus the weight of a cubic foot of water on the earth's surface is taken as 62.4 lb but its density is $62.4/32.2 = 1.94$ slugs/ft^3. In the metric system the mass of a liter of water is 1000 g, or 1 kg, and it occupies on the average 1000 cm^3. Thus its density is $1000/1000 = 1$ g/cm^3, while its weight is $(1000 \text{ g})(980 \text{ cm/sec}^2) = 9.8 \times 10^5$ dyn. These distinctions are necessary to consider properly the work done by a pump in a weightless environment compared to one in which there is a gravitational attraction.

Pressure P is defined as force per unit area, or $P = F/A$. If we have a column of liquid of cross section A then the force of the liquid on the bottom of the column is its weight. The mass of a height h of the liquid is its volume Ah times its mass per unit volume, which was defined as the density ρ, or $m = \rho Ah$. The weight W is this mass times the acceleration of gravity g, and therefore $W = \rho Ahg$. This weight is the force downward of the column on the area A at the bottom, and thus the pressure is

$$P = \rho Ahg/A = \rho gh. \tag{1.1}$$

Note that if this column is open to the atmosphere then on its top is the pressure arising from the column of air above it, decreasing in density with increasing altitude, which extends to the outer vacuum of space. This is the air pressure for a given locality, P_0, and at sea level it is usually taken as 14.7 lb/in.^{2}, or 1.01×10^6 dyn/cm^2, and therefore the pressure on the bottom of the water column is $P = P_0 + \rho gh$.

Air pressure is usually given in terms of the height of a column of mercury that it will support, the mercury being in an evacuated tube. In this situation the force downward of the mercury must equal the force upward of the air pressure. $P_{\text{Hg}} = \rho gh = (13.6 \text{ g/cm}^3)(980 \text{ cm/sec}^2)h = 1.01 \times 10^6 \text{ dyn/cm}^2$, and therefore $h = 76$ cm. In physiological systems pressures are often measured in centimeters of H_2O. Such a measure is 1/13.6 that of a centimeter of Hg.

It is seen by the above discussion that there are two ways of recording pressure: one is the total or *absolute* pressure $P = P_0 + \rho gh$; and the other is the *gauge* pressure $P - P_0 = \rho gh$, which is the pressure above (or below) atmospheric pressure. An example of gauge pressure is the sphygmomanometer, which balances arterial pressure against that of a column of mercury open to the atmosphere. This can be schematically represented in Fig. 1.1.

In Fig. 1.1, pressure on the left-hand side of the base is $P + \rho gh_1$ and on the right-hand side is $P_0 + \rho gh_2$, where P is the pressure of the arterial system and P_0 is the pressure of the atmosphere. The pressure at the base is

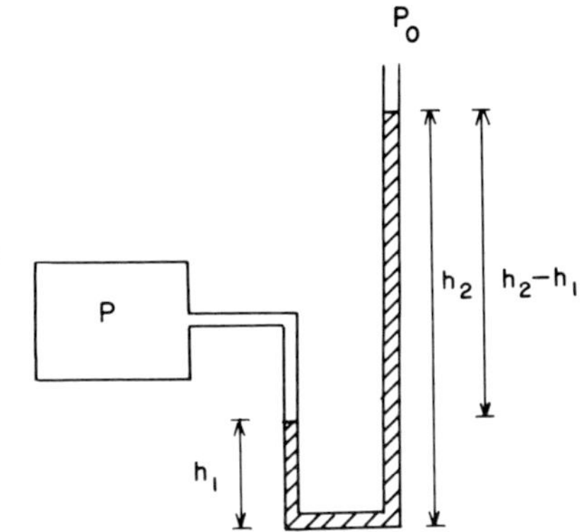

FIG. 1.1 Schematic of pressure relations in a sphygmomanometer.

the same on both sides, and since pressure is equal at all points at the same depth,

$$P + \rho g h_1 = P_0 + \rho g h_2 \qquad \text{and} \qquad P - P_0 = \rho g (h_2 - h_1) \qquad (1.2)$$

The height h_1 is usually calibrated as zero on the scale, and thus the measurement of blood pressure is gauge pressure.

CONTINUITY

When an incompressible fluid flows through either a real tube or a mathematical one we must consider how the assumption of incompressibility may be expressed quantitatively when the tube changes in diameter. In Fig. 1.2 consider two cross sections of areas A_1 and A_2 with fluid flow velocities of v_1 and v_2, respectively. Since distance is the product of velocity and time, in an interval dt we may represent the length of a column of liquid that has flowed past A_1 as $v_1\,dt$. The mass of this liquid is the length times the area of the base times the density, or $\rho A_1 v_1\,dt$. Similarly, the mass that has flowed past A_2 is $\rho A_2 v_2\,dt$. If the fluid is not compressible the density is constant and the mass flowing into the tube must equal the mass flowing out, or

$$\rho A_1 v_1\,dt = \rho A_2 v_2\,dt$$

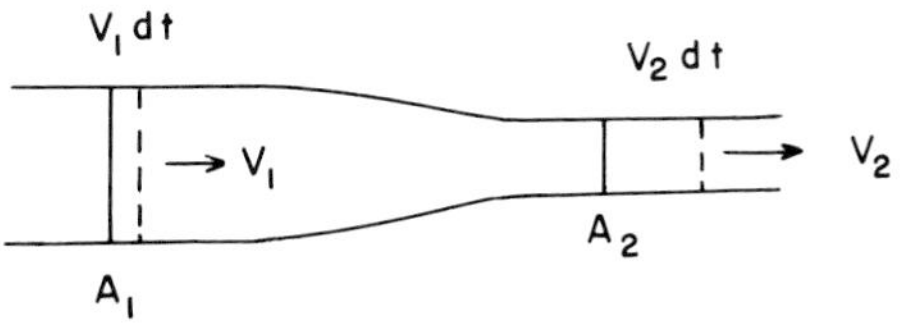

FIG. 1.2 Incompressible fluid flowing in a tube with changing cross-sectional area.

and in equal time

$$A_1 v_1 = A_2 v_2 \tag{1.3}$$

This means that the product Av is a constant along a tube, and that if the tube narrows the velocity of flow increases and if it widens the velocity decreases. For example, a river's rapids occur where it is narrowest while its stillest waters occur where it is broadest. Equation (1.3) is called the *equation of continuity* because, by assuming continuous flow with no density changes, it is simply a statement of the conservation of mass.

BERNOULLI'S EQUATION

When a fluid is in motion there is a change in pressure. Daniel Bernoulli used the conservation of energy as well as the above equation of continuity to derive a general theorem of the total pressure of a nonviscous flowing fluid. Note that we specify now that the fluid be nonviscous in order to derive the simple relationship expressing the change in pressure in terms of the changes in height and velocity. If the fluid were viscous or turbulent we would have to consider energy loss terms, which arise from frictional work of the liquid and are dissipated as heat.

Figure 1.3 shows a situation that is the sum of the previous two diagrams, namely, a change in height of a mass of the fluid and also a change in velocity. The h's represent average heights and the pipe can be of any shape or angle. The principle in the derivation of Bernoulli's equation is that the system is mechanically conservative in that there is no heat loss. Hence the work put into the system goes only into kinetic energy and potential energy, or

$$\Delta W = \Delta \mathrm{KE} + \Delta \mathrm{PE}$$

Note that force F_1 is in the direction of the fluid flow and does positive work, while force F_2 is in the direction opposite to the flow and hence does

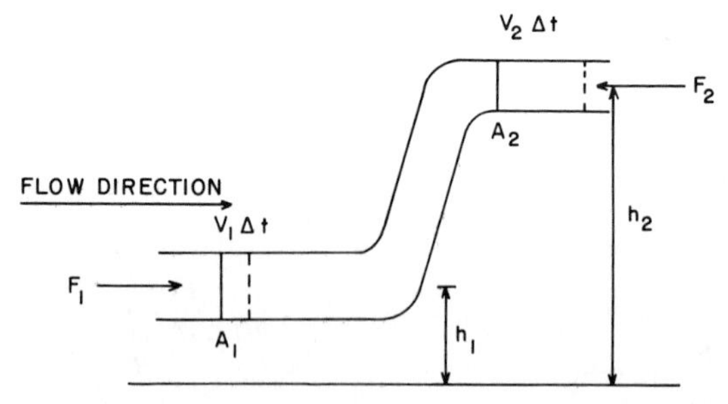

FIG. 1.3 Incompressible fluid flowing in a tube with changing cross-sectional area and changing height.

negative work. In time Δt the left-hand part of the fluid advances $v_1\,\Delta t$ and the right-hand part advances $v_2\,\Delta t$. Since work is $F \cdot d$, the net work done by the system is

$$\Delta W = F_1 v_1\,\Delta t - F_2 v_2\,\Delta t = (P_1 A_1)(v_1\,\Delta t) - (P_2 A_2)(v_2\,\Delta t)$$

where pressure is defined as $P = F/A$. We have seen in the preceding section that in either volume $A_1(v_1\,\Delta t)$ or $A_2(v_2\,\Delta t)$ the mass is the same or

$$\Delta m = \rho A_1 v_1\,\Delta t \tag{1.4}$$

If we also use the equation of continuity,

$$A_1 v_1 = A_2 v_2 \tag{1.3}$$

we may write the change in work as

$$\Delta W = A_1 v_1\,\Delta t(P_1 - P_2) = (\Delta m/\rho)(P_1 - P_2) \tag{1.5}$$

Note that we need not consider the fluid in the tube between A_1 and A_2 because the sum of the KE and PE remains the same at all times for this section. Returning to the conservation of mechanical energy,

$$\Delta W = \Delta\text{KE} + \Delta\text{PE}$$

by substituting into it Eq. (1.5) and the definitions $\Delta\text{KE} = \frac{1}{2}\Delta m v^2$ and $\Delta\text{PE} = \Delta m g h$ we obtain

$$(\Delta m/\rho)(P_1 - P_2) = (\tfrac{1}{2}\Delta m {v_2}^2 - \tfrac{1}{2}\Delta m {v_1}^2) + (\Delta m g h_2 - \Delta m g h_1)$$

Upon cancelling Δm throughout and rearranging, we obtain

$$P_1 + \rho g h_1 + \tfrac{1}{2}\rho {v_1}^2 = P_2 + \rho g h_2 + \tfrac{1}{2}\rho {v_2}^2 \tag{1.6}$$

This is Bernoulli's equation, which is nothing more than a statement of the conservation of energy.

If we continue to assume no friction with the walls of a tube we would expect a pressure decrease where the velocity of fluid flow increases. In Fig. 1.4 we see a design called a Venturi meter, in which the relative heights

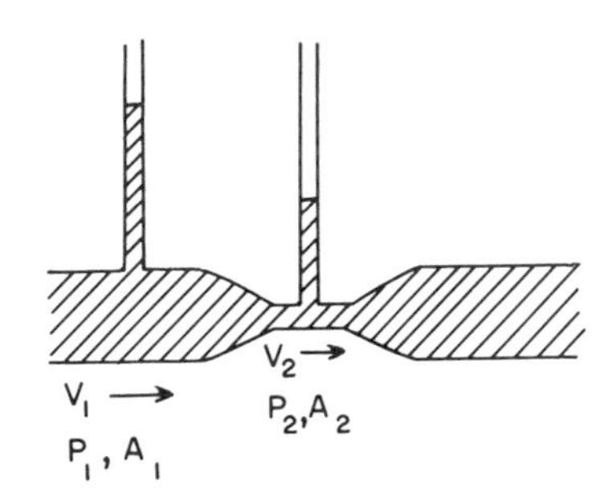

FIG. 1.4 Schematic of the Venturi effect.

of the liquid in the vertical tubes are indicative of the relative velocities of fluid flow through the two areas. In the narrow section the velocity is higher, and therefore the last (KE) term on the right side of Eq. (1.6) is larger than the equivalent term on the left. Therefore, if the height change is negligible, P_2 is less than P_1.

VISCOSITY

Because of viscosity a force must be exerted to cause one layer of fluid to flow past another. The more force required, the greater is the viscosity. For example, one must exert a greater force to stir a pot of molasses than a pot of water.

If we apply a shear stress to a solid it deforms in a manner that is proportional to the magnitude of the stress. If we apply a shear stress to a simple fluid it flows. Let us assume that the fluid is incompressible. This assumption is not a drastic one because in the usual experimental range of low velocity of flow most liquids have this characteristic.

Suppose we have two plates with a liquid between and we exert a shearing force on the upper plate to the right. Because of the viscosity of the liquid we must exert an equal and opposite force on the bottom plate to keep it in position, as in Fig. 1.5. If originally there were a vertical stack of bead markers in the liquid, we would observe the beads become displaced to the right, with the displacements being greater the farther the beads are from the bottom. If each bead corresponds to a layer of liquid then the picture is one of successive layers sliding on one another, like cards in a deck. Such sliding of layers is called *laminar* flow.

Since the resistance to flow is essentially an internal friction that the liquid layers exert on each other, the amount of flow obtained for a given force F exerted will depend upon the area A of the liquid layers displaced. Thus the quantity is the *shearing stress*, defined as force per unit area, or F/A. Note that the F and A vectors in this case are orthogonal whereas in Fig. 1.3 they are parallel. If we had a motion picture of the beads we could measure the distance x that each one travelled in a given time and thus determine the velocity in the x direction, $v = dx/dt$, of each layer. If we had extremely small beads of

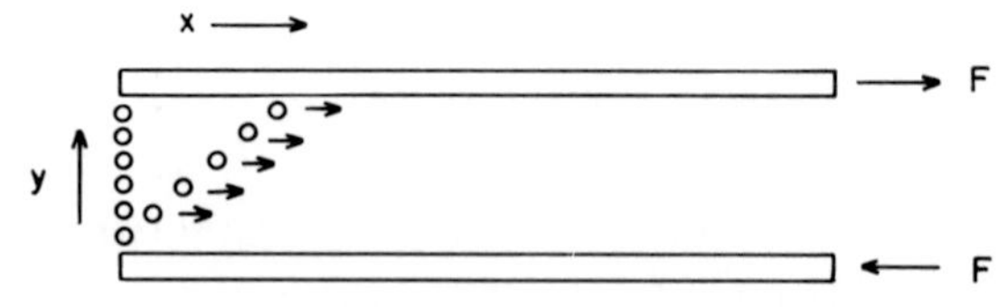

FIG. 1.5 Behavior of markers in fluid at times $t = 0$ and $t = \Delta t$ when the fluid is subjected to shearing force F.

TABLE 1.1

Fluid, at 20°C	η (P)
Benzene	0.0065
Water	0.0100
Blood	0.02–0.04
Glycerol	8.3
Caster oil	9.9
Air	180×10^{-6}

diameter dy we could make a plot of the *velocity gradient*, dv/dy. Note that Fig. 1.5 suggests a linear relationship between the velocity and y distance; thus dv/dy would be a constant. This, however, is generally not the true situation.

If we could take a series of measurements from motion pictures of the movement of the marker beads in a given liquid we would see that the velocity gradient is proportional to the shearing stress, or

$$dv/dy \propto F/A$$

This proportionality is maintained in all Newtonian liquids. Their differences in behavior is contained in the proportionality constant η, called the *coefficient of viscosity*, and the introduction of such a proportionality constant allows us to express the relationship as an equality:

$$F/A = \eta(dv/dy) \tag{1.7}$$

The dimensionality of η is readily determined from the following:

$$\frac{F\ (\text{dyn})}{A\ (\text{cm}^2)} = \eta\left(\frac{\text{dyn/sec}}{\text{cm}^2}\right)\frac{dv}{dy}\frac{(\text{cm sec}^{-1})}{(\text{cm})}$$

In the cgs system, 1 dyn sec cm^{-2} is called one poise (P). Some typical values are given in Table 1.1 for several fluids at 20°C.

The viscosity of liquids decreases with increasing temperature. Table 1.2 illustrates this for the case of water.

TABLE 1.2

η (P)	Temperature (°C)
0.0179	0
0.0131	10
0.0100	20
0.0080	30
0.0065	40

The explanation for this lies in the realm of the study of thermally created microvoids of atomic dimensions that occur in liquids. Motion within a liquid occurs when one molecule pushes another aside. If there is space for the pushed molecule to occupy, the motion will be easier and result in a lower viscosity.

POISEUILLE'S LAW

We now recognize that all liquids have a viscosity and that in the laminar flow region this manifests itself as the sliding of planes of liquid across one another, where the planes near the interface of the solid surface of the container are the slowest. Now roll the lower solid surface of Fig. 1.5 into a tube without the upper plate and let there be a pressure of the fluid that is greater on one end than on the other. If the tube is held externally then this fluid pressure constitutes the shearing force. There is a backward pressure of the fluid ahead of the region under consideration. The difference between the forward and backward pressures would be an accelerating force if it were not opposed by the frictional force of the fluid. Consider the fluid flow in the tube of Fig. 1.6. If the fluid flows from left to right there is a positive pressure P_1 toward the right and a back pressure of P_2 toward the left. If we consider a cylinder of radius r filled with fluid, the unbalanced force causing the fluid to move to the right is $F = P \times A = (P_1 - P_2) \times \pi r^2$. Since this cylinder moves with constant velocity, i.e., it does not accelerate, this unbalanced force must be opposed by an equal and opposite frictional force due to the viscosity of the fluid. The frictional force from Eq. (1.7) is $F_f = \eta A(dv/dy)$, which for a cylinder of length L is

$$F_f = \eta 2\pi r L(dv/dr) \tag{1.8}$$

where y, the distance from the surface, is changed to r in this geometry. (r is zero at the center). Since v decreases as r increases, the gradient dv/dr is negative. Furthermore, F_f due to friction is in the direction opposite to F due to the driving force of the pressure. Since the forces balance we may write

$$\sum F = 0 = F - F_f = (P_1 - P_2) \times \pi r^2 + \eta 2\pi r L(dv/dr)$$

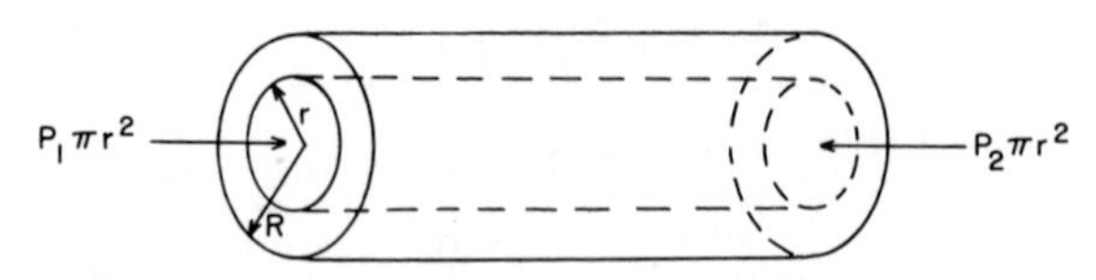

FIG. 1.6 Forces exerted by a fluid flowing in a tube of radius R on a cylindrical segment with radius r.

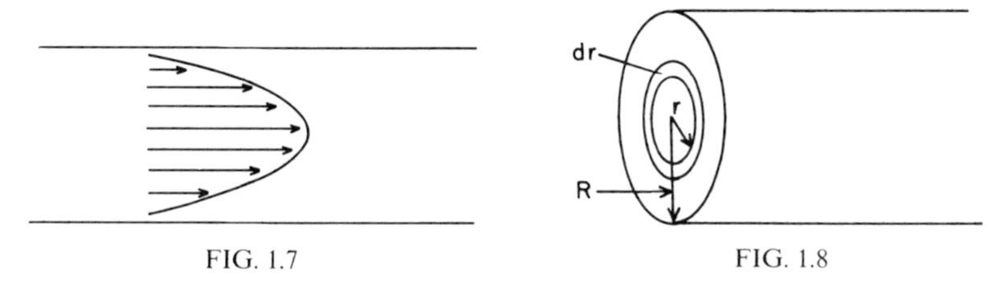

FIG. 1.7 FIG. 1.8

FIG. 1.7 Parabolic profile of velocities of fluid flow in a rigid tube.

FIG. 1.8 Schematic of fluid in a tube, of radius R, within which is an annulus of radius r and width dr.

or

$$-dv = [(P_1 - P_2)/(2\eta L)]r\,dr \tag{1.9}$$

The velocity varies between the limits 0 at $r = R$ (the inner wall of the tube) and v at any r between 0 and R. To find the velocity as a function of distance from the wall of the tube we integrate the above equation between these limits, giving

$$-\int_v^0 dv = [(P_1 - P_2)/(2\eta L)]\int_r^R r\,dr$$

From this we obtain

$$v = [(P_1 - P_2)/(4\eta L)](R^2 - r^2) \tag{1.10}$$

This velocity profile is the equation of a parabola and is sketched in Fig. 1.7. The actual shape of the parabola depends on the parameters of the equation for a particular system.

A useful quantity is the rate of fluid flow through the tube. We may find this by calculating the amount of fluid crossing a given element in the tube per unit time. For this element we may take the annulus shown in Fig. 1.8. If we designate the flow rate by Q, the volume per unit time, then this is equal to the velocity of flow of an incremental annulus times the area of the annulus or

$$dQ = v\,dA = v2\pi r\,dr$$

where v is given by Eq. (1.10). Substituting v and integrating over the entire tube, i.e., from $r = 0$ to $r = R$, we obtain

$$Q = [(P_1 - P_2)/(4\eta L)]2\pi\int_0^R (R^2 - r^2)r\,dr$$

or

$$Q = \pi R^4/(8\eta L)(P_1 - P_2) \tag{1.11}$$

This is known as *Poiseuille's law* because it was first derived by him, and it is in his honor that the unit of viscosity is called the *poise*. For a constant $(P_1 - P_2)$ the flow rate decreases with increasing viscosity and tube length. These results are expected from experience. What could not have been predicted is the R^4 behavior. This means, for example, that if one tube is one-half the diameter of another then, all other things being equal, the flow rate in the former is one-sixteenth as much as the latter's.

It should be noted that Poiseuille's law has the same form as Ohm's law, where Q would be the current, $(P_1 - P_2)$ the voltage difference, and $\pi R^4/(8\eta L)$ the conductivity. Thus the laws of a direct current (dc) network can be directly applied to the fluid flow in that resistances (resistance is the reciprocal of conductivity) in series and parallel add as follows:

$$R_{\text{tot}} = R_1 + R_2 + \cdots + R_n \qquad \text{(series)}$$

$$\frac{1}{R_{\text{tot}}} = \frac{1}{R_1} + \frac{1}{R_2} + \cdots + \frac{1}{R_n} \qquad \text{(parallel)}$$

and also that Kirchhoff's laws are obeyed.

There is a significant difference, however; in a wire the electrical conductivity increases proportionally to the square of the radius, but in a tube the flow rate increases proportionally to the fourth power of the radius.

REYNOLDS' NUMBER

If the beads in Fig. 1.5 were numbered they would maintain their numbering order in laminar flow, although at differing velocities. If the flow rate increases, a value will be reached at which the bead numbers begin to interchange due to local vortices that occur in the fluid. This interchange is called *turbulent flow* and clearly a lot of the work in forcing flow in such a condition goes into frictional work within the fluid.

Referring to Poiseuille's law, Eq. (1.11), we see that for a given fluid in a tube of fixed length and radius the flow rate is a linear function of the pressure. If a fixed fraction of the work goes into turbulent friction a linear relation between flow and pressure still obtains but with a different slope. This is illustrated schematically in Fig. 1.9, where the slopes are equivalent to conductance in the electrical circuit analogy and curves I and II are for two tubes with different radii.

An engineer named Reynolds found that the point at which the slope changed from laminar to turbulent flow has a constant relationship in the ratio of inertial forces to viscous forces, now called Reynolds' number Re, given as follows:

$$Re = \frac{\text{inertial forces}}{\text{viscous forces}} = \frac{\rho \bar{v}^2/r}{\eta \bar{v}/r^2} \tag{1.12}$$

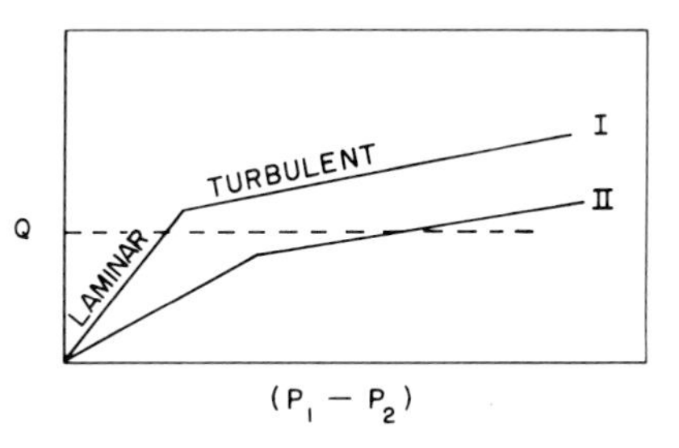

FIG. 1.9 Laminar and turbulent flow curves for tubes of two different radii $r_I > r_{II}$. Q represents a biologically required flow rate. [Adapted from Richardson and Neergaard (1972).]

where ρ is the density, $\bar{v}$ is the average velocity (volume rate of flow/cross-sectional area), r is the radius of the tube, and η is the viscosity. Turbulent flow occurs when $Re > 1000$–1200. Note that Re is sometimes defined in terms of the diameter of a tube and the critical value for turbulence will be $Re > 2000$. Note also that the Reynolds number for the boundary of the regimes varies greatly with other parameters, such as wall surface conditions, etc.

The significance of flow rate and Reynolds' number is seen if we consider a physiological system in which the metabolic requirement is for a constant flow rate. Assume that originally the system is an efficient one in which the radius of the blood vessel permits laminar flow for a given pressure difference. If the blood vessel becomes constricted, i.e., r decreases, then the velocity $\bar{v}$ must increase as the fourth power of the radius to maintain the flow rate. Curve II in Fig. 1.9 indicates the flow rate for the smaller radius according to Poiseuille's law. The dashed line indicates the constant flow rate requirement. It is seen for a constricted tube that the fluid would become turbulent and that the pressure would have to be increased considerably to maintain the flow rate.

LAW OF LAPLACE

In physiological systems the tubes through which fluids flow are not rigid but can deform and recover their shape, as is seen in the bending of a joint or in pressing against a blood vessel. A complete analysis is not simple since neither the radius of the blood vessel nor the blood pressure is constant. (The latter varies with distance and with time because the heart beats instead of exerting a constant pressure.) However, the fundamental principles must be established before refinements can be considered.

Each blood vessel has an internal pressure P_i and an external tissue pressure P_e. The difference between the two is called P_{TM}, the transmural pressure. In Fig. 1.10 there is represented a uniform section of a tube of length l, wall thickness t, transmural pressure P_{TM}, and radius r. Consider an arbitrary plane through the cylinder. The force trying to separate the upper half from the lower half is the pressure P_{TM} times an area A. However,

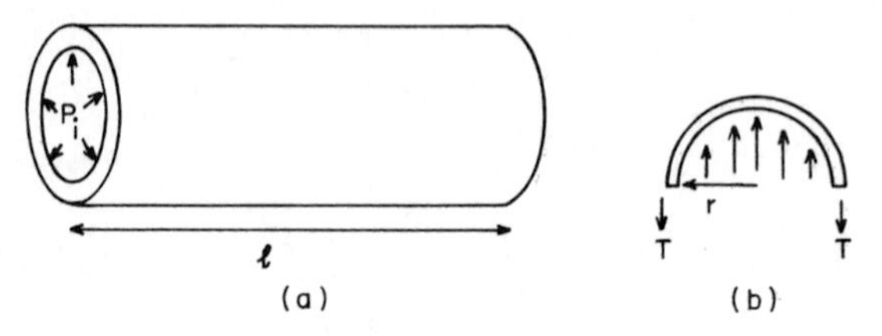

FIG. 1.10 Diagram of (a) transmural pressure and (b) wall tension in a tube.

it is obvious from the sketch in Fig. 1.10b, or easily proven by integration, that the area against which the pressure acts is the diameter of the tube times its length. Thus

$$P_{\mathrm{TM}} A = P_{\mathrm{TM}} 2rl.$$

The force holding the tube together is twice the tension per unit length T times the length, or

$$P_{\mathrm{TM}} 2rl = 2Tl$$

and

$$P_{\mathrm{TM}} = T/r \tag{1.13}$$

This relation is known to physiologists and engineers as the law of Laplace. It shows that, for a pipe of a given wall thickness, the smaller the radius the larger the pressure that can be contained.

Note that if this same exercise is done for a sphere instead of a cylinder the result is

$$P_{\mathrm{TM}} = 2T/r \tag{1.14}$$

a result well known in the study of liquid drops and bubbles. The review by Burton (1962) shows that blood vessel and heart chamber wall thicknesses obey the law of Laplace.

Laplace contributed so much to mathematical physics that it seems a shame to honor his name with such a trivial law. However, since it is in use, we might refer to it as Laplace's zeroth law.

The use of the symbol T above should be clarified. The engineering specifications of a design are given in terms of *tensile stress*. This is the force per unit cross-sectional area. If we denote tensile stress by σ_τ then the connecting relation with T above is

$$\sigma_\tau \times \text{width} \times \text{length} = T \times \text{length},$$

$$\sigma_\tau wl = Tl \qquad \text{and} \qquad \sigma_\tau w = T$$

and Eq. (1.13) in terms of tensile stress is

$$P_{\mathrm{TM}} = \sigma_\tau w/r \tag{1.15}$$

THE RESPIRATORY SYSTEM

The air conduit system in man and other air-breathing mammals exists primarily to supply oxygen to the blood and to allow the blood to dispose of carbon dioxide. The gas exchange takes place between very thin layers of tissue in the lung. There are many small air sacs in the lung called *alveoli*, about 300 million in man, with a surface area of 80 to 90 m^2. In the surface of these alveolar sacs is a network of blood capillaries, about 7 μm in diameter. Since red blood cells are disks about 8 μm in diameter and 2 μm thick, they must squeeze through and thus press directly against the membrane wall for ease of gas exchange. In this way the oxygen does not have to diffuse through the blood plasma to get to the blood cell.

Many of the aveoli are connected to each other in groups called *alveolar septae*, illustrated in Fig. 1.11, and these in turn are connected to tubes called *alveolar ducts*. These ducts are connected to larger tubes called *bronchioles*, which have further alveoli attached to them. The bronchioles are connected to still larger tubes called the *bronchi*, which are ultimately connected to the *trachea*, or windpipe. All of the components, large and small, are elastic and can expand both radially and longitudinally during inspiration, although they do not expand by the same percentage.

The lungs are in the *thoracic cavity*, i.e., in the rib cage and above the diaphragm, and are attached physically only at the trachea near the bronchi. The lungs are not physically attached at any other point and are free to move. The space between the outer part of the lungs and the inner part of the thoracic cavity is called *intrapleural space*. There is a fluid in this space that lubricates a downward sliding of the lungs due to gravity. There is a continual tendency for the pleural capillaries to absorb this intrapleural fluid, which gives the fluid a negative pressure with respect to the atmosphere of -10 to -12 mm-Hg. This holds the lungs tightly to the chest wall while permitting sliding motion, and when the chest wall is expanded by muscular action during inspiration the lungs expand elastically. The lungs have a continual tendency to collapse, i.e., pull away from the chest wall, caused not only by their elasticity but by the surface tension of a fluid that lines the alveoli. This latter effect is controlled by another fluid in the alveoli called the *surfactant* that, much like the action of a detergent, reduces the surface tension of the fluid lining the alveoli. The resulting collapse tendency of the lungs can be measured by the amount of negative pressure in the intrapleural space required to keep the lungs from collapsing. This is measured when the alveolar spaces are open to the atmosphere through the trachea during inspiration. This pressure is normally about -4 mm-Hg with respect to atmospheric pressure. When the lungs are expanded at the end of a deep inspiration the intrapleural pressure may be -9 to -12 mm-Hg. Thus the

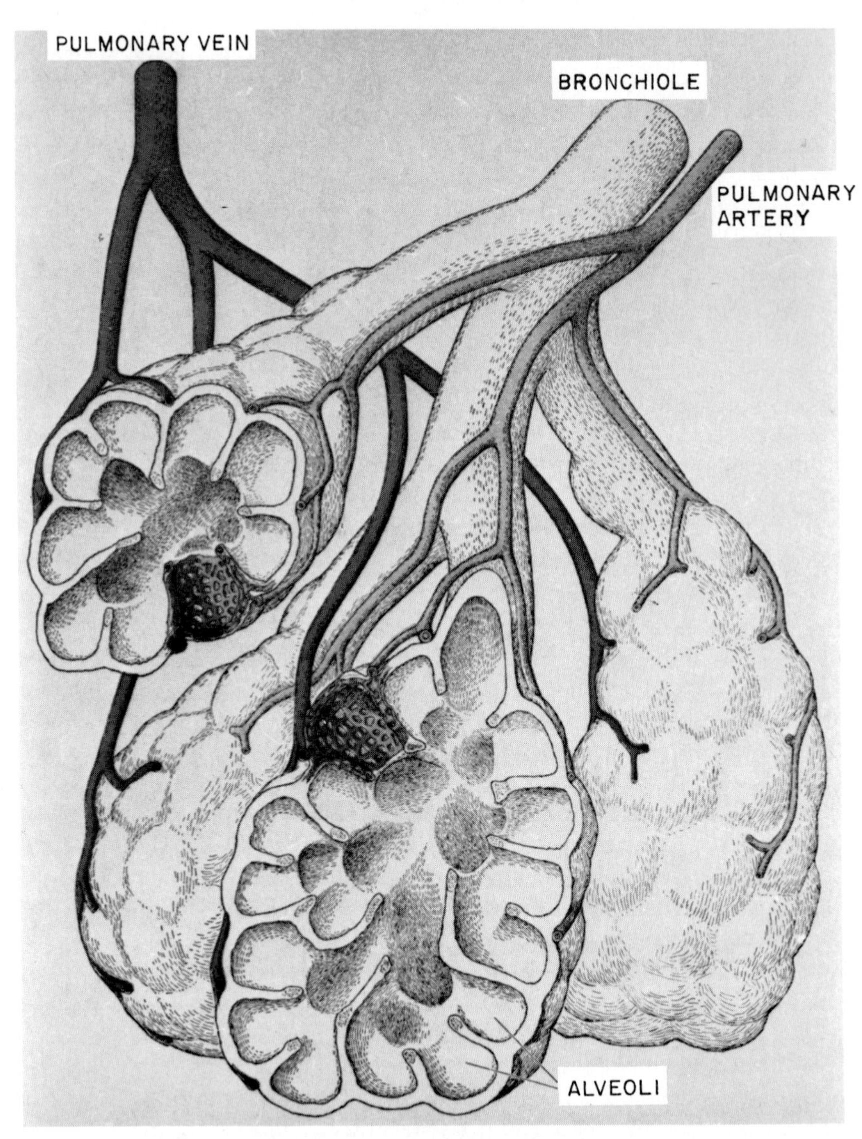

FIG. 1.11 Cross-sectional view of alveolar septae. All of the smaller blood vessels have been omitted except for the capillary networks embedded in two of the alveolar walls. [From J. A. Clements, Surface tension in the lungs, December, p. 120. Copyright © 1962 by Scientific American, Inc. All rights reserved.]

intrapleural fluid pressure of −10 to −12 mm-Hg is a safety factor to keep the lungs from collapsing.

The lungs are expanded by muscular motion, which both lowers the diaphragm and expands the rib cage. When the diaphragm is lowered gravity stretches the lungs vertically. Since work is done against elastic forces in inspiration, expiration is accomplished by the elastic restoring force, which in normal breathing is entirely passive.

WORK OF RESPIRATION

The work of breathing is work done to overcome several types of resisting forces. The first and most obvious is the work against the elastic forces, which is recoverable as heat upon expiration. There is also work done in the sliding of the lungs, as described, and abdominal viscera displaced by the motion of the diaphragm. There is also work done in moving ambient air from the outside to the alveoli. This air flow has two characteristics, viscous and turbulent. With the velocities involved the flow of air through a straight tube can be below the turbulent region. However, the arrangement to the nasal passages is designed deliberately to cause turbulence, both to accomplish more efficient warming and to cause turbulent separation of foreign particles in a way similar to centrifuging. The continuous branching of the air passages also causes turbulence. Another possible energy loss, that arising from accelerated movements of the organs, was shown very early by Rohrer (1925) to be negligible.

The mechanical system of breathing is similar to a bellows with a spring between the handles. Work is done to expand the bellows against the elastic force of the spring, the air resistance of the nozzle, and the viscous forces of the leather of the bellows. Upon removing the external force holding the bellows apart the spring tension will collapse them. Since, by definition, mechanical work = force times distance, force divided by area = pressure, and distance times area = volume, work can therefore be expressed as pressure times volume. Therefore, the work of breathing can be measured by obtaining the pressure–volume relation. This is usually done by measuring the pressure difference between the trachea and the alveoli as a function of volume of the lungs during breathing.

If the pressure difference and volume are obtained at the same times during inspiration and expiration, curves such as the one shown in Fig. 1.12a are observed.

If the pressure and volume at the same times, represented by the dashed lines, are plotted on a P–V diagram, a curve such as the one shown in Fig. 1.12b results. If purely elastic work were involved the dashed line in Fig. 1.12b would result and no hysteresis would be observed. Since in a P–V diagram

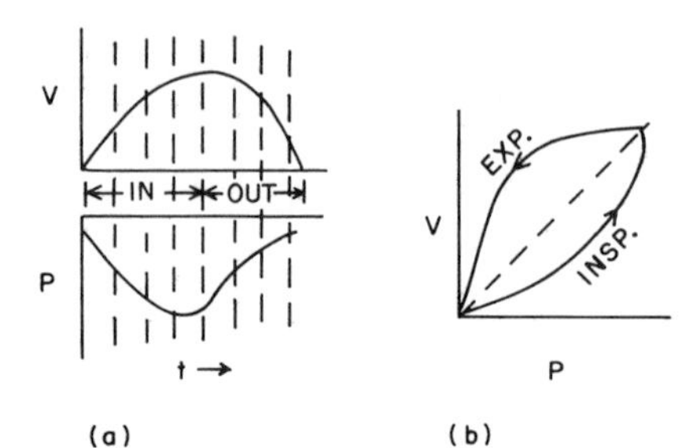

FIG. 1.12 Construction of a P–V diagram for respiration. [Adapted from Fritts and Richards (1960) *in* "Medical Physics" (O. Glasser, ed.), Vol. III. Copyright © 1960 by Year Book Medical Publishers, Inc., Chicago. Used by permission.]

the area under the curve represents the work done, the difference between the areas of inspiration and expiration, i.e., the area of the loop, is the work in breathing that results from nonelastic forces.

A number of studies have been made to determine the relative magnitudes of the contributions to the work of breathing. Only some representative experiments will be mentioned to demonstrate the techniques employed. For example, Bayliss and Robertson (1939) used a cyclic pump to inflate the lungs of anesthetized animals for the purpose of measuring pressure and volume. They attempted to find the magnitude of the air viscosity term by changing the composition of the air (by blending various mixtures of H_2 and O_2) and measuring the area of the resulting hysteresis loop. They measured the relative change of viscosity of the mixture by finding the flow rate through a small glass capillary. Although they observed a relationship, it was not sufficiently consistent to deduce a precise value of the air viscosity term.

The method of Otis *et al.* (1950) illustrates the modern technique of measurement in man. A trained subject is placed in a Drinker apparatus, known commonly as an iron lung, which then does the work of breathing for him by changing the pressure surrounding his rib cage. The subject breathes through a large tube, which offers minimal resistance to air flow, to which a flow meter is attached, i.e., a meter that measures the volume rate of flow dV/dt. In this breathing tube is a shutter that interrupts the flow of air very rapidly and then opens again. It is done so rapidly, in about 0.01 sec, that the breathing subject is unaware of the interruption. While the flow is interrupted a strain gauge manometer measures the pressure. When the air flow is stopped at any time the pressure at the mouth is equal to the trachea–alveolar pressure difference, and the time integration of the air flow dV/dt to the time of interruption is the volume in the lungs. An example of the data obtained in this experiment is shown in Fig. 1.13. These data are plotted by the method of Fig. 1.12b and yield the curve of Fig. 1.14. The slope between the maximum and minimum lung volumes yields the elastic constant K from the relation

$$P_{el} = KV \tag{1.16}$$

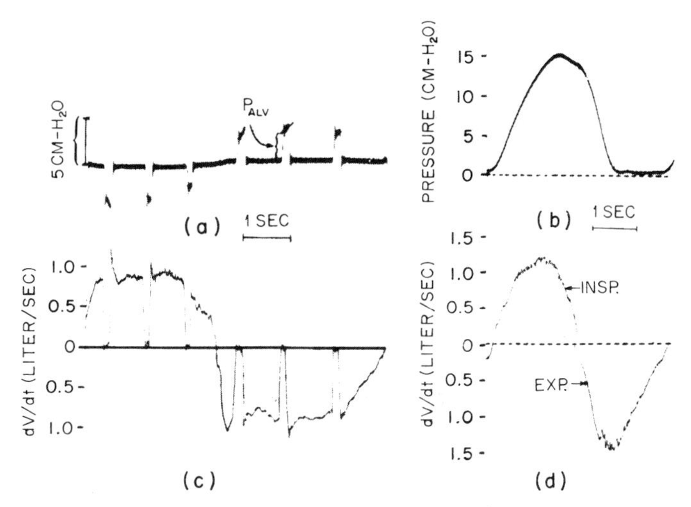

FIG. 1.13 Simultaneous records of (a) pressure at mouth and (b) flow rate versus time. (c) tracing of pressure gradient between mouth and inside of Drinker respirator. (d) Flow rate versus time. [From Otis *et al.* (1950).]

The flow of the air is resisted by both viscous and turbulent forces. From Poiseuille's law we expect the flow rate dV/dt to be proportional to the mouth–aveolar pressure difference. It is also assumed that turbulence loss, since it involves kinetic energy, is governed by a velocity-squared term and can be accounted for by setting such a term proportional to the alveolar pressure. Thus the resistance terms for air flow are

$$P_{\text{alv}} = K_1(dV/dt) + K_2(dV/dt)^2 \tag{1.17}$$

The data of several runs of the type shown in Fig. 1.13 are plotted in Fig. 1.15. The best-fit curve of Eq. (1.17) yields the constants $K_1 = 1.7$ and $K_2 = 1.9$.

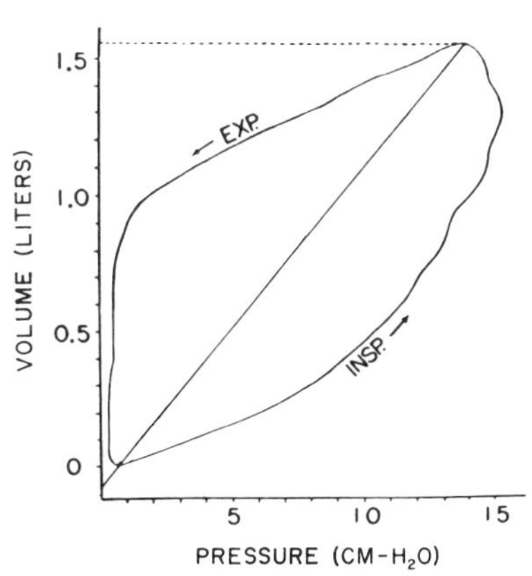

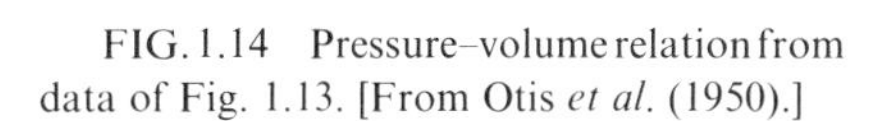
FIG. 1.14 Pressure–volume relation from data of Fig. 1.13. [From Otis *et al.* (1950).]

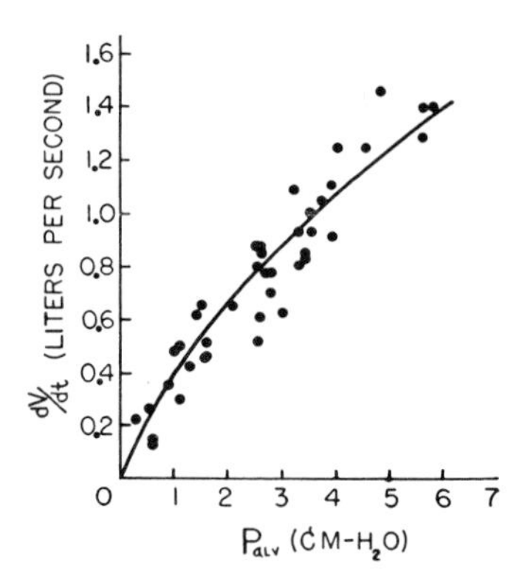

FIG. 1.15 Relationship between instantaneous rate of flow of respired gas and pressure gradient between alveoli and mouth, from data of Fig. 1.12. The curve drawn through the points represents Eq. (1.17) with $K_1 = 1.7$ and $K_2 = 1.19$. [From Otis *et al.* (1950).]

There are minor correction factors to be made because there is a phase lag in pressure between the mouth and the interior of the Drinker apparatus. These investigators therefore modified K_1 to K' and K_2 to K''. A quantitative description of this modification is given in their paper. Thus, the total pressure required for breathing is the sum of the elastic term of Eq. (1.16) and the alveolar pressure terms of Eq. (1.17):

$$P_{\text{tot}} = KV + K'(dV/dt) + K''(dV/dt)^2 \tag{1.18}$$

FREQUENCY OF RESPIRATION

The breathing pattern illustrated in Fig. 1.13 can be approximated by a sine wave as in Fig. 1.16, which obeys the relation

$$dV/dt = A \sin \omega t \tag{1.19}$$

where A is the amplitude and $\omega = 2\pi\nu$, where ν is the frequency. The volume of inspired air, called the tidal volume V_{T}, is the integral over half a cycle of Eq. (1.19):

$$V_{\text{T}} = \int_0^{\pi/\omega} A \sin \omega t \, dt = A(\pi\nu) \tag{1.20}$$

The total work done in inspiration is the integral of the differential expression

$$dW = P\,dV$$

which, upon substituting Eq. (1.18) and (1.19), is

$$dW = KV\,dV + K'A^2 \sin^2 \omega t\,dt + K''A^3 \sin^3 \omega t\,dt \tag{1.21}$$

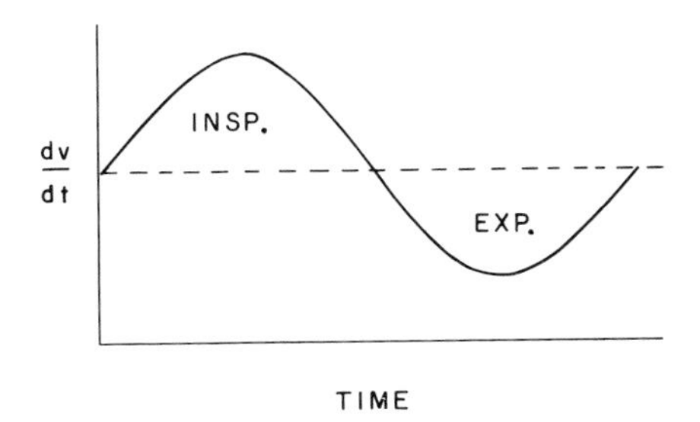

FIG. 1.16 Sinusoidal approximation of the data of Fig. 1.13.

Therefore, the total work of inspiration of volume V_T in time π/ω, i.e., one half-period, is

$$W = K \int_0^{V_T} V\,dV + \int_0^{\pi/\omega} (K'A^2 \sin^2 \omega t + K''A^3 \sin^3 \omega t)\,dt$$

$$= \tfrac{1}{2}KV_T{}^2 + \tfrac{1}{4}K'\pi^2 v V_T{}^2 + \tfrac{2}{3}K''\pi^2 v^2 V_T{}^3 \tag{1.22}$$

(Eq. (1.20) has been used to replace A).

The mean rate of doing work is the work per breath times the frequency of breathing:

$$\text{mean rate of doing work} = \tfrac{1}{2}KvV_T{}^2 + \tfrac{1}{4}K'\pi^2(vV_T)^2 + \tfrac{2}{3}K''\pi^2(vV_T)^3 \tag{1.23}$$

This mean rate is an expression for the total work of breathing per unit time, assuming that the expiration is passive.

Equation (1.23) is plotted in Fig. 1.17 with a suitable approximation for the lungs' dead-space volume, i.e., the volume of unused air such as in the trachea, and the constants evaluated in the experiment. It is interesting to note that the total curve exhibits a minimum—the elastic, viscous, and turbulent contributions are shown to illustrate the origin of the minimum. The optimal frequency of 12–15 breaths per minute for minimum work shown in this figure is the range ordinarily observed in the breathing of resting subjects and clearly confirms the principle of minimum effort by which so many body functions seem to be regulated. However, the principle of minimum effort prevails only when it coincides with the principle of maximum comfort. Thus a person with pleurisy might be willing to sacrifice a little energy for the sake of a more comfortable pattern of breathing.

This technique has been used to study changes during exercise and is readily extended to the study of pathological conditions and of effects of various breathing mixtures and pressure.

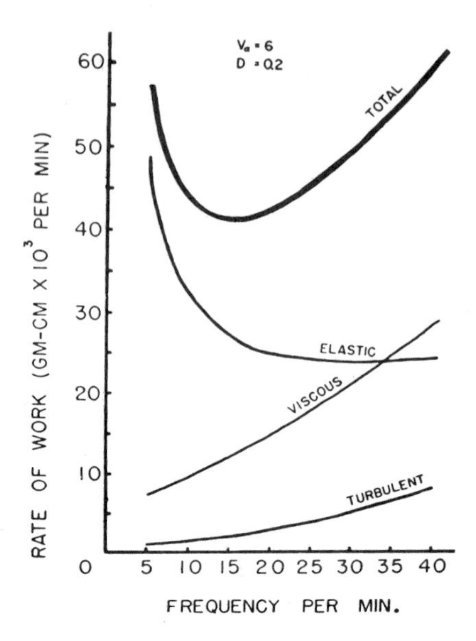

FIG. 1.17 Relationship of elastic, viscous, turbulent, and total work of breathing per minute to frequency of breathing when alveolar ventilation is 6 liter/min and dead space is 200 cm^3. [From Otis *et al.* (1950).]

MICTURITION

Micturition is the process in which the bladder empties itself. It is appropriate for discussion here because urine is essentially water and can be considered as a Newtonian fluid.

The bladder in a simplified sense is a bag with very fine openings connected to tubes called *ureters* through which the urine enters from the kidneys. The tubes are designed so that back pressure from a full bladder will not force urine back into the kidney. At the bottom of the bladder is an opening connected to the discharge tube, called the *urethra.* The bladder, although having some elastic properties, cannot be considered as a purely elastic balloon in which the pressure is proportional to the degree of distension. Rather, there is a long plateau of comfort, as seen in Fig. 1.18, which shows bladder pressure versus volume.

When pressure signals a desire to micturate a complex series of trained reflexes begins. In Figs. 1.19a and 1.19c there is shown a simplified representation of the muscle fibers around the bladder neck and urethra of a relaxed distended bladder. They are apparently arranged so that the more the bladder distends the tighter will be the urethra closure. Figures 1.19b and 1.19d show the same bladder during voiding. Note that the opening achieves a funnel-like

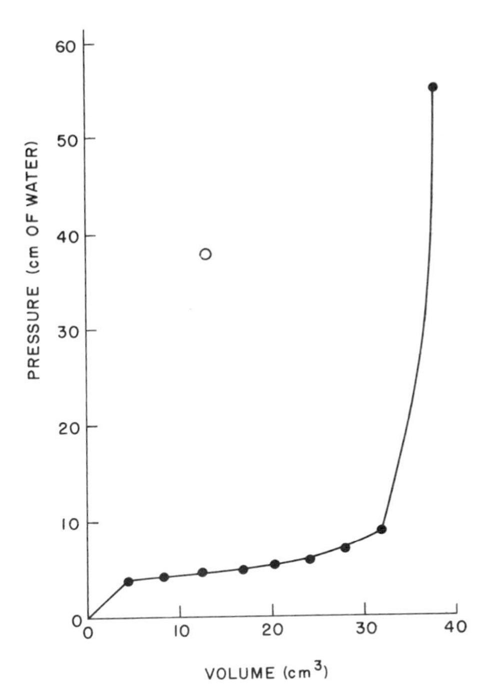

FIG. 1.18 Pressure–volume curve for a cat's bladder. Open circle indicates pressure obtained during a micturition reflex. [Redrawn from Tang and Ruch (1955).]

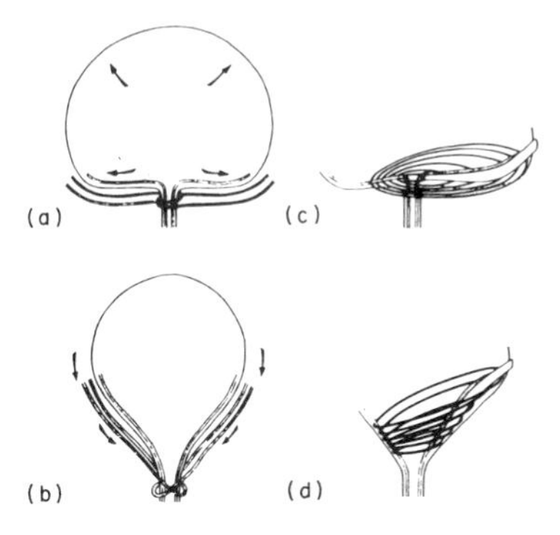

FIG. 1.19 Simplified representation of arrangement of muscle fibers around bladder neck and their mechanism of function. (a) Relaxed distended bladder. (c) Arrangement of muscle fibers around neck in this position. (b) Bladder during voiding. (d) Muscle fibers in neck during voiding. [From E. A. Tanagho *in* "Hydrodynamics of Micturition" (F. Hinman, ed.), 1971. Courtesy of Charles C. Thomas, Publisher, Springfield, Illinois.]

shape due to the contraction of the inner longitudinal muscle coat. This in turn causes a relaxation of the closure muscles. In normal voiding the flow rate starts slowly, increases to a maximum, and then decreases. At the maximum, usually a plateau, the normal voiding pressure is about 40 to 50 cm of water with a flow rate of about 20 cm^3/sec.

The science of urology and micturition is a very broad field and we will concern ourselves solely with the problem of determining if an obstruction exists by measurement of the readily measurable physical parameters: bladder pressure, urine flow rate, stream pressure, and velocity. Although the problem seems to be straightforward it has not yet been solved, and the following discussion will consider some of the various attempts.

One such attempt was based on the intuitively appealing assumption that a stream from an obstructed tube would not cast as far a distance as that from an unobstructed one. It was suggested in the *Southern Medical Journal* by Ballenger *et al.* (1932) that men make note of the distance of their urine stream by periodically making a mark on the side of a barn throughout their lives as a possible technique of indicating prostatic obstruction. Such a systematic form of scientific data collection is usually unavailable to urban dwellers, however. Furthermore, although a prostatic obstruction might be indicated by this technique, other obstructions could not be satisfactorily diagnosed. This is apparent to anyone who has placed his finger over the exit opening of a garden hose or a faucet. If the exit orifice is decreased in area but the pressure is undiminished, Eq. (1.3) shows that the exit stream velocity increases and hence the cast distance of the existing stream will be greater. If there is an obstruction behind the orifice, such as a kink in the hose, the pressure at the orifice will be diminished and therefore the cast distance will be less. Clearly, distance alone is insufficient information.

The approach of Whitaker and Johnson (1966) was to increase the amount of information obtained from the urine stream. Their experimental arrangement is indicated in Figs. 1.20 and 1.21. A pressure sensor was placed against the bladder wall, the stream was directed against a spring-loaded tambour, which measured the force, and the urine then ran onto a weighing device. The force and weight went into a computer, which gave a profile of the velocity

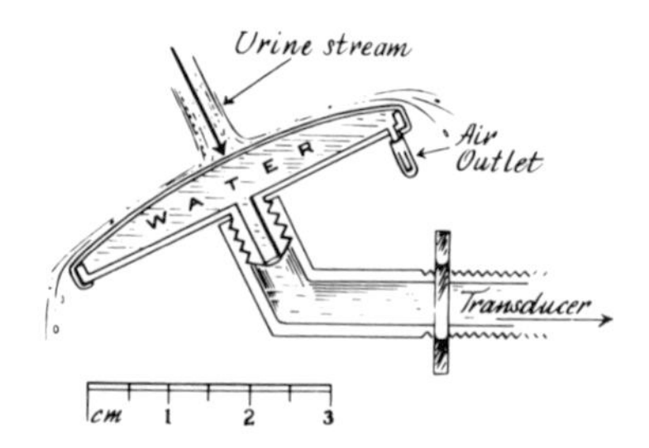

FIG. 1.20 Tambour for measuring momentum transfer of urine stream. [From Whitaker and Johnson (1966.) © 1966 The Williams & Williams Co., Baltimore, Maryland.]

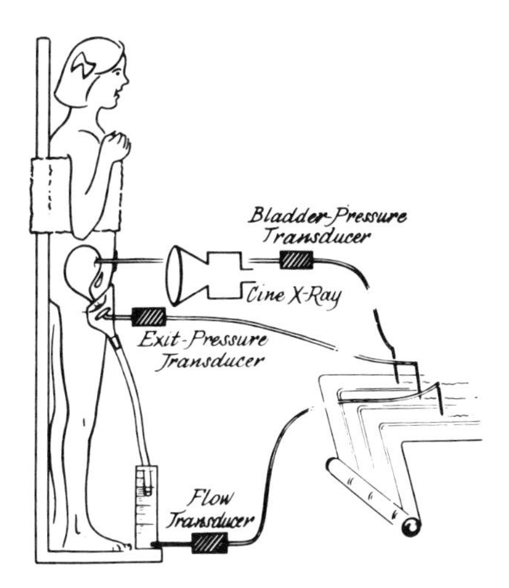

FIG. 1.21 Arrangement for simultaneous recording of bladder pressure, urine momentum transfer, and mass. [From Whitaker and Johnson (1966). © 1966 The Williams & Williams Co., Baltimore, Maryland.]

and momentum transfer as a function of time. Even with this increased information differences in urethras were difficult to separate.

The extraction of information from the urine stream is clearly more complicated than the problem of the hole in the tin can encountered in first-year college physics. We will next consider the urethra as a resistant tube and modify the Bernoulli equation to include a resistance loss term.

URETHRAL RESISTANCE

The urethra is an elastic tube with a varying cross-sectional shape and area. It can stretch to provide a higher flow rate for a higher bladder pressure. Although the flow rate is below the turbulent rate for flow through a glass tube, there is turbulence within the urethra, which arises from its internal design. The purpose of this turbulence is apparently to flush out bacteria. However, the design must be rather clever because surgical changes such as widening of the posterior urethra can cause a permanent localized infection. Experiments by Marberger (1971) with bacterial solution flowing through glass tubes with a variety of strictures show that there is a higher concentration of bacteria in the turbulent vortices. Thus, flushing is inadequate and better design and understanding is required to decide on surgical procedures for urethral alteration (see Hinman's comments (1971) on this subject). However, recognizing that there are such resistance losses, they can be included in the analysis and thereby be quantitatively determined. We shall now modify the Bernoulli equation to accommodate these.

MODIFIED BERNOULLI EQUATION

The Bernoulli equation (derived earlier),

$$P_1 + \rho g h_1 + \tfrac{1}{2}\rho v_1^2 = P_2 + \rho g h_2 + \tfrac{1}{2}\rho v_2^2 \tag{1.6}$$

is a statement of the conservation of energy in a mechanically conservative system, i.e., one in which there are no frictional losses. In situations in which there are energy losses that are irreversible, correction terms must be introduced.

One of these is the frictional loss due to the shearing interaction of the fluid with the wall. If **f** is the frictional or drag force per unit mass and $d\mathbf{s}$ is an element of length along the path of the fluid, the energy loss between points 1 and 2 is

$$-\rho \int_1^2 \mathbf{f} \cdot d\mathbf{s}$$

Another type of loss occurs when the fluid turns a corner, e.g., at a constriction or expansion of the tube. Such losses are called *geometric losses* and, since they involve kinetic energy, are usually considered to involve the square of the velocity. There may be several of these, numbered 1 to N, in the path from point 1 to point 2. Since each involves the velocity at the given point, these losses are written as the sum

$$-\tfrac{1}{2}\rho \sum_{n=1}^{N} K_n v_n^2$$

where the K_n are coefficients for the loss at each of these points.

There is still another term, which is included for completeness, i.e., losses due to the vibration of the walls of the duct caused by the geometric changes in the stream. Since this involves acceleration and deceleration the term is written

$$-\rho \int_1^2 (\partial \mathbf{v}/\partial t) \cdot d\mathbf{s}$$

where $\mathbf{v}$ is velocity of the particular element considered. Subtracting these energy loss terms results in a modified Bernoulli equation, written

$$P_1 + \rho g h_1 + \tfrac{1}{2}\rho v_1^2 - \rho \int_1^2 \left(\frac{\partial \mathbf{v}}{\partial t} + \mathbf{f}\right) \cdot d\mathbf{s} - \tfrac{1}{2}\rho \sum_{n=1}^{N} K_n v_n^2$$
$$= P_2 + \rho g h_2 + \tfrac{1}{2}\rho v_2^2. \tag{1.24}$$

Equation (1.24) is usually simplified for use. Point 1 is taken in the bladder, so $v_1 = 0$. P_2 is taken externally, so P_1 is measured relative to the atmosphere. It is not known how to evaluate the term $\partial \mathbf{v}/\partial t$ so it is dropped. Experimental conditions can be arranged so that $h_1 = h_2$ or else a correction term can be added to P_1. The friction term is considered to be proportional to the square of the velocity at ds and the proportionality factor is taken as $K_f/2$. Equation (1.24) is thus reduced to

$$P_1 - P_2 = \Delta P = \tfrac{1}{2}\rho \sum_{n=1}^{N} K_n {v_n}^2 + \tfrac{1}{2}\rho \int_1^2 K_f v^2 \, ds + \tfrac{1}{2}\rho {v_2}^2$$

$$= \tfrac{1}{2}\rho {v_2}^2 \left[\sum_{n=1}^{N} K_n (v_n/v_2)^2 + \int_1^2 K_f (v/v_2)^2 \, ds + 1 \right], \quad (1.25)$$

If we introduce the equation of continuity

$$Q = A_1 v_1 = A_2 v_2 \quad (1.3)$$

or

$$v_1/v_2 = A_2/A_1$$

Eq. (1.25) can be written

$$\Delta P = \tfrac{1}{2}\rho {v_2}^2 \left[\sum_{n=1}^{N} K_n (A_2/A_n)^2 + \int_1^2 K_f (A_2/A)^2 \, ds + 1 \right] \quad (1.26)$$

The integral and summation are thus only functions of geometry which, if evaluated for a given system, will yield a constant K, so that

$$\Delta P = \tfrac{1}{2}\rho {v_2}^2 (K + 1) \quad (1.27)$$

Since $Q = Av$ we may write Eq. (1.27) as

$$\Delta P/Q^2 = \rho(K + 1)/2{A_2}^2 \qquad \text{or} \qquad \Delta P = RQ^2 \quad (1.28)$$

where $\underline{R}$ is the resistance term.

However, although this relation has not been verified, it is being used as the basis for a number of urological experiments.

Let us consider some of the assumptions in the derivation of Eq. (1.28).

(1) There is a steady state of urethral flow with stable velocity at any given point, and therefore we may assume $\partial \mathbf{v}/\partial t = 0$. This is not a good assumption because the elasticity of the urethra permits the flow rate to increase and then recede.

(2) Friction between the stream and the urethral wall is constant. This term involves the square of the velocity and assumes a constant velocity, which is never the true situation.

TABLE 1.3

Free flow (ml/sec)	Catheter inserted	Free flow (ml/sec)	Catheter inserted
4.2	6.7	15.9	13.2
7.1	8.4	19	14.6
10	10.6	21.9	16.2
13	11.1		

(3) The Bernoulli equation is for a single streamline. This is not the case for turbulent flow in the urethra. This difficulty manifests itself in that Eq. (1.28) does not become Poiseuille's law, Eq. (1.11), for zero friction and turbulence.

(4) The geometry of the urethra is assumed constant during the voiding. This assumption denies the most unique feature of the urethra, namely, the ability to accommodate to changing pressure and flow.

The accommodation of the urethra mentioned above in (4) is clearly shown in Table 1.3 from data of Gleason *et al.* (1967). They inserted a catheter into a large number of patients and filled their empty bladders to a constant amount. They then measured voiding rates with the catheter in the urethra and with it removed. Example data in Table 1.3 show that the catheter interferes only at high flow rates and that the uretha clearly has expanded to accommodate the catheter.

An approach of Ritter *et al.* (1964) was to try for a better separation of velocity and friction factors using engineering data. Their tests showed that in normal young women approximately two-thirds of the pressure drop is converted into velocity of the stream via Bernoulli's principle and one-third is due to frictional losses. They separated the pressure drop into a velocity part and friction part, i.e.,

$$\Delta P_{\text{tot}} = \Delta P_{\text{vel}} + \Delta P_{\text{fric}} \tag{1.29}$$

in which they accepted the kinetic energy relation

$$\Delta P_{\text{vel}} = (\rho/2)v^2 \tag{1.30}$$

However, they used data for flow in tubes from "The Chemical Engineer's Handbook" based on Moody's (1944) calculation that

$$\Delta P_{\text{f}} = 32f LQ^2\rho/(\pi^2 d^5) \tag{1.31}$$

In this equation L is the length and f is Fanning's friction factor, which is related to the roughness of the tube walls. They assumed a reasonable value for f and that turbulent flow was taking place, and evaluated ΔP_{tot}, showing that indeed the friction loss is about one-third or less.

Zatz (1965) started with Eq. (1.29)–(1.31) but combined them. From the relation

$$Q = Av = (\pi d^2/4)v$$

Eq. (1.30) is then

$$\Delta P_{\mathrm{vel}} = 8Q^2\rho/(\pi^2 d^4) \tag{1.32}$$

By putting this and Eq. (1.31) into Eq. (1.29) and factoring, one arrives at the relation

$$\Delta P_{\mathrm{tot}}/Q^2 = [8\rho/(\pi^2 d^5)](d + 4fL) \tag{1.33}$$

where the second term on the right is the urethral resistance. Although this approach also confirmed the fraction of resistance loss, the usefulness for diagnosis has not been adequately demonstrated.

URINARY DROP SPECTROMETER

A more recent approach has utilized the combined talents of physicists, mathematicians, engineers, and urologists. Starting with the assumption that the downstream drops must contain information concerning their origin, Zinner *et al.* (1969) constructed a device, which they call a *urinary drop spectrometer*. The basic element of this device is a series of fiber optics in a double bank. The interruption of the light determines the velocity of each drop and the information is counted by a computer. This is illustrated in Fig. 1.22. A typical histogram of data during a urination is shown in Fig. 1.23 and, since the distance is known, these data are readily converted to a distribution of the time interval between drops, as shown in Fig. 1.24.

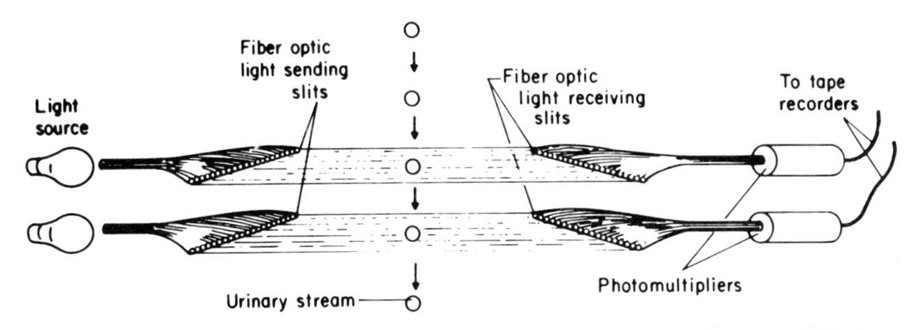

FIG. 1.22 Basic components of urinary drop spectrometer. Drops interrupt light beams between array of fiber optics and their velocities are recorded. [From Zinner *et al.* (1969). © 1969 The Williams & Williams Co., Baltimore, Maryland.]

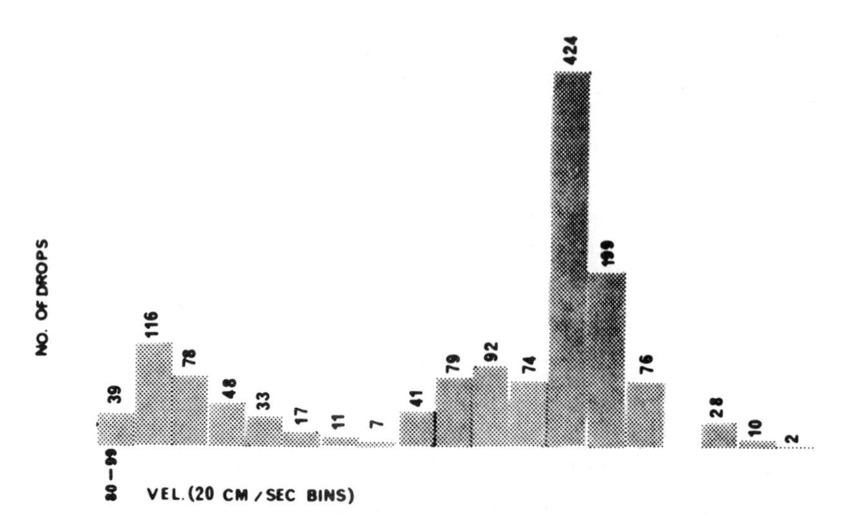

FIG. 1.23 Histogram of velocity distribution of all drops from a single urination of a normal individual. The three peaks, each with a Gaussian distribution, occurred from the initiation of urination, the main body, and at the end or urination, respectively. [From N. R. Zinner and D. C. Harding, *in* "Hydrodynamics of Micturition" (F. Hinman, ed.), 1971. Courtesy of Charles C. Thomas, Publisher, Springfield, Illinois.]

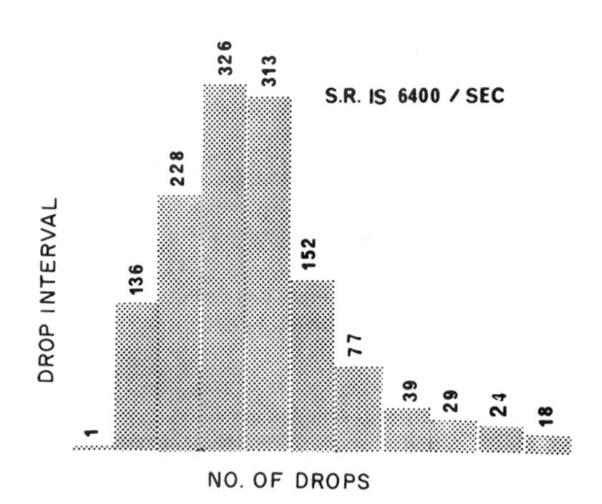

FIG. 1.24 The drop distribution of Fig. 1.23 plotted as a drop interval distribution. This distribution is not Gaussian. [From N. R. Zinner and D. C. Harding, *in* "Hydrodynamics of Micturition" (F. Hinman, ed.), 1971. Courtesy of Charles C. Thomas, Publisher, Springfield, Illinois.]

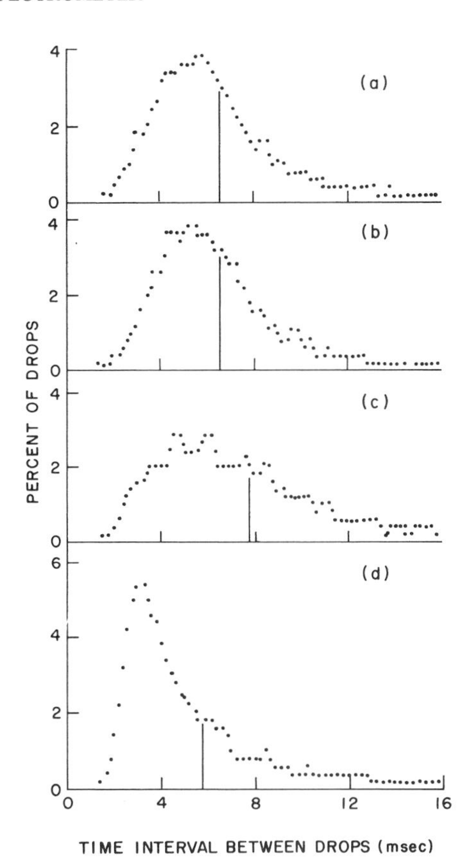

FIG. 1.25 Time interval histograms with urinary drop spectrometer. The vertical lines mark the centroids of distribution. (a) and (b) From the same man on two successive days. (c) From a patient with an enlarged prostate (prostatic hypertrophy). (d) From a patient with an obstruction near the exit orifice (meatal stenosis). [Redrawn from Aiello *et al.* (1974).]

Figures 1.25a and 1.25b show the consistency of the data for the same individual on two consecutive days. Figure 12.5c shows the difference in distribution in a patient with a prostatic obstruction within the urethra, and Fig. 1.25d a patient with an obstruction at the exit orifice.

Clearly, important diagnostic information can be obtained in this way. Why does this type of measurement give better results than the stream analysis approaches? The answer is not yet known because the theory has not kept up with the experiments. In the next section we will summarize the approaches to a theoretical model.

THE CAPILLARY JET STREAM

A capillary jet stream from a circular orifice appears as a standing wave. It collapses, redevelops, collapses again, and eventually breaks up into drops whose spacings are constant and equal to the spacing of the wave. This is illustrated in Fig. 1.26. This phenomenon was first quantified by Lord Rayleigh (1879). His model, based on an earlier study by Savart, postulated that in consequence of surface tension a free cylinder of fluid is not an equilibrium configuration; upon emerging from the containing walls of the capillary, the cylindrical jet suffers an oscillatory disturbance that arises from the surface tension. The formulation leads to a time-dependent boundary value problem whose solution is beyond the scope of this book. The interested reader is referred to the original paper.

The result of the solution of the equation developed by Rayleigh is that the jet would eventually become unstable and break up into drops of uniform diameter of about 1.9 times the jet diameter. The time interval between the drops would be $t = 2\sqrt{2}\pi a/v$, where a is the jet diameter and v the velocity, and the spacing distance or wavelength before the breakup would be $\lambda = 2\sqrt{2}\pi a$. Experiments such as the high-speed photograph of Fig. 1.26 have generally confirmed this model. This approach has been modified by Weber (1931) to include fluid viscosity and air resistance.

The urethral exit orifice is not an open circle, however; it is a closed slit, varying in size with the individual, which opens under pressure. Since it is

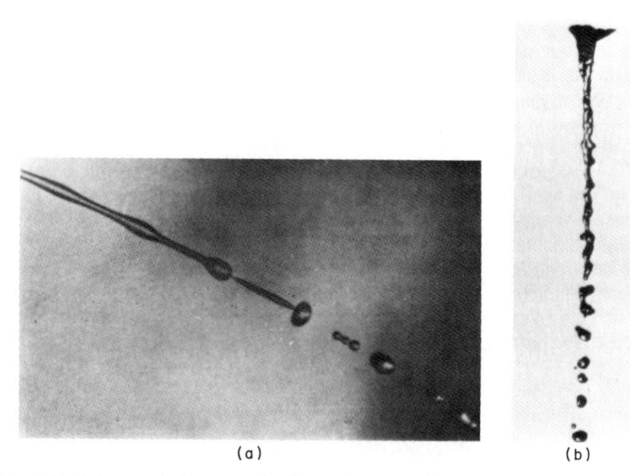

FIG. 1.26 (a) High-speed photograph of cylindrical capillary jet. (b) High-speed photograph of a male urine stream. [From R. C. Ritter and A. M. Sterling, *in* "Hydrodynamics of Micturition" (F. Hinman, ed.), 1971. Courtesy of Charles C. Thomas, Publisher, Springfield, Illinois.]

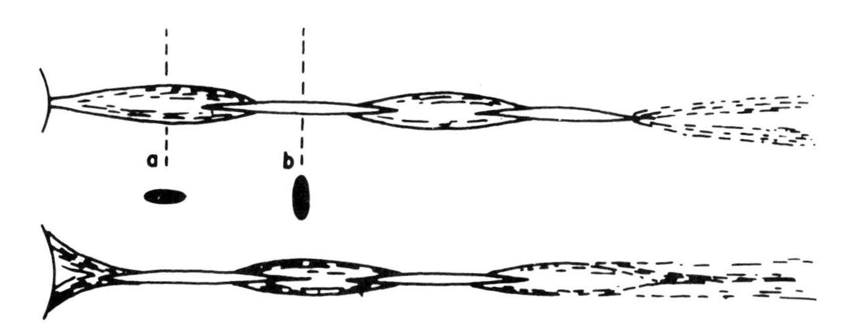

FIG. 1.27 Shape of capillary jet from an elliptical orifice, top and side views. *a* and *b* represent cross-sectional views of the ellipsoidal shape in the two example regions. [From R. C. Ritter and A. M. Sterling, *in* "Hydrodynamics of Micturition" (F. Hinman, ed.), 1971. Courtesy of Charles C. Thomas, Publisher, Springfield, Illinois.]

elastic, the opening is approximately elliptical, with the ratio of the major to minor axis of the ellipse varying with pressure. Figure 1.26 shows the shape of a male urine stream. A stream from a constant elliptical orifice will have an elliptical shape similar to the orifice (see Fig. 1.27). However, in the region of high curvature the surface tension is high and this will exert a force to change the stream shape to a circular one. But, the inertia of the fluid allows the jet to overshoot the circular configuration and go into an elliptical shape whose major axis is perpendicular to that of the original ellipse. It then oscillates back to the original configuration for the same reason. While the jet is undergoing these oscillations, it is moving through space and gives rise to standing waves. Although visual observation of this phenomenon suggests a rotational component to the stream this is an optical illusion.

After several such oscillations axisymmetrical disturbances appear, as seen in Fig. 1.26, and the stream breaks up into drops. The drops are not regular and can be seen to contain oscillatory, and possibly rotational, motion. No device has yet been developed to measure these latter components. Furthermore, the theory assumes laminar flow of the fluid as it approaches the exit orifice, while it is apparent from the earlier discussion of the urethra that the fluid motion is actually turbulent. This seems to be a problem soluble on a high-speed, large-memory computer with a three-dimensional spatial program. But it has not yet been done.

REFERENCES

Aiello, G., Lafrance, P., Ritter, R. C., and Trefil, J. S. (1974). The urinary drop spectrometer, *Physics Today*, September, p. 23.

Ballenger, E. G., Elder, O. F., and McDonald, H. P. (1932). Voiding distance decrease: An important early symptom of prostatic obstruction, *Soutern Med. J.* **25**, 863.

Bayliss, L. E., and Robertson, G. W. (1939). The viscoelastic properties of the lungs, *Quart. J. Exp. Physiol.* **29**, 27.

Bohr, N. (1909). Determination of the surface tension in water by the method of jet vibration, *Roy. Soc. London, Phil. Trans.* **A209**, 281.

Burton, A. C. (1962). Physical principles of circulation phenomena: The physical equilibria of the heart and blood vessels, *in* "Handbook of Physiology," Vol. I, Sect. 2, Circulation. American Physiological Society, Washington, D.C.

Clements, J. A. (1962). *Scientific American*, December, p. 120.

Comroe, J. H. Jr. (1965). "The Physiology of Respiration." Year Book Publ., Chicago, Illinois.

Comroe, J. H., Jr., Forster, R. E., II, DuBois, A. B., Briscoe, W. A., and Carlsen, E. (1963). "The Lung: Clinical Physiology and Pulmonary Function Tests." Year Book Publ., Chicago, Illinois.

Dean, R. B., and Visscher, M. B. (1941). The kinetics of lung ventilation, *Am. J. Physiol.* **134**, 450.

Fritts, H. W., and Richards, D. W. (1960). Respiratory system: External respiration, *in* "Medical Physics" (O. Glasser, ed.), Vol. III. Year Book Publ., Chicago, Illinois.

Gleason, D. M., Bottaccini, M. F., Perling, D., and Lattimer, J. K. (1967). A challenge to current urodynamic thought, *J. Urol.* **97**, 935.

Guyton, A. (1971). "Textbook of Medical Physiology." Saunders, Philadelphia, Pennsylvania.

Hinman, F., Jr. (1971). Infection and urethral function, *in* "The Hydrodynamics of Micturition" (F. Hinman, Jr., ed.). Thomas, Springfield, Illinois.

Holm, H. H. (1964). The hydrodynamics of micturition, *Acta Radiol* [*Diagn.*] *Suppl.* **231**.

Marberger, H. (1971). Flow characteristics, *in* "Hydrodynamics of Micturition" (F. Hinman, ed.). Thomas, Springfield, Illinois.

Mead, J., and Agostoni, E. (W. O. Fenn and H. Rahn, eds.), (1964). Dynamics of breathing, *in* "Handbook of Physiology" Sect. III, Vol. I. Williams and Wilkins, Baltimore, Maryland.

Moody, L. F. (1944). Friction factors for pipe flow, *Trans. ASME* **66**, 671.

Otis, A. B., Fenn, W. O., and Rahn, H. (1950). The mechanics of breathing in man, *J. Appl. Physiol.* **2**, 592.

Otis, A. B., and Proctor, D. F. (1948). Measurement of alveolar pressure in human subjects, *Am. J. Physiol.* **152**, 106.

Perry, J. H. (ed.), (1963). "Chemical Engineers Handbook." McGraw-Hill, New York.

Rayleigh, L. (1879). On the capillary phenomena of jets, *Proc. Roy. Soc.* **29**, 71.

Richardson, I. W., and Neergaard, E. B. (1972). "Physics for Biology and Medicine." Wiley, New York.

Ritter, R. C., and Sterling, A. M. (1971). Exposition of pertinent principles of mechanics and hydromechanics, *in* "Hydrodynamics of Micturition" (F. Himan, Jr., ed.). Thomas, Springfield, Illinois.

Ritter, R. C., Zinner, N. R., and Paquin, A. J. Jr., (1964). Clinical urodynamics II. Analysis of pressure flow relations in the normal female urethra, *J. Urol.* **91**, 161.

Rohrer, F. (1925). "Handbuch der normalen und pathologischen Physiologie," Vol. 2, p. 70. Springer, Berlin.

Tanagho, E. A. (1971). Interpretation of the physiology of micturition, *in* "Hydrodynamics of Micturition" (F. Hinman, ed.). Thomas, Springfield, Illinois.

Tang, P., and Ruch, T. C. (1955). Non-neurogenic basis of bladder tonus, *Am. J. Physiol.* **181**, 249.

Weber, C. (1931). Zum Zerfall eines Flüssigkeitsstrahles, *Z. Angew. Math. Mech.* **11**, 136.

Weibel, E. R. (1963). "Morphometry of the Human Lung." Academic Press, New York.

Whitaker, J., and Johnson, G. S. (1966). Estimation of urinary outflow resistance in children: Simultaneous measurement of bladder pressure, flow rate and exit pressure, *Invest. Urol.* **3**, 379.

Wise, H. M., Jr., Many, M., Birtwell, W. C., Eyrich, T. B., and Maguire, M. (1968). Measurement of urethral resistance, *Invest. Urol.* **5**, 539.

Zatz, L. M. (1965). Combined physiologic and radiologic studies of bladder function in female children with recurrent urinary tract infections, *Invest. Urol.* **3**, 279.

Zinner, N. R., and Harding, D. C. (1971). Velocity of the urinary stream, its significance and a method for its measurement, *in* "Hydrodynamics of Micturition" (F. Hinman, ed.). Thomas, Springfield, Illinois.

Zinner, N. R., Ritter, R. C., Sterling, A. M., and Harding, D. C. (1969). Drop spectrometer: A nondestructive instrument for analyzing hydrodynamic properties of human urination, *J. Urol.* **101**, 914.

CHAPTER 2

Non-Newtonian Fluids: Mucus and Blood

INTRODUCTION

We experience the differences in fluid behavior constantly. One of these differences, viscosity, has been discussed in Chapter 1. We all know the increased force required to stir a jar of honey compared to water, and we say that honey has a higher viscosity. However, although it is easier to stir a jar of mayonnaise than a jar of honey, if we lift out the spoon the honey will virtually all drip from the spoon back into the jar, while none of the mayonnaise will. What physical concepts are required to describe the flow behavior of mayonnaise or that of a raw egg white or tooth-paste?

The study of such materials is called *rheology* (the study of flow), according to which fluids may be put into two general classifications: Newtonian, and non-Newtonian. Actually, all fluids fall into the latter category; the Newtonian treatment is a simplifying assumption that can be applied to only a few fluids in bulk. In the body, only air and water can be treated as Newtonian fluids—all others are non-Newtonian.

The mechanical treatment of non-Newtonian fluid flow was developed by Maxwell in the last century and by Voigt in the early part of the twentieth century. But this study did not grow rapidly until the 1940s, when it began to be stimulated by the needs of the developing plastic industries to control stirring, flow, extrusion, etc. The results of rheology were applied to the paint industry, the baking industry, the ceramics industry, and many others. Here is but one unfortunate example of how the rapid early development of an industry inhibited the later transmission of the wealth of rheological information to the physiologists involved in the study of body fluids. Indeed, only in the last decade has there been significant quantitative application of rheological principles in physiological studies.

In this chapter we will introduce the basic concepts and the results of some experiments on the simpler body fluids. It should be recognized that most body fluids are composed of a variety of complex molecules, each with different bending and twisting forces not only in their main "backbone" but also in their molecular appendages. Each of these has a relaxation time that can vary with temperature, pH, components of solution, and density. Methods for the complete experimental and analytical treatment of such a complex fluid are to be found in any of several books listed in the references at the end of the chapter.

NEWTONIAN FLUIDS

It was shown in Chapter 1 that if a fluid is confined between a stationary and a moving surface the velocity gradient is proportional to the shearing force and the proportionality constant is the viscosity. If, for example, two

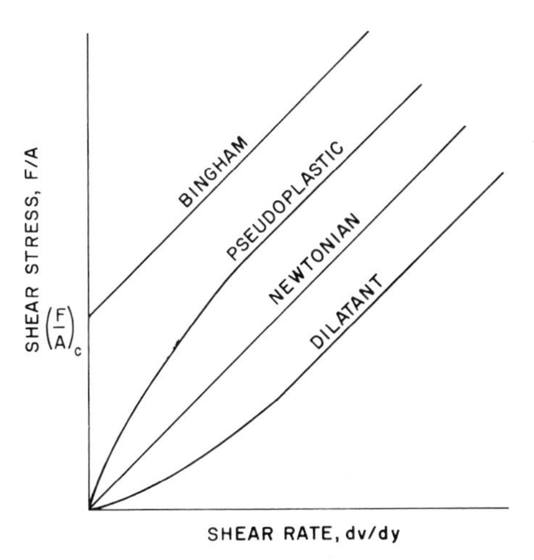

FIG. 2.1 Behavior of shear stress with increasing shear rate for different types of fluids. Final slopes are drawn parallel for simplicity.

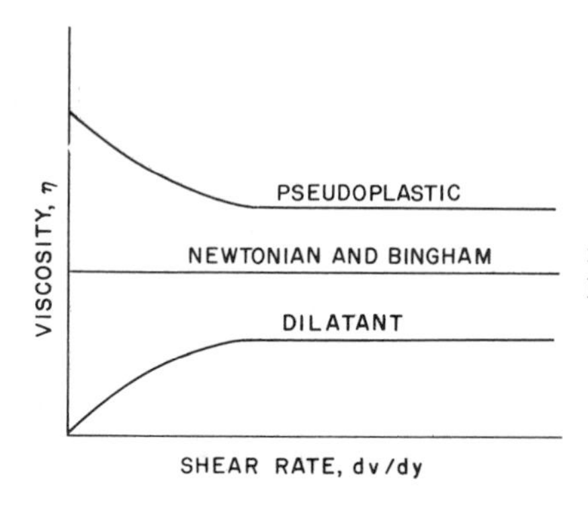

FIG. 2.2 Variation of viscosity with shear rate for the fluids of Fig. 2.1. Constant slopes have been displaced to facilitate viewing.

glass slides have some grease between them and the lower is held while the upper is pushed, the velocity of motion of the upper slide is proportional to the force applied to it. Since the grease will be sheared by the difference in motion of the two slides, the applied force is called the "shearing force." It is obvious that if the slide areas are doubled twice the shearing force will have to be applied to achieve the same velocity. Thus the important parameter is force per unit area, F/A, termed *shear stress* and generally represented in the literature by the symbol σ. It is also evident that the velocity of a grease molecule next to the stationary slide is zero while the velocity of one next to the moving slide is the velocity of the slide. The velocity of any other molecule in between will depend on the molecule's distance from the bottom slide; the greater its distance y the greater its velocity v. Thus there is a velocity gradient, called *shear rate*, dv/dy, for which a frequent symbol is $\dot{\gamma}$. If the viscosity symbol is η the Newtonian relation for laminar flow, as shown in Chapter 1, is

$$F/A = \eta(dv/dy) \tag{1.7}$$

Thus the Newtonian viscosity, while dependent on temperature and pressure, is independent of the rate of shear. It is constant. An example plot of shear stress versus shear rate is shown in Fig. 2.1. The constant viscosity versus shear rate is shown in Fig. 2.2.

NON-NEWTONIAN FLUIDS

Non-Newtonian fluids are those in which the viscosity depends on the shear rate in the fluid, the experimental apparatus used to measure the viscosity, or even the previous history of the fluid. The three most general types can be put into three general classifications.

(1) Fluids for which the rate of shear depends on the shear stress only.

(2) Fluids for which the relation between shear stress and shear rate depends on the time of application of the stress or the previous history.

(3) Fluids that have characteristics of both solids and liquids and exhibit partial elastic recovery after deformation. These are called *viscoelastic* fluids.

We will not consider fluids of category (2) in this chapter because of the complexity of the mathematics. Although in all probability some body fluids have this characteristic, at the present time the writer does not know of any to which this treatment has been applied.

Fluids of category (1) are called *non-Newtonian viscous fluids* and these are subdivided into three distinct types: (a) Bingham plastics; (b) pseudoplastic fluids; and (c) dilatant fluids. These will be discussed in order.

(a) *Bingham plastics* Bingham plastics are those that require a critical shear stress $(F/A)_C$ before they begin to flow, after which they have a constant viscosity. The resulting curve is shown in Fig. 2.1. Examples of this type are mayonnaise, toothpaste, and some slurries.

(b) *Pseudoplastic fluids* Pseudoplastic fluids flow better as the shear rate increases, gradually approaching a constant viscosity, as shown in Fig. 2.1. Fluids of this type are often slurries with asymmetric particles. When motionless, the particles are randomly oriented, but as the flow begins they align themselves in the direction of flow so that their axes are, on the average, at a uniform orientation with the direction of flow. The apparent viscosity decreases as they align themselves and eventually becomes constant. Examples of fluids showing this kind of behavior are solutions of high polymers and, as we will see later, blood.

(*c*) *Dilatant fluids* Dilatant fluids usually have no critical shear stress but, as the shear rate increases, their viscosity increases. This situation occurs in concentrated suspensions. When such a system is at rest there is only sufficient liquid to fill the voids between the particles. At low shear rates the liquids acts as a lubricant between the particles. At higher shear rates the dense packing of the particles is broken up and the particles move apart or *dilate*. There is no longer sufficient liquid to fill the voids and lubricate their motion, so that the applied stress has to be larger. This results in an apparent viscosity increase with increasing rate of shear. An example of a fluid showing this kind of behavior is wet cement. A stick may be moved slowly through it with some ease, but rapid motion is difficult, and the appearance is changed from fluidlike to granular.

VISCOELASTICITY

Many materials have mechanical properties that cannot be adequately defined without simultaneously considering both elastic and viscous effects.

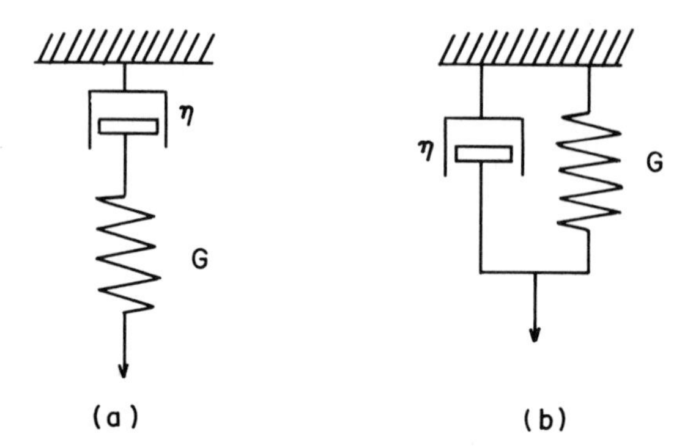

FIG. 2.3 (a) Maxwell element. (b) Voigt element.

There are two basic mechanical mechanisms that can be used to describe such behavior, the Maxwell model and the Voigt model.

In the Maxwell model, the material can respond to an applied stress by two different types of displacements, elastic and flow, and the total displacement is the sum of the two. Figure 2.3a shows a mechanical analog of such a system. The spring is purely elastic with an elastic constant G called the *shear modulus*. This is the usual proportionality constant between stress σ and elastic displacement x_e as in the Hooke's law relation

$$\sigma = Gx_e \tag{2.1}$$

The other mechanical object in Fig. 2.3 is known as a *dashpot*. It is a piston in a cylinder with a small hole for leaking in fluid or air. As a force is applied to it the air flows through the hole and the piston achieves a permanent displacement in the direction of the applied force. When the force is removed the piston no longer moves and a flow displacement x_f is achieved. The instantaneous rate of flow is proportional to the stress and inversely proportional to the viscosity or

$$dx_f/dt = \sigma/\eta \tag{2.2}$$

The total displacement is $x = x_e + x_f$. Since we do not know x_f explicitly, we can add the differentials by taking the derivative of Eq. (2.1). Thus,

$$dx/dt = dx_e/dt + dx_f/dt = (1/G)(d\sigma/dt) + \sigma/\eta \tag{2.3}$$

is the differential equation governing the motion of this system.

By designing an appropriate experiment we can simplify the solution of Eq. (2.3). Suppose we quickly pull on this system to a certain deformation x and then stop. Physically, all of the displacement first occurs in the spring but, with time, the spring pulls the dashpot with diminishing force until the displacement is in the dashpot instead of in the spring. However, since there

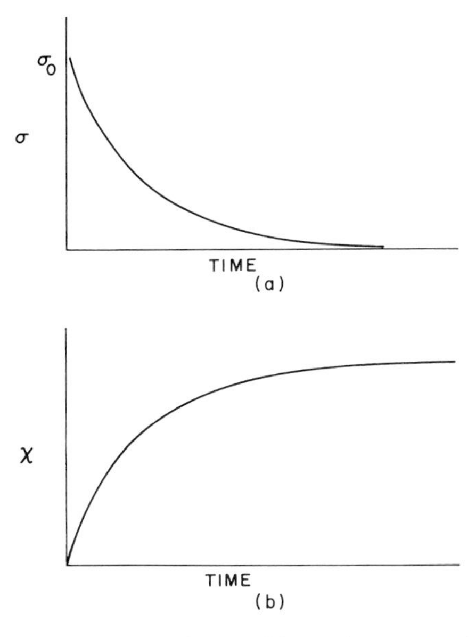

FIG. 2.4 (a) Exponential relaxation of stress with time for a Maxwell element (see Eq. (2.4)). (b) Exponential growth of displacement with time for a Voigt element (see Eq. (2.6)).

is an instantaneous displacement and then no further total displacement we may say that at zero time $dx/dt = 0$. Thus Eq. (2.3) is now

$$(1/G)(d\sigma/dt) + \sigma/\eta = 0$$

whose solution can be written by inspection as (see Appendix D.1)

$$\sigma = \sigma_0 e^{-Gt/\eta} \tag{2.4}$$

and the stress relaxes exponentially with time (Fig. 2.4a). The ratio η/G is called the *relaxation time*. For time scales of a short duration compared to the relaxation time, the system will behave as an ideal elastic body, i.e., the system once displaced will automatically return to its original configuration. For times long compared to the relaxation time the system will behave as an ideal viscous liquid, i.e., the system is displaced slowly and does not return to its original position. Note that in the elastic part the work done on the system is recovered by the system. Thus there is no net work and therefore no energy is absorbed by the system. In the viscous part work, force times distance, is done on the system and it therefore absorbs energy, usually as heat.

The Voigt model is one of retarded elastic response in that a single elastic mechanism operates. However, this response is retarded by a parallel viscous element. This is shown in Fig. 2.3b in which the spring and dashpot are in parallel. In this case we cannot instantaneously displace the system since the dashpot prevents it. Also, there can be no permanent displacement; the stress in the spring returns the system to zero displacement when external stress is removed. It is therefore more representative of a solid in that there is no nonrecoverable flow. In this case the elastic displacement and flow displacement are the same so that the sum of Eqs. (2.1) and (2.2) is

$$\eta(dx/dt) + Gx = \sigma \tag{2.5}$$

If a constant stress is applied at $t = 0$ the solution of Eq. (2.5) is (see Appendix D.3)

$$x = (\sigma/G)(1 - e^{-Gt/\eta}) \tag{2.6}$$

This is an exponential growth, and a typical plot is shown in Fig. 2.4b. If some displacement x_0 is achieved and the stress is removed, then $\sigma = 0$ in Eq. (2.5) and we have

$$\eta(dx/dt) + Gx = 0 \tag{2.7}$$

The solution is

$$x = x_0 e^{-Gt/\eta} \tag{2.8}$$

and the system returns exponentially to its original configuration.

The experimental time scale is an important factor in the Voigt model also. If a constant stress is applied for a time duration that is very short compared to the relaxation time η/G then only the first part of the deformation–time curve is observed. For short time duration the exponent of Eq. (2.6) is small and can be expanded to two terms only,

$$x = \frac{\sigma}{G}\left\{1 - \left[\left(1 - \frac{G}{\eta}t\right)\right]\right\}, \qquad x = \frac{\sigma}{\eta}t$$

Thus the early portion has a constant slope σ/η and the system behaves like a simple fluid with viscosity η. If the experimental time scale is long compared to the relaxation time then the dashpot will extend and relax without any noticeable retardation and the system will seem to be purely elastic.

Many complex systems can be represented by various parallel and series combinations of the Maxwell and Voigt elements, and even a critical shear stress can be included when necessary. The procedure, then, is to examine flow data and to make different models, varying the constants to see whether the data can be matched.

For example, consider a series combination of a Maxwell and Voigt viscoelastic system as in Fig. 2.5, with elastic constants G_1, G_2 and viscosities η_1,

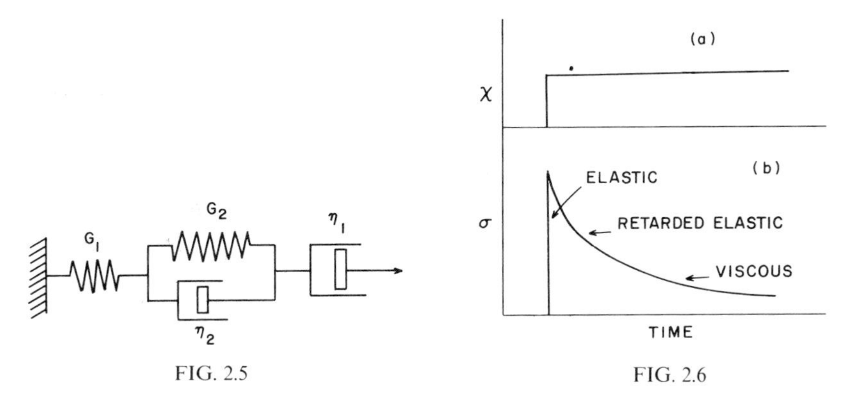

FIG. 2.5 FIG. 2.6

FIG. 2.5 Series combination of a Maxwell and a Voigt viscoelastic element.

FIG. 2.6 (a) Time dependence of displacement of the Maxwell–Voigt combination of Fig. 2.5. (b) Time dependence of a sudden stress placed on this combination.

η_2. The differential equation is a combination of Eqs. (2.3) and (2.5). Suppose the end of the system at the right is suddenly displaced a distance x and held there. This is illustrated in Fig. 2.6a. If we examine the time change of the displacement, we see that initially there is an elastic response governed by spring constant G_1; the dashpots have not had time to yield. If the viscosity of the dashpot in the Voigt element η_2 is less than that of η_1, the next type of motion is the retarded elastic response of the Voigt element governed by η_2 and G_2. Finally, as time progresses the dashpot of the Maxwell element

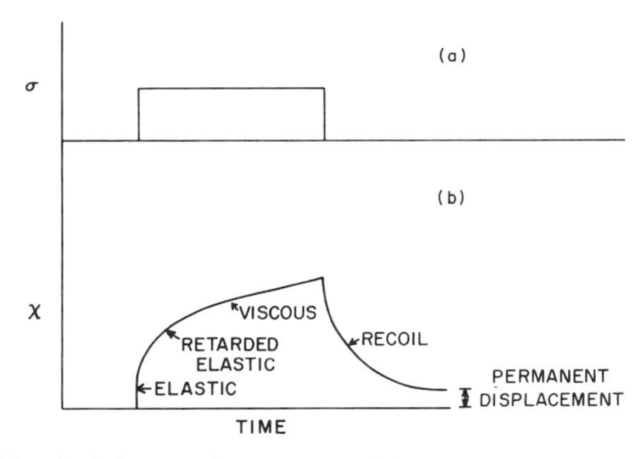

FIG. 2.7 Hypothetical curves of a creep test result for the series Maxwell–Voigt elements of Fig. 2.5. (a) Stress versus time. (b) Displacement versus time.

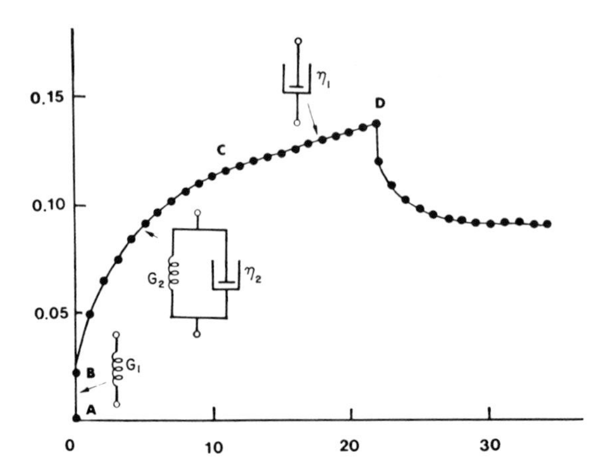

FIG. 2.8 Creep curve for sputum (25°C). The abscissa is time (min), the ordinate compliance (cm^2/dyn). A–B, elastic region; B–C, Voigt viscoelastic region, C–D, viscous region. [From S. S. Davis, *in* "Rheology of Biological Systems" (H. L. Gabelnick and M. Litt, eds.), 1973. Courtesy of Charles C. Thomas, Publisher, Springfield, Illinois.]

flows and relieves the internal stress. Thus, although overlap occurs, different time regions are predominantly governed by different modes of relaxation, and the constants can be obtained from the curve. These regions are shown in Fig. 2.6b. This type of experiment is called *stress relaxation.*

Another type of experiment is called the *creep test*. In this, the system is subjected to a constant force and the extension is measured as a function of time. This is illustrated in Fig. 2.7. Note that in this experiment the relaxation of the dashpot in the Maxwell element, η_1, undergoes a permanent displacement. Thus when the stress is removed the springs recoil and the system shortens, but not to the original length.

An example of data obeying this model is shown in Fig. 2.8, which is a creep curve for sputum. This curve was analyzed by a similar model and the elastic and viscosity parameters were determined. Thus with an appropriate model and experimentally determined constants the mechanical response of this system is completely described.

An immediate application of these studies is an external technique for the measurement of the effectiveness of various remedies for the reduction of sputum consistency, i.e., cough medicines. Unfortunately, sputum is a coarse mixture of varying concentrations and a miniaturization technique was not available for small samples. However, standardized samples of cow musin are available and were used. Figure 2.9 shows the results of such a standard test performed externally on mucin samples. Note that some remedies are no more effective than water.

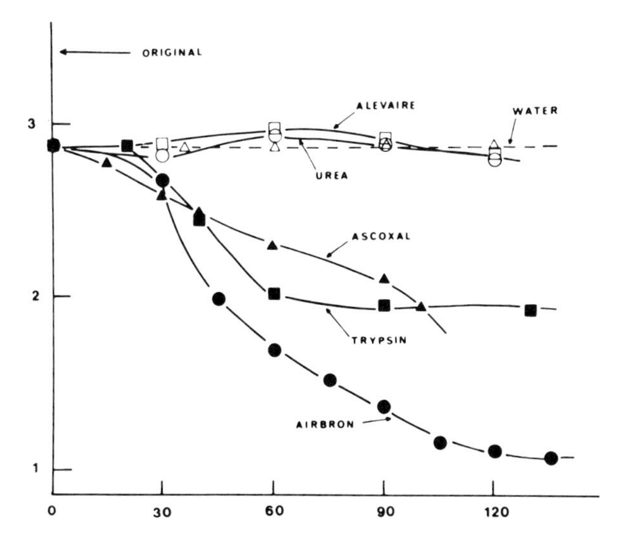

FIG. 2.9 Effect of mucolytic agents on apparent viscosities of bovine gastric mucin (shear rate = 370 $\sec^{-1}$). The abscissa is contact time (min), the ordinate apparent viscosity (P). [From S. S. Davis, *in* "Rheology of Biological Systems" (H. L. Gabelnick and M. Litt, eds), 1973. Courtesy of Charles C. Thomas, Publisher, Springfield, Illinois.]

SINUSOIDAL STRESS

The mechanical models have equivalent electrical analogs. These are shown in Fig. 2.10. The application of a sinusoidal stress to a mechanical system results in the same equations as the application of a sinusoidal voltage to an electrical system. The stress σ can be represented by the horizontal component of a rotating vector of length σ_0 and an angular velocity ω. The deformation, called γ for an arbitrary direction, will be given by the horizontal component of the rotating γ vector, and the phase angle between the σ_0 and γ vectors will be a constant angle δ. We can define *elastic reactance*

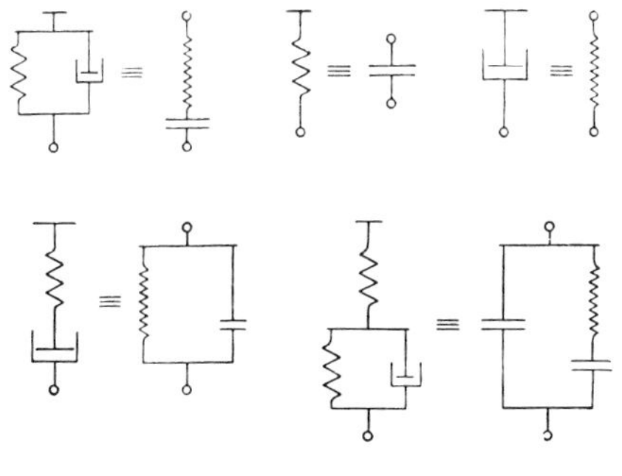

FIG. 2.10 Equivalence of mechanical and electrical networks. Note that elasticity is equivalent to capacitance because both store energy, and viscosity is equivalent to resistance because both dissipate energy. Stress is equivalent to potential difference and flow to electric current. [From Alfrey (1948).]

X as analogous to capacitive reactance in the electric analog, and R can be defined by analogy to η. Thus

$$X = -G/\omega$$

and the viscoelastic impedance Z is such that

$$Z = R + iX \qquad \text{and} \qquad |Z| = (R^2 + X^2)^{1/2}$$

A purely elastic material has a deformation strain in phase with the stress, and a purely viscous material has strain 90° out of phase with stress.

Continuing the analogy with the electrical system, we can express the dynamic relations between stress and strain using complex functions. Define the following:

$$\sigma^* = \sigma_0 e^{i\omega t} = \sigma_0(\cos \omega t + i \sin \omega t); \tag{2.9}$$

$$\gamma^* = \gamma_0 e^{i(\omega t - \delta)} = \gamma_0[\cos(\omega t - \delta) + i \sin(\omega t - \delta)];$$
$$G^* = \sigma^*/\gamma^* = (\sigma_0/\gamma_0)e^{i\delta} = G_0 e^{i\delta} = G_0(\cos \delta + i \sin \delta) \tag{2.10}$$

Thus G^*, the dynamic modulus, is also a complex number, which can be expressed as a sum of real and imaginary components as

$$G^* = G' + iG'' \tag{2.11}$$

where

$$G' = G_0 \cos \delta \quad \text{and} \quad G'' = G_0 \sin \delta$$

It is seen that if stress and strain amplitude and phase angle are measured, G' and G'' can be evaluated. What are these quantities? G' is the component that represents the in-phase relation between stress and strain, i.e., the elastic part. Since the elastic part stores energy, which is recovered in the next half-cycle, it is called the *storage modulus*. G'' represents the 90° out-of-phase relation between stress and strain and represents the viscous or energy-absorbing part. It is therefore called the *loss modulus*.

BEHAVIOR OF POLYMER SOLUTIONS

If an arbitrary magnitude of G^* is calculated as a function of frequency for a single Maxwell element, G' and G'' plots appear as in Fig. 2.11. The shape of the G' curve arises because at low frequencies in a Maxwell system the spring does not stretch, since the dashpot has time to respond. G' increases and eventually saturates because at high frequencies the dashpot has no time to respond and all of the response is elastic. The shape of the G'' curve goes down at high frequencies for the reason already stated—the dashpot does not have time to respond, and hence there is no viscous flow. It

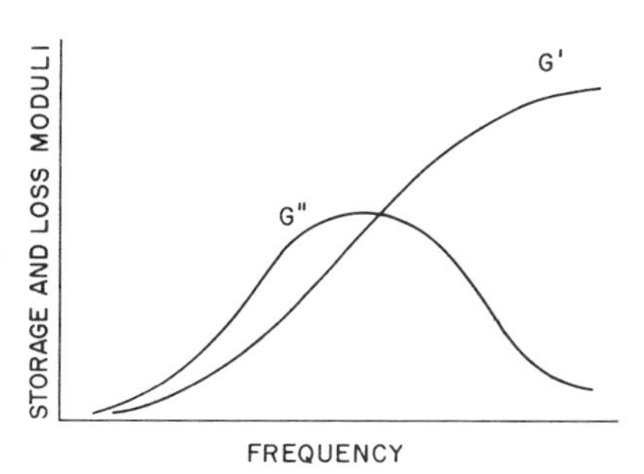

FIG. 2.11 Storage modulus G' (in-phase) and loss modulus G'' (out-of-phase) of a single Maxwell element as a function of frequency.

is also low at low frequencies because, although flow takes place, it is so slow that the rate of energy dissipation is small.

Figure 2.11 illustrates the behavior with respect to frequency of the dynamic moduli of a single Maxwell element. The theoretical response of a system of three Maxwell elements with different constants is shown in Fig. 2.12. These models have been used to assist in the analysis of polymer solutions. The springs correspond to the bending and stretching of molecular segments and the dashpots to the slipping or sliding of molecules or their chain segments past each other.

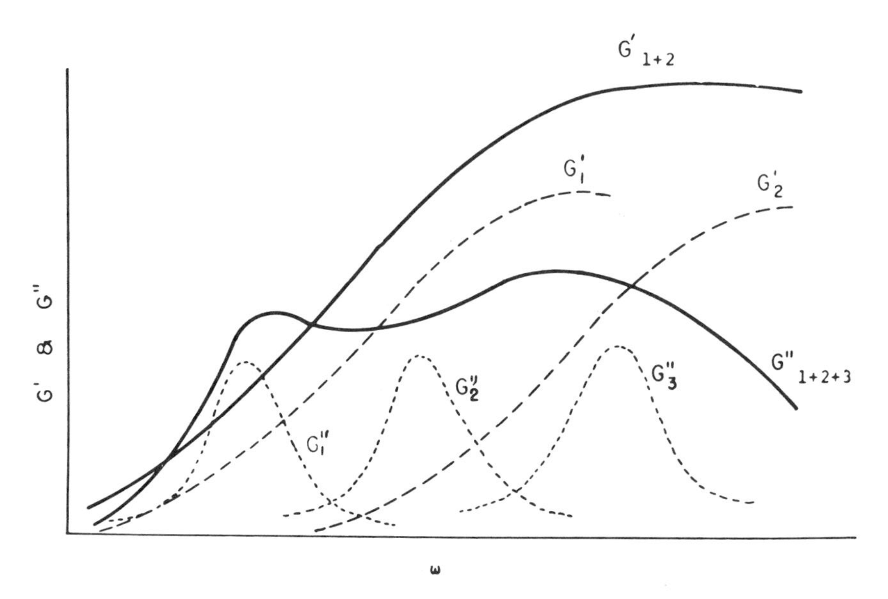

FIG. 2.12 Response of a system of three Maxwell elements. The solid curves are the observed sums of the elements, each with their own elastic and viscosity constants. [From R. J. Lutz, M. Litt and L. W. Chakrin, *in* "Rheology of Biological Systems" (H. L. Gabelnick and M. Litt, eds.), 1973. Courtesy of Charles C. Thomas, Publisher, Springfield, Illinois.]

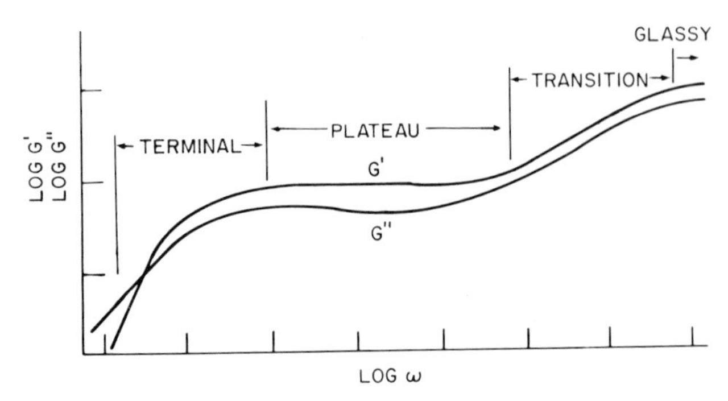

FIG. 2.13 Typical viscoelastic behavior of a high molecular weight polymer solution. [From R. J. Lutz, M. Litt, and L. W. Chakrin, *in* "Rheology of Biological Systems" (H. L. Gabelnick and M. Litt, eds.), 1973. Courtesy of Charles C. Thomas, Publisher, Springfield, Illinois.]

Polymer solutions generally yield curves of G' and G'' as in Fig. 2.13. The interpretation of the four zones are, in general terms, the following.

(1) The terminal zone at low frequencies is the slippage of molecules past each other or the equivalent rearrangement of the "backbones" of the molecules in a viscous-like behavior.

(2) The plateau occurs because of molecular entanglements in which the points of entanglement behave as if they were cross-linked points for which there is no relative motion. Thus the same strain is observed for a given stress regardless of frequency, resulting in a constant modulus.

(3) The transition region is intermediate between the rubberlike character of the plateau and the high glassy region. It involves configurational rearrangements of chains with varying degrees of motion.

(4) The glassy region is at such high frequency that there is no motion of sliding or rotation except some elasticity of the side chains.

Let us now consider how this type of experiment can assist in the understanding of a molecular solution of biological importance. We will briefly review some work on the mucin molecule by Lutz *et al.* (1973). The structure of the mucin molecule is shown in Fig. 2.14. It is basically a polypeptide backbone with short polysaccharide chain side groups. The molecule is slightly acidic and so increasing the pH of the aqueous solution should increase the negative charge on the molecule. At high pH, then, the molecules should be highly charged and therefore repel one another, thereby keeping their distance apart as great as possible. Thus the region of flexibility from entanglements should decrease, resulting in a diminished plateau region, and the terminal viscous region due to molecular sliding should shift to

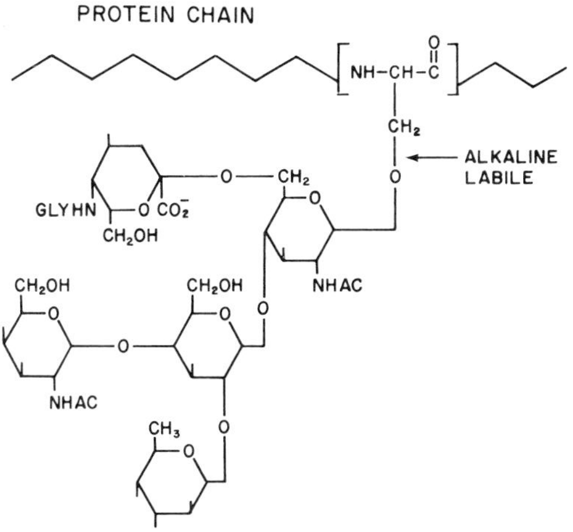

FIG. 2.14 Proposed structure for pig submaxillary mucin. [After Carlson (1968).]

higher frequencies. These phenomena are shown in Fig. 2.15. Each of the charged molecules is surrounded by an electric double layer, which makes its hydrodynamic cross section larger the higher the charge on the molecule. If ions are added to the solution, e.g., NaCl, the thickness of the double layer is decreased and correspondingly the range of interaction of the molecule is decreased. We would therefore expect a smaller plateau for the higher ionic concentration. This is shown in Fig. 2.16. Greater interaction under any charge condition is to be expected if the concentration of mucin is increased, which is also seen in the curves of Fig. 2.16.

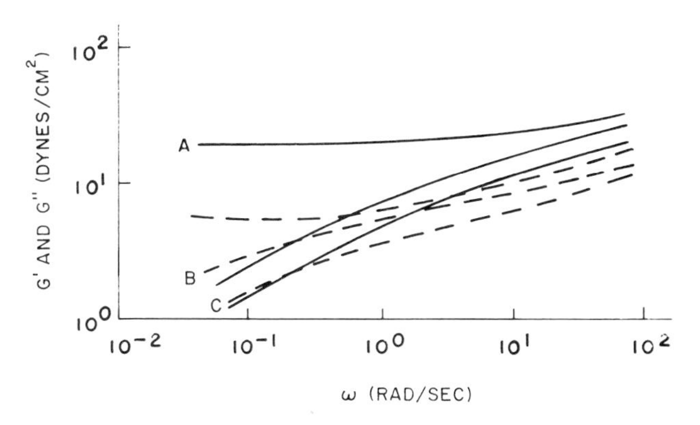

FIG. 2.15 Storage modulus (——), loss modulus (– – –) curves for 2 wt % reconstituted mucus at pH values of A, 5.7; B 6.9; C, 8.4. [Redrawn from Lutz *et al.* (1973).]

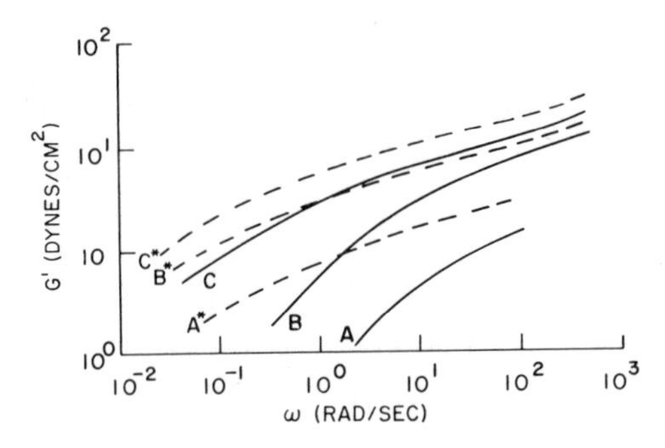

FIG. 2.16 Storage modulus versus frequency for reconstituted mucus. For pH 8.4: ---, 0 M-NaCl; ———, 0.15 M-NaCl; A and A*, 2 wt/ % mucin; B and B*, 3.5 wt/ % mucin; C and C*, 5 wt/ % mucin. [Redrawn from Lutz *et al.* (1973).]

From this example, which is one of many, we see that modern viscoelastic techniques not only can give a description of the mechanical properties of such a material but can contribute useful information on the fundamental properties of the interacting species. Similar studies are taking place on changes in these properties under pathological conditions.

SOME RHEOLOGICAL PROPERTIES OF BLOOD

Blood is composed of red blood cells, *erythrocytes*, in a medium of protein molecules and other cells, the basic fluid being plasma. The volume concentration of erythrocytes, called *hematocrit*, in a healthy individual is about 45%. The erythrocytes are essentially cylindrical disks about 8 μm in diameter with depressions in the centers on both sides. A cross-sectional view is seen in Fig. 2.17, and a photograph of normal and sickled cells is seen in Fig. 2.18. One of the uses of this depressed center is for the storage of excess membrane so that the cell can distort elastically. The capillaries are about 7 μm in diameter and the erythrocytes distort to pass through, as seen in Fig. 2.19. The purpose of this distortion is to present a greater diffusion area for gas exchange between the erythrocyte and the capillary wall.

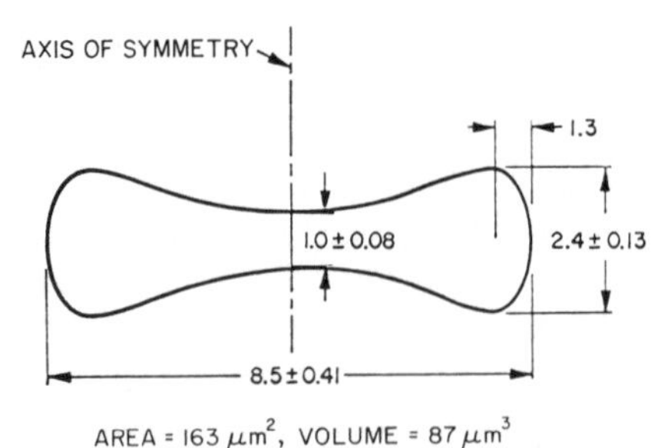

FIG. 2.17 Geometry of human red blood cell by Ponder. [Reprinted from Y. C. Fung, *Fed. Proc.* **25**, 1761 (1966).]

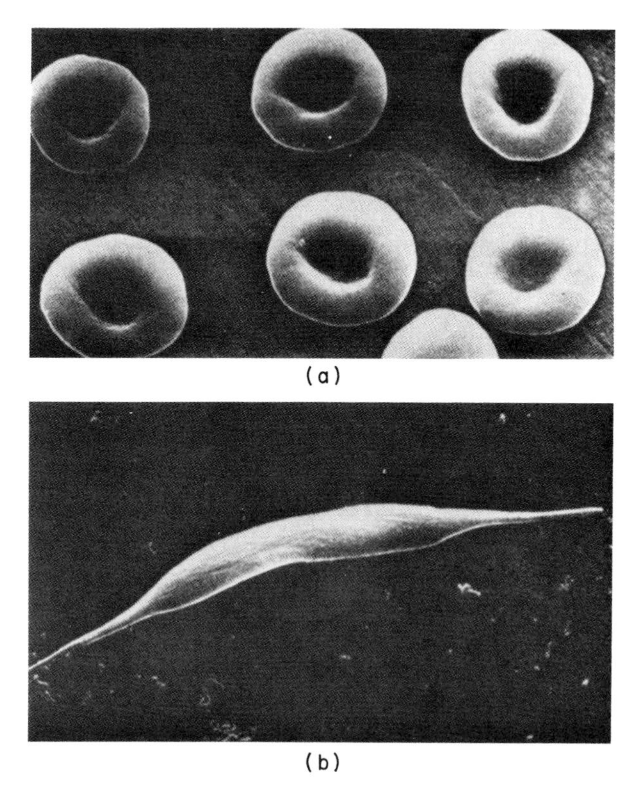

FIG. 2.18 (a) Normal red blood cells. (b) Red blood cell in sickle cell anemia. [Photograph by Baker (1975).]

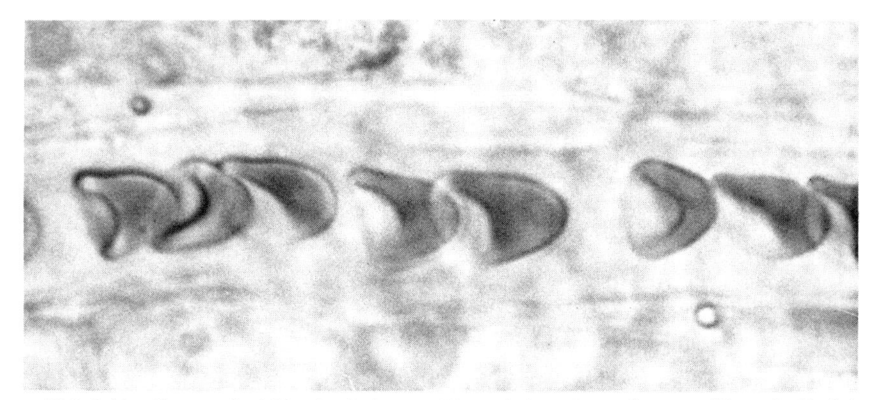

FIG. 2.19 Shapes of red blood cells in a capillary of about 7-μm diameter. [From R. Skalak and P. I. Branemark, *Science* **164**, 717 (1964). Copyright 1964 by the American Association for the Advancement of Science.]

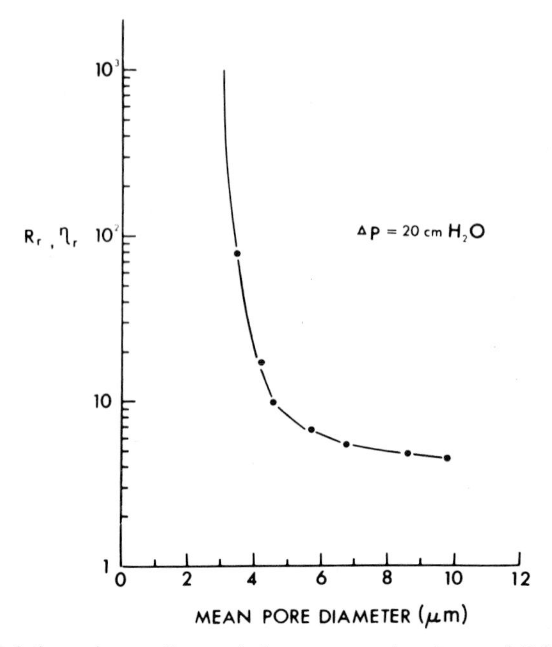

FIG. 2.20 Relative resistance R_r, or relative apparent viscosity η_r, of 40% red blood cells in Ringer solution through polycarbonate sieves of various mean pore diameters. [From Chien *et al.*, *in* "Rheology of Biological Systems" (H. L. Gabelnick and M. Litt, eds.), 1973. Courtesy of Charles C. Thomas, Publisher, Springfield, Illinois.]

In a viscosity measurement the more the cell must deform, the larger is its resistance to flow, resulting therefore in a greater apparent viscosity. This viscosity effect of pore diameter is shown in Fig. 2.20, in which the red blood cells have been removed from the whole blood and placed in Ringer's solution; this solution tends to duplicate the pH and ionic content of the *in vivo* environment. Erythrocytes age, and in about 70 days the membrane hardens. This prevents them from passing through the fine pores of the spleen and they become trapped there and are dissolved. The sickle cell has less membrane elasticity and becomes trapped earlier; the result is a deficiency of erythrocytes, i.e., anemia.

When whole blood is motionless or subject to low shear rates, red blood cells cluster, much like stacks of poker chips. When the shear rate increases these clusters, called *rouleaux*, separate and the cells move separately. A photograph of this is shown in Fig. 2.21. However, because of their high volume concentration, they interfere with each other, and so their flexibility is an important feature of their flow characteristic. The membrane flexibility can be eliminated with aldehyde fixation so that the cells become hardened. The comparison of viscosity versus shear rate of normal and hardened cells

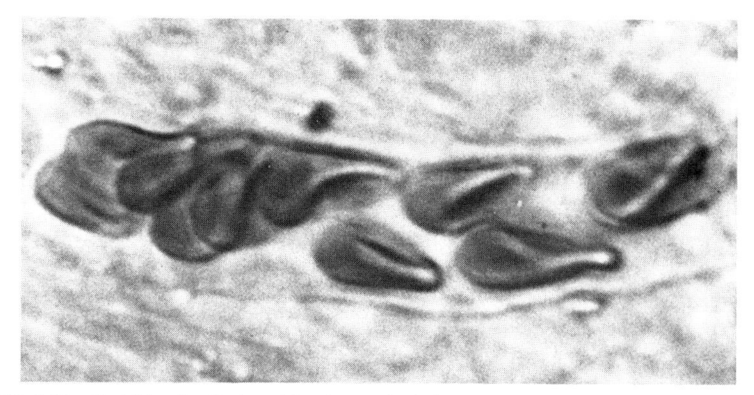

FIG. 2.21 Red blood cells in a blood vessel of about 12-μm diameter illustrating on the left a rouleaux and on the right the separation that occurs with increasing shear rate. [From R. Skalak and P. I. Branemark, *Science* **164**, 717 (1969). Copyright 1964 by the American Association for the Advancement of Science.]

in plasma is shown in Fig. 2.22. It is seen that normal blood in the non-Newtonian classification is pseudoplastic and that hardened cells make a dilatant fluid. Recall that pseudoplastic fluids approach a constant viscosity because of the alignment of their asymmetric particles. Careful observations have shown that the plane of red blood cells approaches an angle of $<20^\circ$ from the flow direction, while hardened cells tend to tumble. Figure 2.23 illustrates results of an orientation measurement.

Blood rheology measurement seems to have great potential as a diagnostic tool. Changes have been reported for conditions of myocardial infarction, coronary occlusion, arteriosclerosis, diabetes, etc. One example of change in

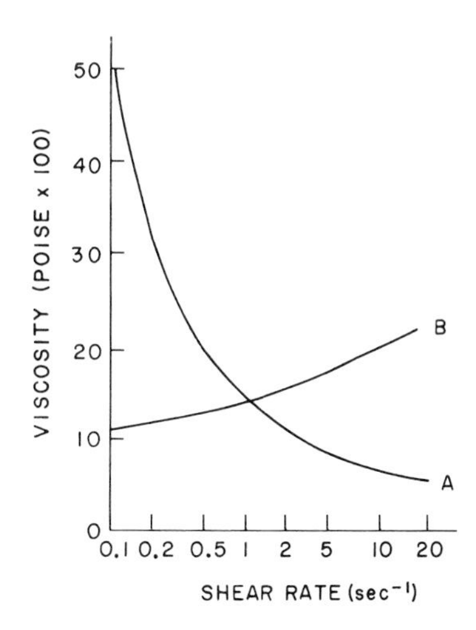

FIG. 2.22 Viscosity versus shear rate for 40% red blood cells in plasma. A, normal; B, hardened. [Redrawn from Wells and Goldstone, *in* "Rheology of Biological Systems" (H. L. Gabelnick and M. Litt, eds.), 1973. Courtesy of Charles C. Thomas, Publisher, Springfield, Illinois.]

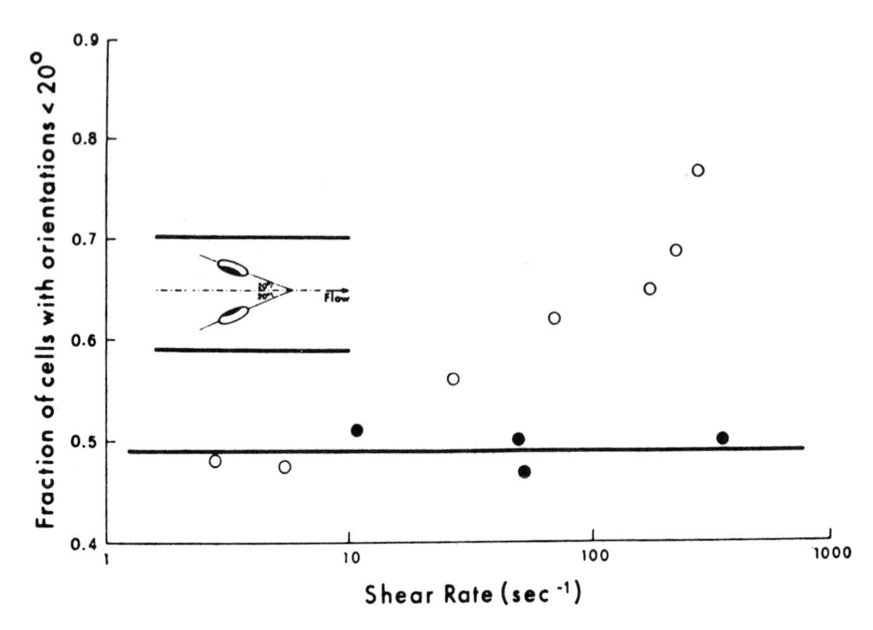

FIG. 2.23 Fraction of red blood cells with orientation of their plane <20° versus shear rate. ○, normal cells; ●, hardened cells; —, theoretical behavior of rigid disks. [From Chien *et al.*, *in* "Rheology of Biological Systems" (H. L. Gabelnick and M. Litt, eds.), 1973. Courtesy of Charles C. Thomas, Publisher, Springfield, Illinois.]

viscosity is illustrated in Fig. 2.24, in which Dintenfass *et al.* (1966) show the general behavior of viscosity versus shear rate of whole blood for groups of normal individuals (lower curve) and those suffering from coronary occlusion, myocardial infarction, and arterial thrombosis (upper curve). Note that the divergence of the two curves is greatest at low velocity gradients. This indicates that it is the characteristics of the aggregation of the red blood cells that have changed the most in pathological conditions. It is not yet known whether these viscosity changes are symptoms or are contributory. Dintenfass (1969) suggests that a vicious circle is established, which is illustrated in Fig. 2.25. In this circle a pathological condition causes a viscosity change, which can then aggravate the condition or create an associated complication.

It was seen in Fig. 2.24 that the lower the shear rate, the greater the divergence of blood viscosity between the mean viscosities of groups of normal and abnormal individuals. A convenient experiment for obtaining very low shear rates in blood involves the erythrocyte sedimentation rate (ESR). A recent experiment by V. Riley (1976) illustrates the usefulness of such a technique not only as a possible diagnostic tool but also as an indicator of

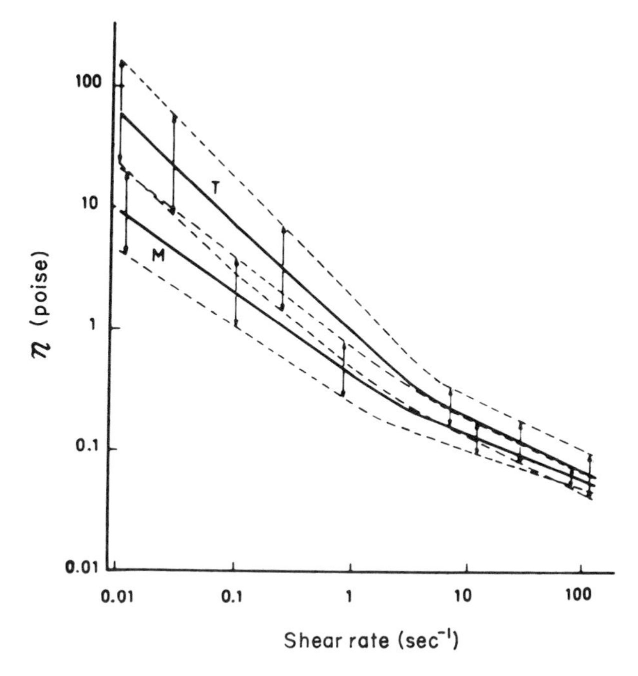

FIG. 2.24 Blood viscosity versus shear rate in normal controls (M) and in patients (T) suffering from coronary occlusion, myocardial infarction, and arterial thrombosis. ——, means; – – –, standard deviations. [From L. Dintenfass, *Am. Heart J.* **77**, 139 (1969) based on data by Dintenfass *et al.* (1966).]

an unknown fraction in the blood which may be the origin of the effect. The erythrocyte sedimentation rate for a large sample of healthy females (over 3000) was measured, and the frequency distribution of the percentage versus rate is shown by the wider shading of Fig. 2.26. This distribution was corrected for hematocrit variations, and the corrected distribution is the narrow area. The frequency distribution, corrected and uncorrected for hematocrit, for patients with early stages of breast malignancies, are shown by the data points. There is a clear separation from the distribution curve for normal women. Clearly, this is statistical, and a given patient may have a sedimentation rate that appears to be normal. Its use as a diagnostic tool is therefore still limited. However, an increase in plasma fibrinogen, as well as foreign proteins, has been found to occur in a significant number of cancer patients. Thus the question remains, whether the sedimentation rate is altered because of specific foreign proteins produced by the malignancy or by the secondary accumulation of fibrinogen. Much work in this area remains to be done.

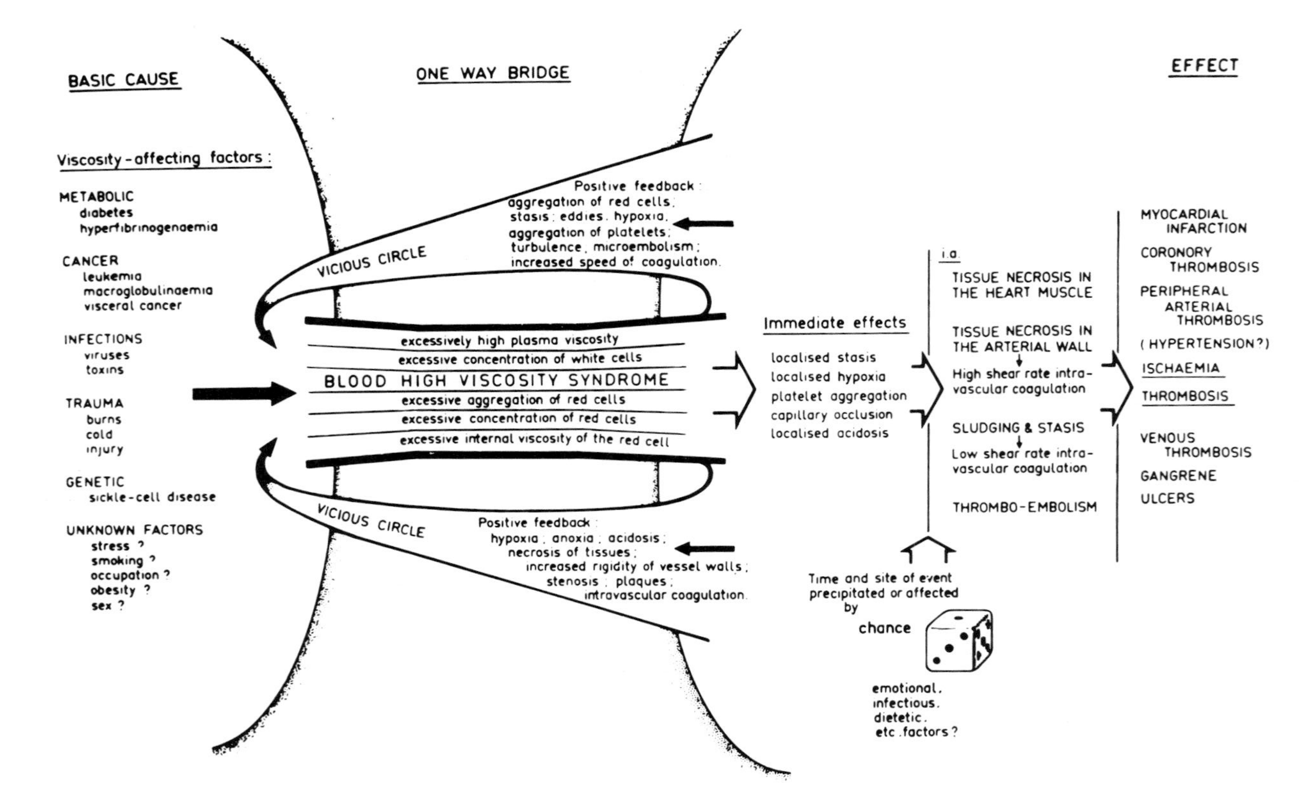

FIG. 2.25 Proposed one-way bridge of the blood high viscosity syndrome which illustrates the role of high blood viscosity in the development of ischemia, infarction, and thrombosis. [From L. Dintenfass, Blood rheology in the pathogenesis of the coronary heart disease, *Am. Heart J.* **17**, 139 (1969).]

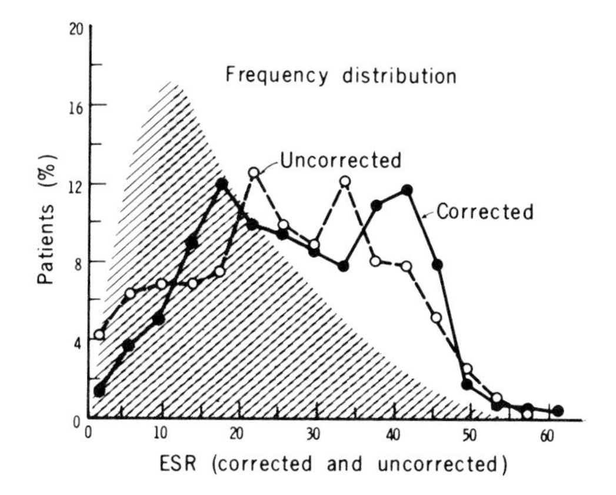

FIG. 2.26 Wider shading is the corrected distribution of erythrocyte sedimentation rate versus percent of normal group. Narrow area is the same data corrected for hematocrit differences. Data points are (○) uncorrected and (●) corrected (for hematocrit differences) distributions of sedimentation rates for a group with early stage breast cancer. [From V. Riley, *Science* **191**, 86 (1976). Copyright 1976 by the American Association for the Advancement of Science.]

An example of another diagnostic test is the resistance of red cell membrane walls to shear stress. In this experiment the percent of killed blood cells, determinable by microscope counting of the number of ruptured membranes, is plotted against shear stress. Figure 2.27 shows an example of this type of experiment. It is seen that dramatic changes in membrane strength are detectable.

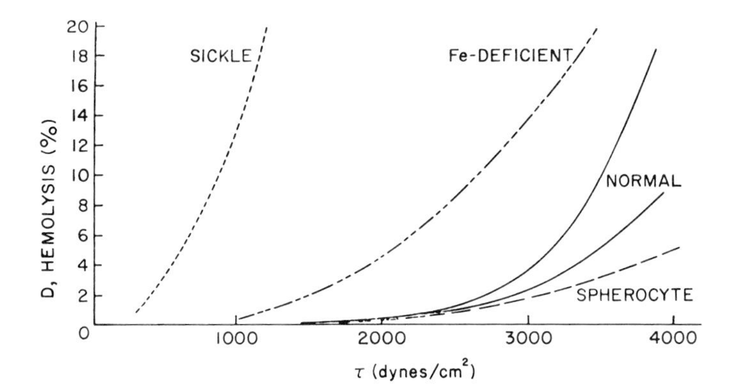

FIG. 2.27 Comparison of shear destruction of normal and abnormal red blood cells. Note that τ is used in the figure to denote stress. [From R. N. MacCallum *et al.*, *in* "Rheology of Biological Systems" (H. L. Gabelnick and M. Litt, eds.), 1973. Courtesy of Charles C. Thomas, Publisher, Springfield, Illinois.]

The study of rheology of blood continues on essentially two levels: the phenomenological, i.e., clinical observation; and the hydrodynamic, i.e., mathematical interpretation. In this section we have described only the former because, although intensive theoretical work is going on, the mathematical treatment of the hydrodynamics of particles in fluids is beyond the scope of this book. The reader interested in a theoretical discussion of the viscosity of suspensions might start with the papers by Vand (1948).

REFERENCES

Alfrey, T., Jr. (1948). "Mechanical Behavior of High Polymers." Wiley (Interscience), New York.

Baker, R. F. (1975). *Chem. Eng. News*, Jan. 6, p. 16.

Bayliss, L. E. (1952). Rheology of blood and lymph in deformation and flow, *in* "Biological Systems" (A. Frey-Wyssling, ed.). North-Holland Publ., Amsterdam.

Bayliss, L. E. (1962). The rheology of blood, *in* "Handbook of Physiology: Circulation" (W. F. Hamilton, ed.), Sect. 2, Vol. 1. American Physiological Soc., Washington, D.C.

Carlson, O. (1968). *Proc. Int. Conf. Cystic Fibrosis of the Pancreas* (*Mucoviscidosis*), 4*th*, Part II, p. 304. Karger, Basel, Switzerland.

Chien, S., Usami, S., Jan, K. M., and Skalak, R. (1973). Macrorheological and microrheological correlation of blood flow in the macrocirculation and microcirculation, *in* "Rheology of Biological Systems" (H. L. Gabelnick and M. Litt, eds.). Thomas, Springfield, Illinois.

Cokelet, G. R. (1972). The rheology of human blood, *in* "Biomechanics" (Y. G. Fung, N. Perrane, and M. Anliker, eds.). Prentice Hall, Englewood Cliffs, New Jersey.

Conway, B. E., and Dobry-Duclaus, A. (1960). Viscosity of suspensions of electrically charged particles and solutions of polymeric electrolytes, *in* "Rheology: Theory and Applications" (F. R. Eirich, ed.), Vol. III. Academic Press, New York.

Davis, S. S. (1973). Rheological examination of sputum and saliva and the effect of drugs, *in* "Rheology of Biological Systems" (H. L. Gabelnick and M. Litt, eds.). Thomas, Springfield, Illinois.

Dintenfass, L. (1969). Blood rheology in pathogenesis of the coronary heart diseases, *Am. Heart J.* **77**, 139.

Dintenfass, L., Julian, D. G., and Miller, G. (1966). Viscosity of blood in normal subjects and in patients suffering from coronary occlusion and arterial thrombosis, *Am. Heart J.* **71**, 587.

Eirich, F. R. (ed.) (1960). "Rheology: Theory and Applications." Academic Press, New York.

Fung, Y. C. (1966). Theoretical consideration of the elasticity of red blood cells and small blood vessels, *Fed. Proc.* **25**, 1761.

Gabelnick, H. L., and Litt, M. (1973). "Rheology of Biological Systems." Thomas, Springfield, Illinois.

Goldsmith, H. L. (1971). Deformation of human red cells in tube flow, *Biorheology* **7**, 235.

Gregerson, M. I., Bryant, C. A., Hammerle, W. E., Usami, S., and Chien, S. (1967). Flow characteristics of human erythrocytes through polycarbonate sieves, *Science* **157**, 825.

Ham, C. D., and Barnett, B. (1973). Measurements of the rheological properties of biological fluids, *in* "Rheology of Biological Systems" (H. L. Gabelnick and M. Litt, eds.). Thomas, Springfield, Illinois.

Lutz, R. J., Litt, M., and Chakrin, L. W. (1973). Physical–chemical factors in mucus rheology, *in* "Rheology of Biological Systems" (H. L. Gabelnick and M. Litt, eds.). Thomas, Springfield, Illinois.

MacCallum, R. N., O'Bannon, W., Hellums, J. D., Alfrey, C. P., and Lynch, W. C. (1973). Viscometric instruments for studies of red blood cell damage, *in* "Rheology of Biological Systems" (H. L. Gabelnick and M. Litt, eds.). Thomas, Springfield, Illinois.

Reiner, M. (1960). "Deformation, Strain and Flow." Lewis, London.

Riley, V. (1976). Breast cancer patients: Substance in blood causing acceleration of erythrocyte sedimentation rate, *Science* **191**, 86.

Rouse, P. E., Jr. (1953). A Theory of linear viscoelastic properties of dilute solutions of coiling polymers, *J. Chem. Phys.* **21**, 1272.

Schmid-Schonbern, H., Wells, R., and Goldstone, J. (1969). Influence of deformability of human red cells upon blood viscosity, *Circ. Res.* **25**, 131.

Scott-Blair, G. W. (1949). "A General Survey of Applied Rheology," 2nd ed. Pitman, London.

Skalak, R., and Branemark, P. I. (1969). Deformation of red blood cells in capillaries, *Science* **164**, 717.

Vand, V. (1948). Viscosity of solutions and suspensions, I, II, *J. Phys. Colloid Chem.* **52**, 277, 300.

Wells, R. (1970). Syndromes of hyperviscosity, *N. Engl. J. Med.* **283**, 183.

Wells, R., and Goldstone, J. (1973). Rheology of the red cell and capillary blood flow, *in* "Rheology of Biological Systems" (H. L. Gabelnick and M. Litt, eds.). Thomas, Springfield, Illinois.

Wilkinson, W. L. (1960). "Non-Newtonian Fluids." Pergamon, Oxford.

CHAPTER 3

The Nerve Impulse: Action Potential and Transmission

INTRODUCTION

The organization of the nervous system is extremely complex. The system is composed of nerve cells called *neurons*, but these have types of almost every imaginable extreme, and within a type it has been said that no two neurons are exactly alike. Neurons, however, do have similar parts. Like all other cells of the body the neuron is bounded by a *plasma membrane*, about 80 Å thick, although more detailed studies have indicated that the membrane consists of three different layers. The cell body or *soma* is the part that contains the *nucleus* of the cell, and both the soma and the nucleus vary in size with the size of the neuron; the soma may be from 6 to 100 μm in diameter. The space between the nucleus and the membrane contains a material called *cytoplasm*, which varies in consistency from a gel near the membrane to a liquid near the nucleus and contains small structures and particles. Within the cytoplasm surrounding the nucleus is a network of very small tubes called the *endoplasmic reticulum* and many small particles called *ribosomes*, which are the principal sites of protein synthesis. Very small units in the cells are present, called *mitochondria*. These may number from a few hundred to many thousand and come in a variety of shapes and sizes, the sizes ranging from tenths of microns to several microns. It is believed that when oxygen and nutrients come in contact with enzymes in a mitochondrion they combine to form water and carbon dioxide. The liberated energy is used to form *adenosine triphosphate*, ATP. The ATP diffuses through the cell, releasing its stored energy when needed. Other cellular structures have been identified but their functions are beyond the scope of this chapter.

Some neurons have a signal transmitting tube extending from the soma, called an *axon*. This is filled with cytoplasm and is surrounded by the membrane. Some axons have another type of cell tightly surrounding the membrane, called *Schwann cells*, and this encasement is called *myelin*. Axons that are so myelinated have regular gaps in the myelin sheath; the gaps are called *nodes of Ranvier* and are about 1 μm wide. We shall see later how these gaps play a special role in rapid impulse transmission. Axons may divide into smaller branches to distribute signals to more than one terminal. The terminals, or signal junctions, have the general name of *synapses*. These have a variety of shapes, such as knobs, buttons, end-feet, claws, and even more elaborate mossy types. Two terminals face one another and the transmission of a signal between them is chemical in nature. Current research is most active in the synapse area because the signal transmission along the axon is believed to be reasonably well understood. We will confine the discussion of this chapter to the research that led to the present model of nerve impulse transmission along a single axon.

The detailed mechanism by which a nerve axon transmits an impulse has been studied for many decades. It has long been known to be electrical in nature, but when a nerve is examined for electrical transmission characteristics it is found wanting. The obvious connection is to the underwater cable studied by Lord Kelvin in 1856. Instead of a wire there is an electrolyte in the interior, with a relatively high resistivity of about 50 Ω cm. The insulation of the membrane at the surface of the axon has a capacitance of about 1 μf/cm^2 and thus has a large ability to store electrical energy. However, the insulation is very leaky, with a resistance of only a few thousand Ω cm. If the axon is considered as a passive transmission cable with these constants, the losses are very great; its core and surface leakage are many orders of magnitude greater than that of commercial cable, and its sheath capacity about 10^6 times larger. A weak signal applied to an axon fades out within a millimeter. This type of calculation is based on the Kelvin equation, which will be developed later in the chapter. If a small electrical signal is imparted to an isolated nerve fiber it does indeed behave in this manner. However, if a signal of about 100 mV is imparted rapidly across the membrane with an internal electrode (Fig 3.1) a current density of only 10^{-8} C/cm^2 sec will cause a large signal (Fig. 3.2). This signal, called the *action*

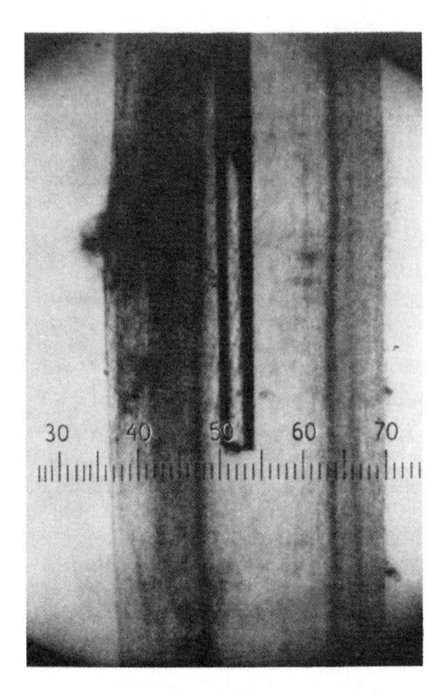

FIG. 3.1 Photomicrograph of silver electrode coated with silver chloride inside a squid axon. 1 scale division = 33 μm. [From Hodgkin and Huxley (1939).]

potential, is propagated along the nerve fiber at constant amplitude and velocity. Following this impulse there is a time span, called the *refractory period*, during which the nerve may not carry another signal. This time is the order of milliseconds. As we review the subsequent studies of the detailed mechanism, we will understand why this is also called the *repolarization period*. In a resting nerve the inside is negative with respect to the external medium. This is called the *resting potential.*

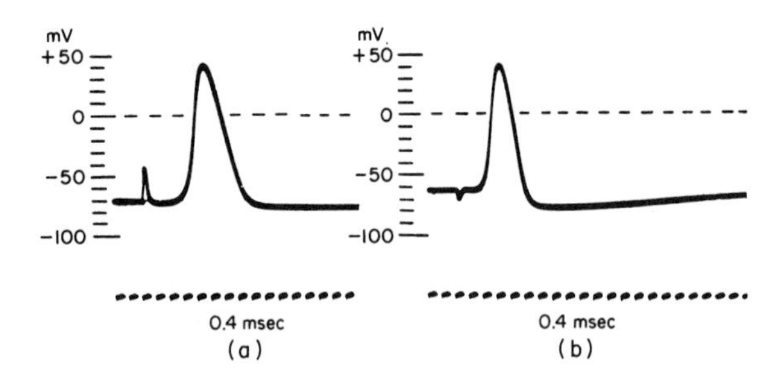

FIG. 3.2 Action potential recorded between inside and outside of (a) intact and (b) isolated squid axon. The vertical scale indicates the potential of the internal electrode in millivolts, the sea water outside being taken at zero potential. [Unpublished work by A. L. Hodgkin and R. D. Keynes quoted by Hodgkin (1958).]

In order to understand the processes that go on in the nerve membrane we must first develop the laws of diffusion and membrane potential.

FICK'S LAWS OF DIFFUSION

If there is a higher concentration of a movable substance in one region of space than another, thermal motion will cause the substance to migrate to the deficient region. Such a flow of material is called *diffusion*. Consider a system of uncharged atoms or molecules in a one-dimensional tube, as shown in Fig. 3.3. If the jump frequency of atoms is equal in both directions there will be a net flow away from the concentrated region because more atoms will move away per unit time than will move back. Such a flow is called *flux*, which is the number of particles that move across a unit cross-sectional area per unit time; flux therefore has dimensions of $L^{-2}T^{-1}$. (Note that in dimensional notation, L denotes length and T time.) It is more convenient to count the particles in terms of moles instead of individual particles and we will use *mole flux* with symbol n later on. Mole flux has the dimensions $ML^{-2}T^{-1}$. (In dimensional notation M denotes mass.)

In Fig. 3.3 the cross-sectional area A has dimensions L^2 and the dimensions of concentration C is mass per unit volume or ML^{-3}. The rate of mass transfer $\Delta m/\Delta t$ from x_1 in the $+x$ direction is proportional to the area A and the concentration C_1 at position x_1 or

$$\left.\frac{\Delta m}{\Delta t}\right|_{+x} = \beta AC_1,$$

where β is the proportionality constant. For the negative direction of flow

$$\left.\frac{\Delta m}{\Delta t}\right|_{-x} = \beta AC_2$$

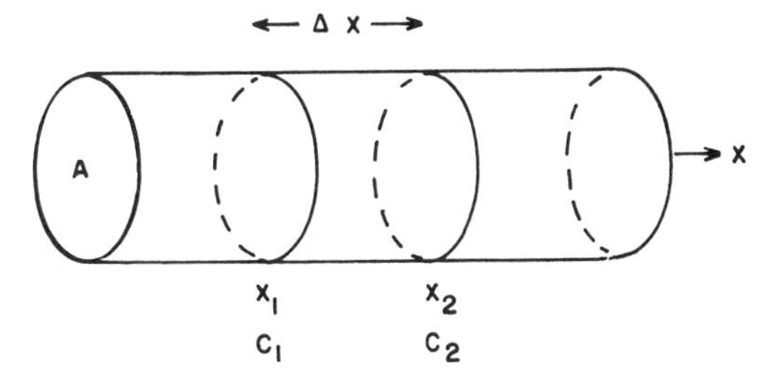

FIG. 3.3 Cylinder of cross section A.

and the net mass transfer in the $+x$ direction is therefore

$$\left.\frac{\Delta m}{\Delta t}\right|_{\text{net}} = \left.\frac{\Delta m}{\Delta t}\right|_{+x} - \left.\frac{\Delta m}{\Delta t}\right|_{-x} = \beta A(C_1 - C_2)$$

We can express the concentration change with distance as $\Delta C/\Delta x$, called the *concentration gradient*, and write the net mass flow as

$$\left.\frac{\Delta m}{\Delta t}\right|_{\text{net}} = \beta A \frac{\Delta C}{\Delta x}$$

As Δx and Δt approach zero this becomes

$$dm/dt = (\beta/dx)A\, dC$$

or

$$dm/dt = PA\, dC \tag{3.1}$$

where β/dx is called the *permeability* P. Since A has dimensions L^2 and C dimensions ML^{-3} we may write

$$(M/T) = P(L^2)(M/L^3)$$

and thus permeability has dimensions LT^{-1}, i.e., dimensions of velocity. The diffusion equation, Eq. (3.1), is called Fick's first law and is often written in terms of the concentration gradient dC/dx:

$$dm/dt = DA(dC/dx) \tag{3.2}$$

The constant D is called the *diffusion coefficient*, and since

$$D/dx = P \tag{3.3}$$

it has dimensions of $LT^{-1}L = L^2T^{-1}$. The diffusion coefficient contains such factors as the jump rate (usually a Boltzmann factor), entropy change, number of atoms involved in a single jump, etc.

In Fig. 3.3, if $C_1 > C_2$ there is a negative concentration gradient, i.e., the concentration decreases with increasing x. If we express the time rate of mass flow per unit area as the mass current J, so that

$$J = (1/A)(dm/dt)$$

we use the negative gradient in substituting Eq. (3.2) and write

$$J = -D(dC/dx) \tag{3.4}$$

as another form of Fick's first law.

Equation (3.2) can be expressed in an alternative form in the following way. From Eq. (3.2) write the net mass transfer per unit time, i.e., J crossing plane 1 of Fig. 3.3 into the volume between planes 1 and 2, as

$$\left.\frac{1}{A}\frac{dm}{dt}\right|_{x_1\ (\text{net})} = D\frac{dC_1}{dx}$$

and the net J leaving the volume across plane 2 as

$$\left.\frac{1}{A}\frac{dm}{dt}\right|_{x_2\ (\text{net})} = D\frac{dC_2}{dx}$$

Thus the net flux into the volume between x_1 and x_2 is

$$(1/A)(dm/dt) = D[(dC_1/dx) - (dC_2/dx)]$$

Take the derivative of both sides with respect to x, giving

$$\frac{1}{A}\frac{dm}{dxdt} = D\frac{d}{dx}\left(\frac{dC_1 - dC_2}{dx}\right)$$

Since, by definition of concentration as mass per unit volume,

$$\frac{dm}{Adx} \equiv dC$$

we may write

$$dC/dt = D(d^2C/dx^2) \tag{3.5}$$

Equation (3.5) is known as Fick's second law of diffusion.

DIFFUSION OF IONS

If we have n moles of ions per unit volume (of mass m as required by the form of the diffusion equation) with valence Z as the diffusing species, and F is the Faraday constant (see Section A.6 of Appendix A), then FZ of current is moved per mole of the ion, and Fick's first law (Eq. (3.4)) may be written as current density due to diffusion arising from a concentration gradient:

$$J_{\text{diff}} = -FZD(\partial n/\partial x) \tag{3.6}$$

where the minus sign indicates that the current density for positive ions is the opposite of the concentration gradient. Note that Z is the ion valence and F is the Faraday constant: $F = 96{,}500$ C/mole/valence. It is also now necessary to use ∂x (the *partial* derivative) instead of dx because n may be a function of time as well as space.

Since we are dealing with ions, an electric field in addition to the concentration gradient can play a role in the diffusion rate. Consider a situation in which there is a field but no concentration gradient. The field will accelerate an ion but collisions will decelerate it. Thus there will be an average drift velocity v and the current density due to the field will be equal to the charge times the velocity:

$$J_{\text{field}} = FZnv \tag{3.7}$$

Mobility, whose symbol is u, is defined as velocity v per unit electric field E for a unit valence, or

$$u = v/(ZE) \tag{3.8}$$

To eliminate the unknown drift velocity substitute v from Eq. (3.8) into Eq. (3.7) and obtain

$$J_{\text{field}} = FZ^2unE \tag{3.9}$$

Note that v and E have the same sign in Eq. (3.8), and hence u is always positive. Also note that the diffusion current density (Eq. (3.6)) contains Z and therefore changes sign when the sign of the ion changes, but that the field current density (Eq. (3.9)) does not change sign since it contains Z^2. However, a negative charge moving in a given direction gives the same current as a positive ion moving in the opposite direction. When a concentration gradient and an electric field are both present the total current density is the sum of the two,

$$J = FZ^2unE - FZD(\partial n/\partial x) \tag{3.10}$$

At this point we assume that since the same type of jump process must occur for the ions, whether they move by concentration gradient or by electric field, the diffusion coefficient D is proportional to the mobility u. Note that D/u has the dimensions of potential:

$$\frac{D}{u} = \frac{L^2T^{-1}}{qTM^{-1}} = \frac{M(L/T)^2}{q} = \frac{\text{energy}}{\text{charge}} = \text{potential}$$

We further assume that the energy distribution of the ions follows a Maxwell–Boltzmann distribution. In such a distribution, if there are N_0 particles, the fraction of them $N(U_1)$ having energy U_1 above the lowest possible, or ground state, energy is given by

$$N(U_1)/N_0 = e^{-U_1/RT}$$

where R is the gas constant and T the absolute temperature.

To apply this to the present case note that if V is the electric potential (energy per coulomb of charge) then

$$FZV = \frac{\text{coulombs}}{\text{mole-valence}} \times \text{valence} \times \frac{\text{energy}}{\text{coulomb}} = \frac{\text{energy}}{\text{mole}}$$

For simplicity define the fraction of ions with energy FZV as $N(FZV)/N \equiv n(FZV)$; we may therefore write

$$n(FZV) = e^{-FZV/RT} \tag{3.11}$$

In steady state there is no net current flow and therefore $J = 0$ and Eq. (3.10) becomes

$$ZunE = D(\partial n/\partial x) \tag{3.10'}$$

Note that

$$\partial n/\partial x = (\partial n/\partial V)/(\partial V/\partial x)$$

and therefore taking the derivative of Eq. (3.11) with respect to V gives

$$\partial n/\partial x = (-FZ/RT)(\partial V/\partial x)e^{-FZV/RT}$$

Upon substituting this expression into the right-hand side of Eq. (3.10') and Eq. (3.11) into the left-hand side we obtain

$$ZuEe^{-FZV/RT} = D(-FZ/RT)(\partial V/\partial x)e^{-FZV/RT}$$

or

$$uE = D(F/RT)(-\partial V/\partial x)$$

Recall that by definition the electric field E is related to the electric potential V by $E = -\partial V/\partial x$ and therefore

$$uE = D(F/RT)E$$

or

$$D = (RT/F)u \tag{3.12}$$

This equation is sometimes called *Einstein's relation.* Upon substituting this for D in Eq. (3.10) we obtain

$$J = ZuRT[(F/RT)ZnE - (\partial n/\partial x)] \tag{3.10''}$$

which is known as the *Nernst–Planck equation.* This equation holds for each type of ion and, if more than one is present, the total current is the sum of the individual currents.

DONNAN EQUILIBRIUM

At this point we may obtain a physical understanding of the origin of membrane resting potentials. Suppose that we have a membrane which separates two ionic solutions, one NaCl and the other NaP, where P^- is a large protein ion which cannot pass through the membrane, whereas the Na^+ and Cl^- ions can pass through the membrane. If C_1 is the concentration of NaCl and C_2 the concentration of NaP, initially the system will look like this:

Outside			Inside	
Na^+	Cl^-	⋮	Na^+	P^-
C_1	C_1	⋮	C_2	C_2

Cl^- will tend to diffuse from the outside to the inside and Na^+ will initially go with it to maintain electrical neutrality. At equilibrium a certain amount X of Na^+ and Cl^- will have passed through the membrane so that the system will look like this:

Na^+	Cl^-	⋮	Na^+	P^-	Cl^-
$C_1 - X$	$C_1 - X$	⋮	$C_2 + X$	C_2	X

Note that X is much smaller than C_1 or C_2 and that there is electrical neutrality in each compartment. At equilibrium, since Na^+ and Cl^- are the moving species, there is no change in the Gibbs free energy when a few ions are moved from one side of the membrane to the other. Therefore the chemical potentials (see Section A.4) of NaCl on both sides of the membrane must be equal. Thus, for dilute solutions Eq. (A.26), derived in Section A.5, is

$$V = (RT/F)\ln(C_2/C_1) \tag{A.26}$$

This is known as the *Nernst equation*, and can be written in terms of the ionic concentrations; for Na^+, $Z = +1$; and for Cl^-, $Z = -1$.

If "i" and "o" stand for inside and outside, respectively, Eq. (A.26) may be written

$$RT\ln(Na_i^+/Na_o^+) = FV \tag{3.13}$$

$$RT\ln(Cl_i^-/Cl_o^-) = -FV$$

where V is the potential difference between the inside and the outside of the membrane. Dividing Eqs. (3.13) and rearranging gives

$$Na_i^+/Na_o^+ = Cl_o^-/Cl_i^- \tag{3.14}$$

which is called a *Donnan equilibrium*.

The origin of the potential of Eq. (3.13) across the membrane is a result of the ions trying to diffuse away from the high concentration regions across the membrane but being prevented by their attractive charge to each other. The Na^+, in excess concentration on the right side of the membrane, will tend to diffuse to the left, while the Cl^- on the left will tend to diffuse to the right. Such diffusion cannot take place since a complete separation of the ions would take an enormous amount of energy. However, a slight charge separation at the membrane can take place to balance the concentration gradient. Thus the entire fluid inside the membrane has a slightly negative potential with respect to the outside fluid. This is known as the *resting potential*, i.e., no current is flowing and the magnitude in volts V is given in Eq. (A.28) as

$$V = 61 \times 10^{-3} \log(C_2/C_1) \tag{3.13'}$$

where the sign of the voltage depends on the choice of C_2 and C_1, as indicated in Eq. (3.13). Thus, knowledge of the relative concentrations inside and outside of a membrane of an ion to which the membrane is permeable will immediately yield the equilibrium or resting potential.

GOLDMAN POTENTIAL

For ions moving in a concentration gradient with or against an electric field, the ion fluxes are not proportional to the concentration gradients alone but must be multiplied by a factor which, due to the electric field, lowers or raises an ion's possibility of passage.

Goldman (1943) assumed that the field E across the membrane is a constant, i.e., $E = -V/W$ (instead of $E = -\partial V/\partial x$), where W is the width of the membrane. This is the same expression as that for a parallel plate capacitor. In this case Eq. (3.10″) has a simple solution. Note first that Goldman's assumption has been put on firmer ground. It has been shown by Arndt *et al.* (1970) that the constant field solution is an exact solution provided that (1) there is electroneutrality on each side of the membrane, and (2) the total numbers of ions of the same valence on both sides are equal. On this basis, since Na^+, K^+, and Cl^- will be shown to be the important ions in nerve membranes, the Goldman assumption is valid.

We may write Eq. (3.10″) for the ions of interest as

$$\begin{aligned} J_{Na} &= -u_{Na}(FV/W)n_{Na} - u_{Na}RT(\partial n_{Na}/\partial x) \\ J_K &= -u_K(FV/W)n_K - u_K RT(\partial n_K/\partial x) \\ J_{Cl} &= -u_{Cl}(FV/W)n_{Cl} + u_{Cl}RT(\partial n_{Cl}/\partial x) \end{aligned} \tag{3.15}$$

These equations are simple first-order linear differential equations. If we write for the first one

$$\alpha = -RTu_{\mathrm{Na}} \quad \text{and} \quad \beta = -u_{\mathrm{Na}}FV/W$$

we may write each equation as

$$(dn/dx) + (\beta/\alpha)n = J/\alpha$$

This equation is of the form $dy/dx + Py = Q$ which is readily solved by the integrating factor method (Appendix D.3) to give

$$n = e^{-(\beta/\alpha)x}(J/\alpha)(\alpha/\beta)e^{(\beta/\alpha)x} + Ce^{-(\beta/\alpha)x}, \qquad n = (J/\beta) + Ce^{-(\beta/\alpha)x}$$

where C is a constant of integration. With the following boundary conditions

$$x = 0 \quad \text{(inside)} \qquad n = n_{\mathrm{i}}$$

$$x = W \quad \text{(outside)} \qquad n = n_{\mathrm{o}}$$

and upon rearranging we obtain for the Na^+ current

$$J_{\mathrm{Na}} = -u_{\mathrm{Na}}\frac{FV}{W}\frac{n_{\mathrm{Na_o}} - n_{\mathrm{Na_i}}e^{-FV/RT}}{1 - e^{-FV/RT}} \tag{3.16}$$

Let us now express this current in terms of a membrane permeability. Recall from our definition of permeability that it is

$$P = D/dx \tag{3.3}$$

and that the diffusion coefficient in the free solution is

$$D = (RT/F)u \tag{3.12}$$

Combining these we obtain $u = FWP/RT$, since $dx = W$. We can thereby eliminate u from Eq. (3.16) to obtain

$$J_{\mathrm{Na}} = -\frac{F^2V}{RT}P\frac{n_{\mathrm{Na_o}} - n_{\mathrm{Na_i}}e^{-FV/RT}}{1 - e^{-FV/RT}} \tag{3.16'}$$

This equation expresses the relation between the current through the membrane and the applied field. If the current is zero, Eq. (3.16′) reduces to the resting Nernst potential of Eq. (3.13).

NERVE RESTING POTENTIAL

The squid *Loligo forbesi* has a giant nerve fiber about 1 mm in diameter, about 50 times larger than a comparable nerve in man. When isolated from the animal the fiber continues to conduct electrical impulses for several

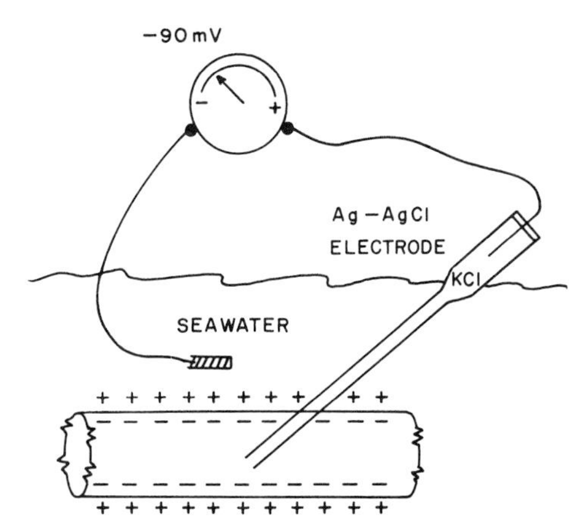

FIG. 3.4 Measurement of the membrane potential of a nerve fiber using a microelectrode. [Redrawn from Guyton (1971).]

hours. The fibers are large enough to permit the insertion of an electrode wire, which can be used to control internal potential, or a fine glass capillary filled with an electrolyte, which can be used either as a voltage probe or as a pipette to add or extract fluid (see Figs. 3.1 and 3.4). Analysis of the internal fluid for ion concentrations compared to seawater and blood are shown in Table 3.1. Note that there are other organic components but, since they are not important to our present purposes, they will not be discussed. The only anion listed is Cl^-, but charge balance is also maintained by organic anions.

Examination of Table 3.1 indicates that the resting membrane is permeable to potassium ions but not to sodium ions. If we make an immediate crude calculation from the Nernst potential for the nerve due to potassium ion concentration difference between the nerve interior and blood, from Eq. (3.13), using Eq. (3.13′),

$$V = (RT/F)\ln(K_o^+/K_i^+) = -61 \times 10^{-3} \log(400/20)$$

TABLE 3.1

	Ion concentration (mmole/kg-H_2O)		
Substance	Nerve interior	Blood	Seawater
H_2O	865	870	966
K	400	20	10
Na	50	440	460
Cl	40	560	540
Cu	0.4	10	10

we obtain $V = -79$ mV, with the outside positive with respect to the inside.

In the squid nerve (axon) the normal resting potential is -70 mV against the external fluid, which is reasonably close to that against seawater from the Nernst potential. Correction factors derived by the refinement of the resting potential and the inclusion of relative transport rate of ions across the membrane, which includes the permeabilities discussed in the Goldman potential, have yielded values even closer to the true potentials.

ACTION POTENTIAL

Two measurements substantiate the model of the activation of nerve impulse. First, radioactive tracer measurements have shown that there is continuous transport of Na^+ ions across the membrane in both directions. This indicates that there is not a static impermeability of Na^+ ions; rather, there is an active transport process, which, at the expense of metabolic energy, pumps the Na^+ ions from the inside. The second observation is that upon excitation of a portion of the nerve the potential in the interior suddenly reverses from the resting value of -70 mV to $+40$ mV with respect to the outside, as shown in Fig. 3.2, followed by a refractory or recovery period.

The Goldman potential was used to unravel the series of events that take place. Returning to Eq. (3.16′), the total current is the sum of the ionic currents. Define the inside concentrations for sodium, potassium, and chlorine, as Na_i, K_i, and Cl_i, respectively, and correspondingly for the outside concentrations, Na_o, K_o, and Cl_o. Also, let the permeabilities for each be P_{Na}, P_K, and P_{Cl}. We may write the currents due to each ion from Eq. (3.16′), (note the opposite order for the negative Cl^- ion):

$$J_K = -\frac{F^2V}{RT} P_K \left(\frac{K_o - K_i e^{-FV/RT}}{1 - e^{-FV/RT}} \right)$$

$$J_{Na} = -\frac{F^2V}{RT} P_{Na} \left(\frac{Na_o - Na_i e^{-FV/RT}}{1 - e^{-FV/RT}} \right)$$

$$J_{Cl} = -\frac{F^2V}{RT} P_{Cl} \left(\frac{Cl_i - Cl_o e^{-FV/RT}}{1 - e^{-FV/RT}} \right)$$

The total current is

$$\begin{aligned} J_{tot} &= J_K + J_{Na} + J_{Cl} \\ &= -\frac{F^2V}{RT(1 - e^{-FV/RT})} [(P_K K_i + P_{Na} Na_i + P_{Cl} Cl_o) \\ &\quad - (P_K K_o + P_{Na} Na_o + P_{Cl} Cl_i) e^{-FV/RT}] \end{aligned} \tag{3.17}$$

When $J_{tot} = 0$ in equilibrium, Eq. (3.17) can be written as the resting potential V_r where

$$V_r = \frac{RT}{F} \ln\left(\frac{P_K K_i + P_{Na} Na_i + P_{Cl} Cl_o}{P_K K_o + P_{Na} Na_o + P_{Cl} Cl_i}\right) \tag{3.18}$$

which is sometimes referred to as the *Goldman potential of the nerve*. If the permeability of any two of the ions is small, Eq. (3.18) reduces to the Nernst equation.

In a series of experiments, Hodgkin and Katz (1949) varied the external concentrations of the pertinent ions and observed the magnitude of the three phases of the nerve spike potential (Fig. 3.2): the resting potential, the spike or action potential, and the refractory or recovery part. For the various concentrations they then found which permeabilities of Eq. (3.18) consistently matched the observed potentials. Their experiments obtained the results as ratios, shown in Table 3.2.

TABLE 3.2

	P_K :	P_{Na} :	P_{Cl}
Resting	1	0.04	0.45
Peak of spike (action potential)	1	20	0.45
Refractory	1.8	0	0.45

Since the Cl permeability remains constant it is reasonable to assume that it plays a passive role and it traverses the membrane to maintain a Donnan equilibrium. Table 3.2 also indicates that the action potential apparently arises from a sudden permeability change which lets Na^+ rush in, thereby changing the voltage. In the refractory period no Na^+ goes in so it is apparently being pumped out. Since K^+ goes in during this period, it must also have gone out during the action potential. These important experiments clearly indicated that the role of the cations during the spike should be studied in more detail and, in particular, the change in membrane permeability.

Magnitudes of ion diffusion can be estimated. If the change in membrane potential is about 110 mV and the capacitance of the membrane, as measured earlier, is 1 μf/cm^2 both at rest and at the peak of the action potential, the charge transport is about 110×10^{-9} C transferred across 1 cm^2 of membrane. Keynes and Lewis (1951) have shown by radioactive tracer techniques that the net Na^+ transport is about twice this value, more than enough to substantiate the model.

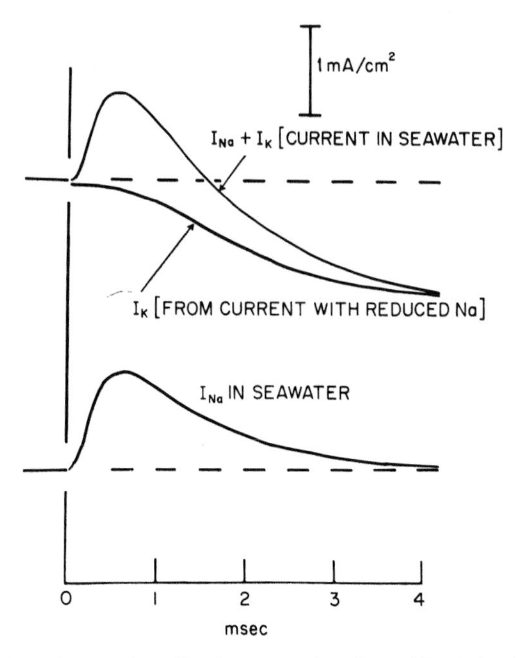

FIG. 3.5 Method of separation of ionic currents into I_{Na} and I_K. A depolarization potential of 56 mV was applied at $t = 0$. [From Hodgkin and Huxley (1952b).]

The study on the efflux of K^+ was made by Hodgkin and Huxley (1952b) in a series of careful experiments. An axon potential was controlled by an internal electrode connected to a feedback amplifier. This is called the *voltage clamp.*

Figure 3.5 illustrates how they separated the Na^+ and K^+ currents. After a sudden 56-mV decrease in membrane potential the total current I as a function of time was recorded with the nerve membrane in seawater. Then the membrane was put into a solution with virtually no Na^+ ions and the excitation performed again. This gave the K^+ current as a function of time. Subtraction of the K^+ current from total current gave the Na^+ current.

This experiment was performed with varying concentrations of Na^+ with consistent results for the outward diffusion of K^+ ions. A simple explanation is now fairly obvious. When the nerve membrane is depolarized it undergoes a rise in its permeability to sodium ions. They rush in and, had the electrode not been draining their current, the inside of the nerve would become more positive, giving rise to the action potential. In support of this idea it was found that if an inert cation such as choline is substituted for the sodium externally the inward current disappears and is replaced by a small pulse of outward current. This latter effect occurs because under these con-

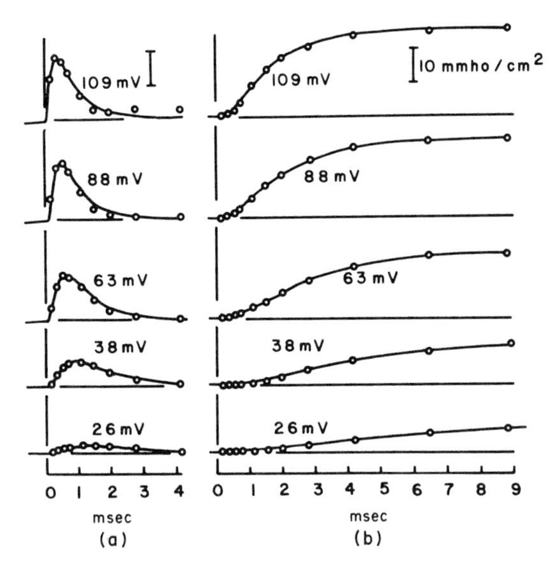

FIG. 3.6 Changes in (a) sodium and (b) potassium conductance associated with different depolarizations at 6°C. The numbers give the depolarizations used. The circles are the experimental points and the smooth curves are the plotted solutions of equations used to describe these changes. [Redrawn from Hodgkin and Huxley (1952d).]

ditions there are more sodium ions inside than outside, and the outward current arises from the tendency of the sodium ions in the axoplasm to move outward to the lower sodium concentration.

One further test of the role of sodium motion was made. If the model is correct, the potential difference between the inside and outside of the membrane is given by the Nernst relation

$$V_{Na} = (RT/F)\ln(Na_o/Na_i) \tag{3.13}$$

If the potential between the inside and outside is controlled by the voltage clamp method to buck this Nernst potential there should be no current flow. This was found to be the case by testing various external concentrations of sodium over a factor of ten.

Referring now to Fig. 3.5, it is seen that the initial Na^+ current rises rapidly and then begins to decrease, approaching zero in about 4 msec. The second component is due to outward flux of K^+, which has been confirmed by radioactive tracer analysis. This current starts slowly and traces an S-shaped curve, which eventually causes the inside to lose more positive charge than it gained from the sodium ion entry. The magnitude of the Na and K currents depends on the stimulating potential (Fig. 3.6). Further evidence has shown that the K^+ current varies with the external K^+ concentration, although not as steeply as predicted by the Nernst equation.

It was later shown that accumulation of K^+ ions on the outer layer of the membrane caused deviation from a Nernst behavior related to the K^+ concentration in the external solution.

REFRACTORY PERIOD

The action potential has been shown to occur when the permeability of the membrane to Na^+ and K^+ changes. The nerve then requires a certain period of time to restore its resting potential. What happens during this recovery period? Clearly the sodium ion must be pumped out. The potassium ion can be pumped back in or simply respond passively and diffuse inward as the sodium is pumped out. Such a pumping operation requires metabolic energy and if such metabolism is inhibited the pump will cease to work. Experiments have been performed with inhibitors of oxidative metabolism such as cyanide, azide, and dinitrophenol, and an example using the latter will be discussed.

A squid axon is charged with radioactive ^{24}Na and flushed externally with seawater, which is continually sampled. The first part of Fig. 3.7 shows an exponential decrease of ^{24}Na from the interior of the axon (the efflux scale is logarithmic) as expected in a normal diffusion process. After 100 min the seawater outside was replaced by a similar solution which contained dinitrophenol. It is seen that the rate of efflux declines to nearly zero and when the dinitrophenol is removed the system recovers and continues its former exponential excretion of sodium ions. Clearly there is an oxidative

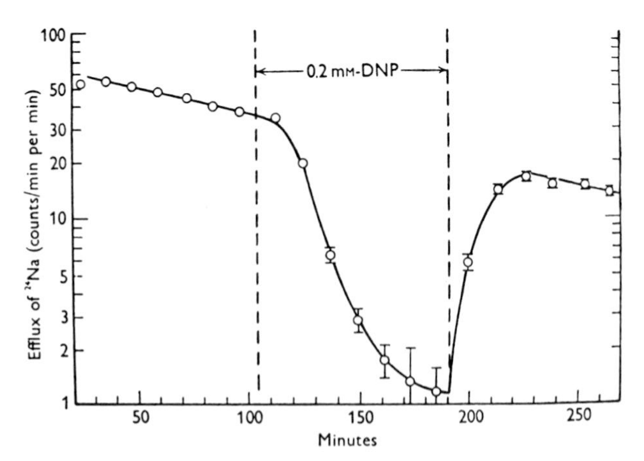

FIG. 3.7 Sodium efflux from a *sepia* axon during treatment with dinitrophenol. At the beginning and end of the experiment the axon was in artificial seawater. The abscissa is time after end of stimulation in ^{24}Na sea water, the ordinate time rate at which ^{24}Na leaves axon. [From Hodgkin and Keynes (1955).]

metabolic pumping mechanism which continuously pumps out sodium in the resting nerve. A further fact on the continuousness of the pumping action was found by Connelly and Cranefield (1953). They observed that the giant squid axon had a resting oxygen consumption rate of 70 mm^3/g hr and that continuous stimulation at a rate of 200 Hz (cycles per second) raised this consumption rate by only 10%.

ANALYTIC TREATMENT OF MEMBRANE CONDUCTION

The membrane current can be considered as composed of three terms: the Na^+ current, the K^+ current, and that due to the leakage of other ions, L:

$$I_m = I_{Na} + I_K + I_L$$

The voltage clamp experiments showed that the current voltage relations obeyed are

$$I_{Na} = g_{Na}(V - V_{Na}), \qquad I_K = g_K(V - V_K), \qquad \text{and} \qquad I_L = \bar{g}_L(V - V_L)$$

where V_{Na} and V_K are the equilibrium potentials for Na and K ions. V_L is the potential at which the leakage current due to chloride and other ions is zero and the g's are the conductances in mho/cm^2.

POTASSIUM CONDUCTANCE

Consider now, as did Hodgkin and Huxley (1952d) potassium data for the action potential current and the return flow shown in Fig. 3.8. They found that if g_K is considered a variable the decay curve followed a first-order equation, i.e., and exponential decrease with time, while a fourth-order

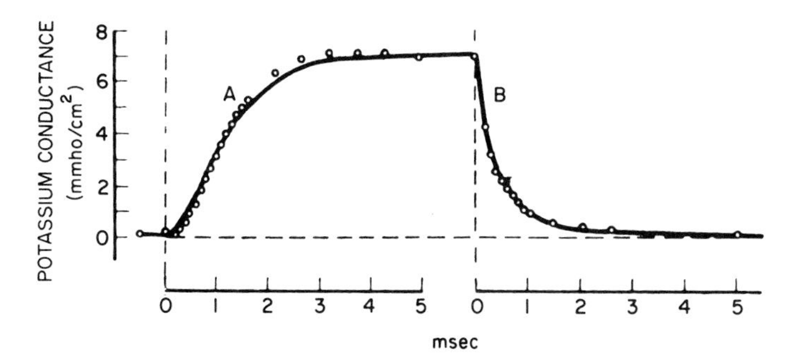

FIG. 3.8 Curve A records the rise of potassium conductance associated with depolarization of 25 mV, curve B the fall of potassium conductance associated with repolarization to the resting potential. The circles are the experimental points, and the smooth curves are the plotted solutions from theoretical equations. [From Hodgkin and Huxley (1952d).]

equation was required to describe the rising conductance. For simplification, they assumed that g_K was proportional to the fourth power of a variable that obeys a first-order equation. Thus the rise of potassium conductance from zero to a final value would be described by the functional dependence $[1 - \exp(-t/\tau)]^4$ and the fall would have the form $\exp(-4t/\tau)$. Other powers were tried in the curve fitting but the fourth-order gave the best fit.

They then assumed that there were two terms governing the potassium conductance g_K,

$$g_K = \bar{g}_K n^4 \tag{3.19}$$

where $\bar{g}_K$ is a constant with the dimensions of conductivity/cm^2 and n is dimensionless. n is a variable between 0 and 1 and might be the fraction of carriers of K^+ inside the membrane, while $n - 1$ would be the fraction of K^+ carriers outside of the membrane. Thus if K^+ movement requires four carriers whose movements are independent of each other, g_K would vary with n^4.

A further assumption is that the movement of these carriers is controlled by rate constants α_n and β_n, which vary with voltage but not with time. Here α_n determines the rate of transfer from outside to inside and β_n determines the rate in the opposite direction (see Section A.7 and Eq. (A.33)). We may write this time rate of change of n as

$$dn/dt = \alpha_n(1 - n) - \beta_n n \tag{3.20}$$

Note that for the condition of resting, i.e., $dn/dt = 0$, Eq. (3.20) yields a resting value of n_0 given by

$$n_0 = \frac{\alpha_{no}}{\alpha_{no} + \beta_{no}}$$

If the voltage is suddenly changed, α_n and β_n take values corresponding to the new voltage, e.g., if the carrier has a negative charge α_n increases and β_n decreases when the membrane is depolarized. The solution of Eq. (3.20) is obtained by simple integration (Section D.1) as

$$n = n_\infty - (n_\infty - n_0)e^{-t/\tau_n} \tag{3.20a}$$

where

$$n = n_0 = \frac{\alpha_{no}}{\alpha_{no} + \beta_{no}} \quad \text{at} \quad t = 0,$$

$$n_\infty = \frac{\alpha_n}{\alpha_n + \beta_n} \quad \text{at} \quad t = \infty,$$

$$\tau_n = \frac{1}{\alpha_n + \beta_n}$$

This solution can be put into the form of Eq. (3.19) for comparison with experiment by substituting $g_{K_0} = \bar{g}_K n_0^4$ and $g_{K\infty} = \bar{g}_K n_\infty^4$ and Eq. (3.20a) into Eq. (3.19) and obtaining

$$g_K = \{(g_{K_\infty})^{1/4} - [(g_{K_\infty})^{1/4} - (g_{K_0})^{1/4}](e^{-t/\tau_n})\}^4 \quad (3.21)$$

where g_{K_∞} is the final value of g_K after the potential is changed and g_{K_0} is the conductance at $t = 0$. Curve fitting was done for each of the curves of Figs. 3.6 and 3.8; the solid lines indicate good agreement with experiment. Details of this curve fitting procedure are given by Hodgkin and Huxley (1952d).

SODIUM CONDUCTANCE

Hodgkin and Huxley (1952d) found that the Na^+ conductance data of Fig. 3.6 could be matched either by a variable that obeys a second-order differential equation or by two variables each of which obeys a first-order differential equation. They chose the latter because the visualization of a mechanism is simpler and the mathematics of curve fitting easier. They assumed that there are two independent events, m and h, which must occur to permit Na^+ conductance, and they found that data could be matched by the product of one event times the third power of the other event:

$$g_{Na} = \bar{g}_{Na} m^3 h \quad (3.22)$$

An example model proposed by them is that the sodium conductance is proportional to three sites inside the membrane, which are occupied simultaneously by three activating molecules but not blocked by one type of inactivating molecule (or the converse). Then m represents the fraction of activating molecules present and $1 - m$ the fraction absent. Similarly, h represents the fraction of inactivating molecules not present and $1 - h$ the fraction present. As in the potassium case m and h have values between 0 and 1. As before, the time rate of change can be written

$$dm/dt = \alpha_m(1 - m) - \beta_m m, \qquad dh/dt = \alpha_n(1 - h) - \beta_h h \quad (3.23)$$

The solutions of these equations satisfying the boundary conditions $m = m_0$ and $h = h_0$ at $t = 0$ are

$$m = m_\infty - (m_\infty - m_0)e^{-t/\tau_m} \quad (3.24)$$

$$h = h_\infty - (h_\infty - h_0)e^{-t/\tau_h} \quad (3.25)$$

where

$$m_\infty = \alpha_m/(\alpha_m + \beta_m) \quad \text{and} \quad \tau_m = 1/(\alpha_m + \beta_m)$$

$$h_\infty = \alpha_h/(\alpha_h + \beta_h) \quad \text{and} \quad \tau_h = 1/(\alpha_h + \beta_h)$$

In the case of Na^+ the conductance of sodium is very small compared with its value at the peak for depolarizations greater than 30 mV. Therefore, m_0 is neglected for values greater than 30 mV. Also, inactivation is nearly complete under these conditions, so that h_∞ is negligible. With these approximations the value of sodium conductance with time may be written

$$g_{Na} = \bar{g}_{Na} m_\infty{}^3 h_0 (1 - e^{-t/\tau_m})^3 e^{-t/\tau_h} \tag{3.26}$$

The data of Fig. 3.6 were used to obtain the constants of this equation, and the resulting theoretical curves are shown as the solid lines of this figure. When the two processes governed by Eqs. (3.21) and (3.26) are added together they give the shape of the action potential of Fig. 3.2. (The comparison of the theoretical to experimental moving impulse is shown in Fig. 3.13.)

THE ROLE OF CALCIUM

It has been known from the work of Gordon and Welsh (1948) that if the concentration of calcium is reduced, nerves fire spontaneously. This immediately raises the possibility that the blocking objects in the membrane pores, as required by the empirical equations of the previous section, are calcium ions. Thus three calcium ions around or within a pore could block the motion of sodium ions through the pore. Upon stimulation of the membrane either physically or electrically some organic carrier nearby could attach itself to a calcium ion and remove it. When all three had been removed and a fourth separate event occurred the sodium ions could rush through the pore.

The removal of the calcium ions need not be a transport mechanism. For example, three of the lipid molecules that form the membrane pore may have calcium ions attached to their pore end and may partially rotate under pressure or electric field stimulus. Such a motion could reduce the calcium ion blockage without necessarily removing the ions themselves. Further work on the blocking mechanism continues and will be discussed later in the chapter.

LOCAL CIRCUIT MODEL

When a nerve is stimulated above some threshold value either mechanically or electrically an action potential results. This action potential is the conversion of the negative internal resting potential to a positive value. The idea of the *local circuit* theory is as follows. Suppose the nerve stimulation has occurred in a region and Na^+ ions have rushed in and made the local interior positive (see Fig. 3.9b.) These changes reduce the membrane potential just ahead of the active region by drawing charge out of the membrane

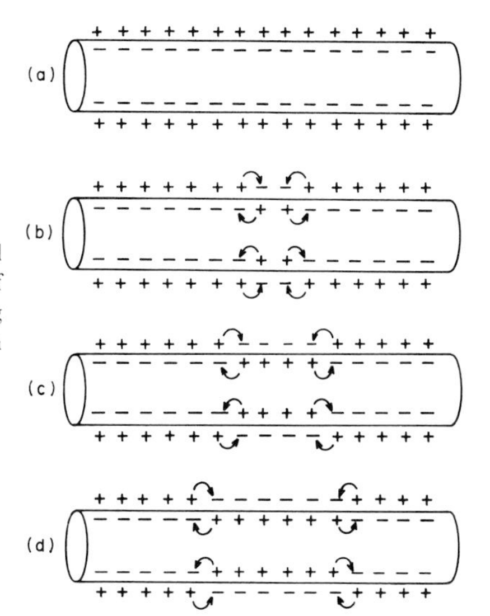

FIG. 3.9 Diagrams based on the local circuit theory illustrating the propagation of the action potential in both directions along a conducting nerve fiber. [Redrawn from Guyton (1971).]

capacitance. When the potential difference in a region adjacent to the reversed part has been reduced by about 20 mV that point becomes sodium permeable and can stimulate an adjacent region, as in Fig. 3.9c. In this way a wave of depolarization propagates along the nerve fiber (Figs. 3.9b, 3.9c, and 3.9d). This wave of depolarization is called *passive spreading*. There is a local ion current flow, as indicated by the arrows, at the edges of the depolarizing membrane, which are called *local circuits*.

MYELINATED NERVE FIBERS

Our discussion to this point has concerned the squid axon, which does not have an insulating sheath of myelin and is therefore called a *nonmyelinated* nerve fiber.

As will be seen in the next section, conduction along such a fiber is comparatively slow. Most of the human nerve fibers are covered with myelin, as mentioned in the Introduction. The myelin sheath is interrupted periodically by gaps, called *nodes of Ranvier*, which are schematically shown in Fig. 3.10. The myelin sheath is a good insulator of reasonable thickness, so that the capacitance in a membrane with myelin is much less than one without myelin. The excitable part of the membrane is exposed at the nodes of Ranvier. The current flows between the nodes as indicated in Fig. 3.10. This current flow, called *saltatory conduction*, is one in which the activity jumps from one node to the next, and the passive spreading occurs much

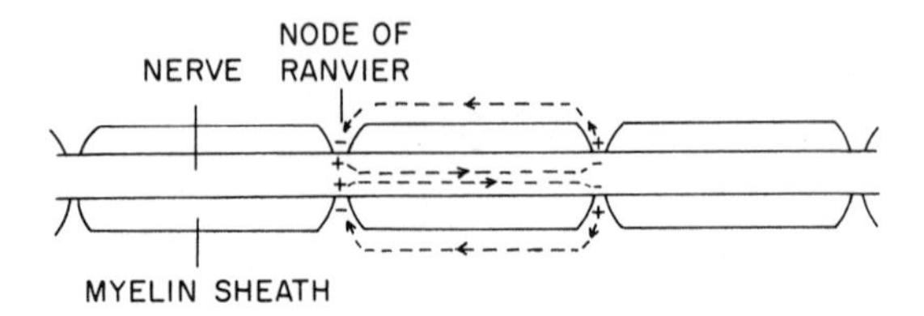

FIG. 3.10 Diagram illustrating the structure of myelinated nerve fiber and saltatory conduction.

faster and with less expenditure of energy than in nonmyelinated fibers. See reviews by Frankenhaeuser (1952) and Tasaki (1952) for details of this mechanism.

IMPULSE PROPAGATION

The local circuit model can be represented by the circuit diagram of Fig. 3.11, in which the current is flowing inside in the longitudinal x direction but is leaking out with currents i_m through the membrane resistances r_m. By Ohm's law the longitudinal current along the axon interior is given as

$$i_{\text{long}} = -(1/r_i)(dV/dx) \tag{3.27}$$

where the negative sign means that the direction of the electric current is positive along a falling (negative) gradient of potential. Let r_i be the core resistance per unit length of axon and V be the internal potential relative to the outside. If the axon is in a large volume of conducting fluid the outside resistance is negligible and the external potential can be taken as zero.

If we further identify R_i as the specific resistivity of the axon and a as its radius then

$$r_i = R_i/\pi a^2 \tag{3.28}$$

The leakage current across the membrane i_m is then

$$i_m = -di_{\text{long}}/dx$$

and upon substituting Eq. (3.27) we obtain

$$i_m = (1/r_i)(d^2V/dx^2)$$

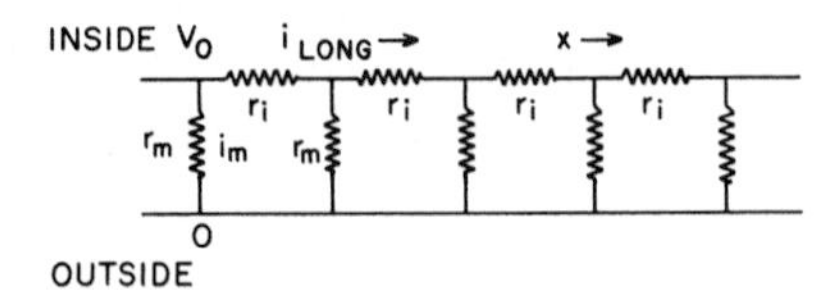

FIG. 3.11 Diagram illustrating resistances and currents in local circuit theory of membrane conductance.

Then substitute Eq. (3.28) and obtain

$$i_m = (\pi a^2/R_i)(d^2V/dx^2) \tag{3.29}$$

The density of the membrane current I_m per unit surface area $2\pi a$ is obtained from Eq. (3.29):

$$I_m = i_m/2\pi a = (a/2R_i)(d^2V/dx^2) \tag{3.30}$$

This is known as the Kelvin equation for a leaking cable.† If a wave of voltage pulse $V(x, t)$ passes along the cable with a constant velocity θ, then $\theta t = x$ and the wave equation for the propagation of the voltage pulse can be written

$$d^2V/dx^2 = (1/\theta^2)(d^2V/dt^2)$$

(See Appendix B for a derivation of the wave equation.) Substitution of the wave equation into Eq. (3.30) yields

$$I_m = (a/2R_i)(1/\theta^2)(d^2V/dt^2) \tag{3.31}$$

We will use this equation in a later substitution.

Across the membrane the total leakage current is the sum of four terms. If V is the membrane potential and V_{Na} the resting Nernst potential for sodium, then the current I_{Na} due to Na is

$$I_{Na} = g_{Na}(V - V_{Na})$$

† One can examine the loss rate of signal in an ordinary transmission cable at this point, by means of the Kelvin equation:

$$-di_{long}/dx = i_m = V/r_m = (1/r_i)\, d^2V/dx^2$$

and therefore $(r_m/r_i)\, d^2V/dx^2 = V$. The solution of this differential equation, which is readily seen by substitution, is

$$V = A \exp[-x/(r_m/r_i)^{1/2}] + B \exp[x/(r_m/r_i)^{1/2}]$$

but, since $V = 0$ at $x = \infty$, the second term must be zero, i.e., $B = 0$. Let $V = V_0$ at $x = 0$ and the particular solution is

$$V = V_0 \exp[-x/(r_m/r_i)^{1/2}]$$

and therefore the strength of the signal decreases exponentially.

Consider the constants stated at the beginning of this chapter for a squid axon; the interior specific resitivity $R_i \simeq 50\ \Omega$ cm and membrane specific resistivity $R_m \simeq 5 \times 10^3\ \Omega$ cm. The resistivities vary as the cross-sectional area and the circumference, respectively, i.e., $r_i = R_i(4/\pi D^2)$ and $r_m = R_m/\pi D$, where D is the axon diameter. If the axon diameter is, for example, 0.01 cm the voltage would fall to one-tenth of its value in about 1 cm.

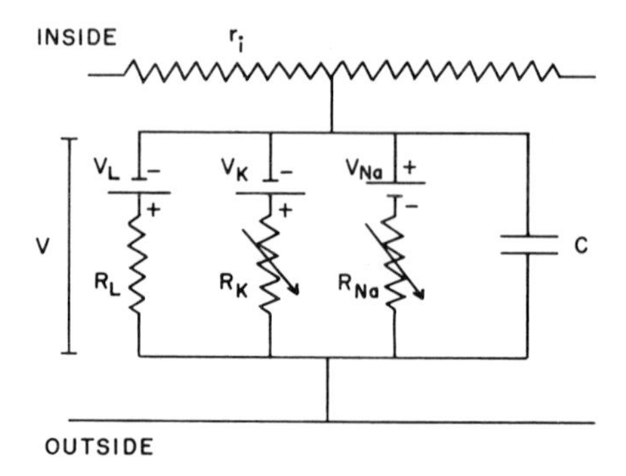

FIG. 3.12 Section of Fig. 3.11 showing the replacement of an r_m element with the experimental findings. C is the membrane capacitance, V_K and R_K the potassium battery and resistance, V_{Na} and R_{Na} the sodium battery and resistance, V_L and R_L the leakage battery and resistance contributed by ions passing through the pores that do not change during activity. V_L is small and R_L large; they are therefore neglected in the calculation.

Similarly,

$$I_K = g_K(V - V_K)$$

and, assuming also an undetermined leakage current I_L,

$$I_L = \bar{g}_L(V - V_L)$$

The reversal of the ions across the membrane also causes a capacitance change, so our circuit must be redrawn to include this. Figure 3.12 shows one of the replacements for the elements r_m of Fig. 3.11.

The total current through the membrane is

$$I_m = C(dV/dt) + (V - V_K)\bar{g}_K n^4 + (V - V_{Na})\bar{g}_{Na} m^3 h + (V - V_L)\bar{g}_L \quad (3.32)$$

where C is the capacitance of the membrane. Equating Eqs. (3.31) and (3.32) gives the nerve current equation:

$$(a/2R\theta^2)(d^2V/dt^2) = C(dV/dt) + (V - V_K)\bar{g}_K n^4 + (V - V_{Na})\bar{g}_{Na} m^3 h + (V - V_L)\bar{g}_L \quad (3.33)$$

In this equation θ, the velocity, is unknown, but it is a constant. Note that this work was done before computers were available. The procedure then was to guess a value for θ and start a trial solution. It was found that V goes to ∞ depending upon whether θ was chosen too high or too low. The properly chosen value of $\theta = 18.8$ m/sec corresponded well with experimental observation of 21.2 m/sec. The theoretical shape shown in Fig. 3.13a also agrees well with the experimental wave form of Fig. 3.13b.

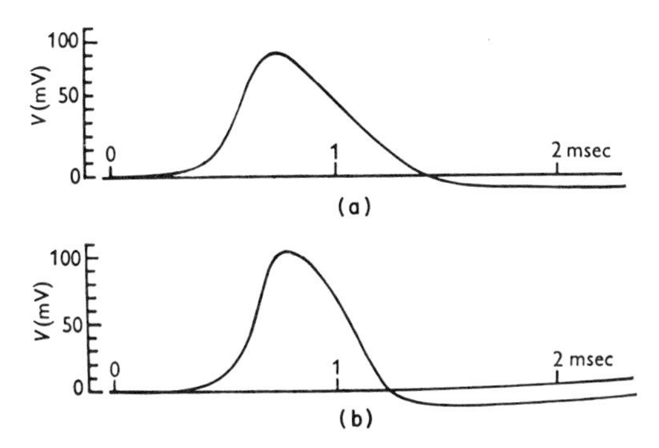

FIG. 3.13 (a) Theoretical propagated action potential, based on Eq. (3.33), from which the calculated velocity is 18.8 m/sec. (b) Experimental propagated action potential; the experimental velocity is 21.2 m/sec. [From Hodgkin and Huxley (1952d).]

VELOCITY AND FREQUENCY

If we examine Eq. (3.33) it is seen that for an initial velocity before much leakage has taken place the approximate equation is

$$d^2V/dt^2 = (2R\theta^2C/a)(dV/dt) \tag{3.34}$$

where $2R\theta^2C/a$ is a constant to be evaluated. The same constant could be obtained for a variety of values of θ^2, C and a, and it is therefore evident that the velocity squared increases proportionately to the radius of the membrane and inversely proportionately with its capacitance. Because the squid axon is large and has a velocity of about 20 m/sec, consider the difficulty in explaining the speed of response of an optic nerve. Its radius can be smaller than 1 μm whereas that of the squid axon which gave $\theta = 20$ m/sec had a radius of 238 μm. The ratio of the velocities in Eq. (3.34) goes as the ratio of the square roots of the radii. Thus the velocity in the optic nerve is about one fifteenth of that of the axon, or about 1 m/sec. It was recognition of this unreasonably low value that led to the postulate of saltatory conduction in myelinated nerves, discussed above.

The different nerve sensations such as touch, vision, etc., arise from a signal to the brain that an action potential has occurred. Since the size of the pulse has been shown to be a constant for a given nerve membrane, relative intensities arise from the frequency of the pulse, as shown in Fig. 3.14. This is most readily comprehended in vision, in which one quantum may cause an action potential in the optic nerve from a given rod, and for which, because of the repolarization time of a few milliseconds for an action potential, several hundred pulses per second are possible, i.e., several hundred quanta per second striking a single rod could be signaled to the brain,

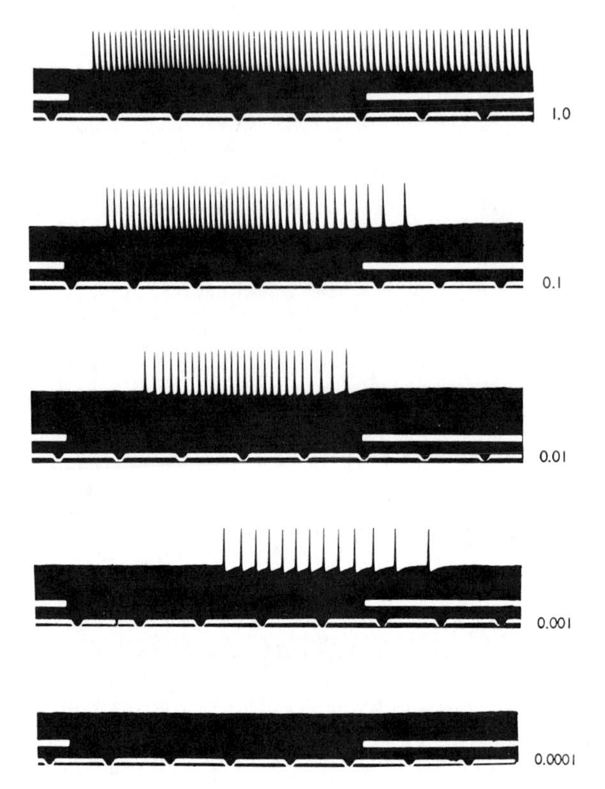

FIG. 3.14 Impulses set up in optic nerve of *Limulus* by 1-sec flash of light (relative intensity shown at right). The lower white line marks 0.2-sec intervals. The gap in the central white line gives the period for which the eye was illuminated (the same in these five cases). [From Hartline (1934).]

giving a sensation of intensity. There have been tests of the discrimination of mechanical frequency of our sense of touch. For example, the finger placed against a dull toothed gear wheel can tell that it has teeth up to a rotation speed in which the teeth are passing the contact point of the finger about 100–200 times per second. At frequencies greater than this the wheel feels smooth.

POISONS, BARBITURATES, AND LOCAL ANESTHETICS

The availability of the squid axon, as well as that of other sea animals, and the techniques of ionic current flow discussed here have led to what are now rather standard methods of measurement of varieties of chemicals that are known to affect the nervous system. Some general categories of chemical effect on the nervous system can be classified. (1) The sodium pump is made inactive, (2) the sodium influx is reduced, (3) the rate of sodium influx is reduced, and (4) the steady state current is reduced. These items have subgroups, and there are drugs that change the shape of the action potential by introducing other disturbances in the curve. We will review briefly examples of the above phenomena.

An example of the inactivation of the sodium pump has already been shown in Fig. 3.7 for dinitrophenol, and it is known that cyanide behaves in a similar manner. The effect is believed to arise from the interference with the oxidative metabolism required to supply the energy for the sodium pump.

The reduction of sodium influx has been studied in detail by a number of investigators (see the review by Narahashi (1971)) for the chemical TTX (tetrodotoxin). This chemical has been identified as the active ingredient in the poison of the puffer fish. Although the poison was known to the ancient Egyptians and was recorded in the first Chinese pharmacopea, it draws most attention in Japan, where the fish is a delicacy. Even though only state-trained and licensed puffer fish cleaners may prepare them (the poison is in the ovary and liver, which must not be punctured) there are some 150 cases of accidental death each year. Also, because of its highly specific action at extremely low concentrations (10^{-8} mole) it is widely used as a tool for a variety of neurophysiological studies.

TTX exists in cation forms and the cation concentration varies with the pH of the solution—the lower the pH the higher the cation concentration. By using the voltage clamp technique, Naharashi *et al.* (1969) found that, although the rate of rise of the action potential was unaffected by TTX, the magnitude of the Na^+ current was reduced. They used the voltage clamp technique and plotted a form of the conductivity relationship that we used earlier. That is,

$$g_P = I_p/(E - E_p)$$

where g_P is the peak transient sodium conductance, I_p the peak current, E the membrane potential, and E_p is the membrane potential where I_p reverses its polarity, i.e., the equilibrium potential for I_p. Typical data are shown in Fig. 3.15, where I_{ss} is the late steady state potassium current (see Fig. 3.5). It is seen that there is a dramatic effect on the action potential; 30×10^{-9} M of TTX reduces the peak transient current by 35% at pH 9 and by 73% for pH 7. It is also seen that there is a negligible effect of TTX on the steady state potassium current. In a similar experiment the TTX was injected inside of the axon, but it exhibited no effect on either I_p or I_{ss}. The conclusion is that the membrane is asymmetric, and that in the cation form TTX can block the outside of the channels through which the Na^+ ions pass and is a poison because the binding energy is too large to be washed free of the membrane.

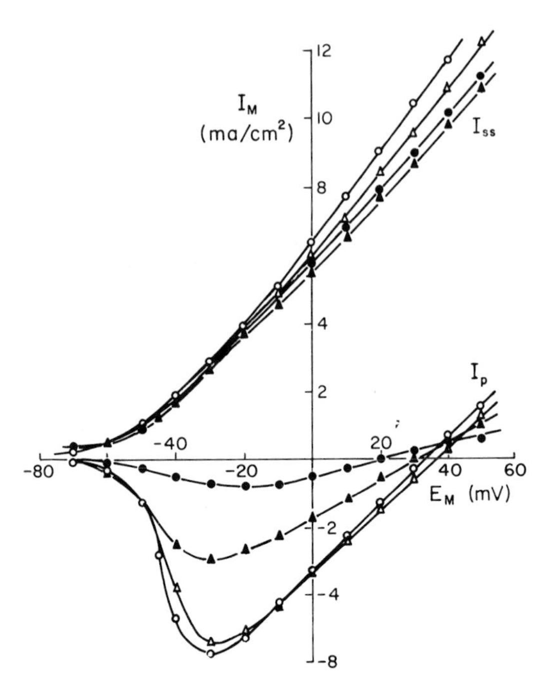

FIG. 3.15 Current–voltage relation at peak transient of action potential (I_p) and steady state currents (I_{ss}) for squid axon at: (○) pH 7 without TTX; (●) pH 7 with 30 nM TTX; (△) pH 9 without TTX; (▲) pH 9 with 30 nM TTX. [From Narahashi *et al.* (1969). © 1969, The Williams & Williams Co., Baltimore, Maryland.]

Barbiturates such as pentobarbital exhibit three effects: they reduce the peak current, they reduce the rise time, and they reduce the steady state current [see Blaustein (1968)]. Barbiturates dissociate into anions and a pH effect has been reported by Narahashi *et al.* (1971). The blocking potency has been found to be higher with lower pH values. This suggests that pentobarbital exerts blocking action in the uncharged form. The blocking action occurs for either internal or external application, but the potency of internal blocking is 6 to 10 times higher than that for external application. These results suggest that the blocking site is inside the membrane and that pentobarbital applied externally can penetrate the membrane in the uncharged form.

Nearly all local anesthetics studied reduce the peak and the steady state current but increase the rise time. An example of the effect on the peak and the steady state current of procaine is shown in Fig. 3.16, and the effect on the time to reach the peak is shown in Fig. 3.17. The blockages occur when the anesthetic is on either side of the membrane, probably because the uncharged amines are able to penetrate the membrane.

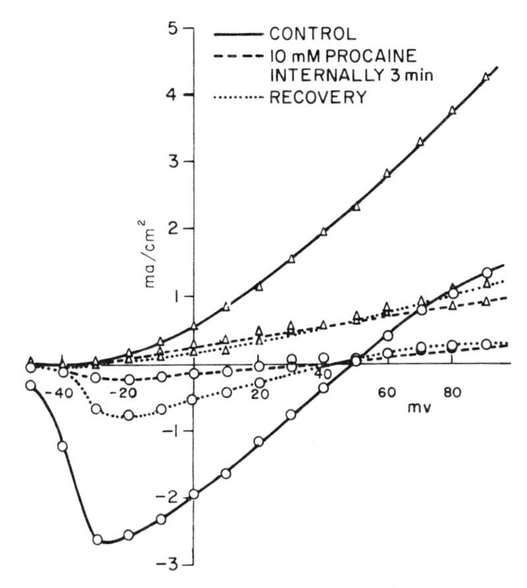

FIG. 3.16 Current–voltage relations for the early transient current (circles) and the late steady-state current (triangles) in a perfused voltage-clamped giant squid axon before and 3-min after application of 10^{-2} mole procaine internally and after washing with normal internal solution. [From Narahashi *et al.* (1967).]

An enormous amount of data concerning the effect of drugs on the action potential of excitable membranes has been accumulated (see the review by Seeman (1972)) in an attempt to understand the nature of the sodium channels. A direct study of the channels is difficult because of their sparcity. In 1 μm^2 of nerve membrane there are about 3×10^6 lipid molecules, while the number of sodium channels is estimated to be only between 2 and 500.

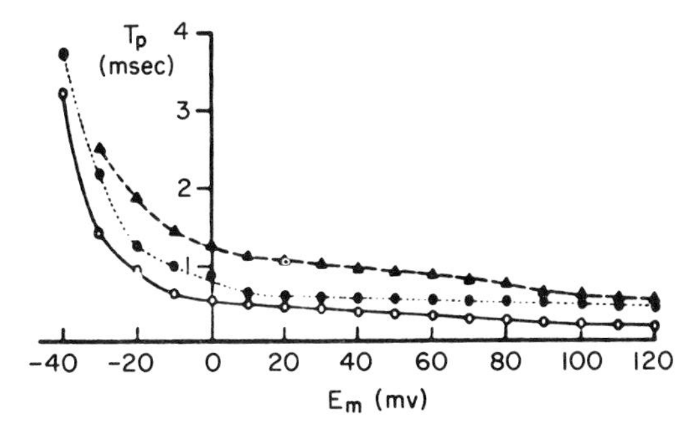

FIG. 3.17 The time for the transient current in a squid axon to reach its peak (T_p) plotted as a function of membrane potential before and during application of 10^{-2} mole procaine internally, and after washing with normal internal medium. ———, control; – – –, 10 m*M* procaine internally 3 min; · · ·, recovery. [From Narahashi *et al.* (1967).]

The channel size is estimated to be a narrow slit with a cross section of about 3 Å × 5 Å. Its small size cross section constitutes a selectivity filter that determines which ions can flow.

It is seen by the above discussion of the effects of a few drugs that there are possibly several different mechanisms of channel blocking, although it has been shown that the effects of anaesthetic mixtures is close to being additive. The charged state of the tertiary amines is the most potent form, which suggests that the blocking effect occurs by chemical bonding to a specific receptor. However, the fact that such a wide range of amines act as anaesthetics would suggest a lack of selectivity of specific receptors. In contrast to this behavior, TTX, which blocks so strongly, will not block at all if very minor changes are made in its structure.

The existence of so much data and the absence of a universally accepted theory has naturally led to the proposal of a variety of models. A summary of, with references to, these models is contained in the article by Lee (1976). The nature of the pore in the membrane, that is, the sodium channel, is an item of active current research, but because the matter is not settled a discussion of the proposed models and their relative merits is beyond the scope of this book.

The effects of drugs on the body are not restricted to their action on sodium currents in excitable membranes. For example, Moore *et al.* (1964) examined the effect of ethyl alcohol in the medium surrounding the squid axon. They had to add at least 4% alcohol by volume before a detectable difference was observed in the action potential. Such a quantity is many times the plasma alcohol content in a drunken human. It is therefore apparent that there is an effect of drugs on the synapses, on other cells of the body, and on the brain itself. These effects are also the object of current intensive investigation.

LITHIUM AND MANIC DEPRESSION

Manic depression is a psychotic condition that manifests itself in extremes of enthusiasm, often accompanied by disorganized behavior, followed by severe depression. Treatment of such a condition by the ingestion of moderate amounts of a lithium salt such as lithium carbonate has had a controversial history. Initial positive results in the control of the manic phase were followed by negative reports and the issue was taken up by the news media. A decade ago a major magazine accused the drug industry of not cooperating with doctors because lithium salts were too inexpensive for profit-making industry to handle and, of course, were not proprietary. A clearer picture of the true situation is now emerging and will be described briefly.

Early in the study of the action potential, Huxley and Stämpfli (1951a, b) used external ions other than sodium to measure the action potential. The lithium ion has the same charge as the sodium ion, and a squid axon gave a similar action potential when placed in a Ringer solution with lithium substituted for sodium. However, with repeated excitations the action potential decreased and, in about 15 min, ceased altogether. Later and more detailed work by Gardner and Kerkut (1968) on snail neurons showed that there is some mechanism in the interior of the neuron that can distinguish between sodium and lithium ions. Sodium apparently plays two roles: to carry the inward current of the action potential, and to stimulate the active transport system that pumps the excess sodium out again. While lithium causes an adequate action potential, it does not activate the sodium pump. Or, if it does, the pump does not transport lithium. Thus, from a biophysical approach lithium did not appear to be beneficial in a disorder of the nervous system.

One could not hope to excise a nerve fiber of a manic depressive patient for experimentation, and it was apparently difficult to detect such a psychosis in a squid. However, the work of a number of investigators such as McConaghey and Maizels (1962) has shown that the membrane of red blood cells behaves much like that of nerve membrane. That is, the ion diffusion, Donnan equilibrium, sodium pump, etc., are quite comparable and, therefore, a blood sample could be used as an analog of a nerve membrane. Maizels (1968) showed, in an experiment similar to that of McConaghey and Maizels mentioned above, that although lithium could be taken into the cell it was not pumped out again. These negative findings in the role of lithium are not final, however. Glen *et al.* (1972) measured the efflux of sodium from normal red blood cells incubated in various concentrations of lithium and potassium. Their results are shown in Fig. 3.18. In this figure it is seen that lithium at concentrations of 3 mM and 6 mM causes a significant stimulation of sodium efflux when the concentration of potassium is 0 or 2 mM, but the effect of lithium is not apparent in the presence of 6 mM potassium, which itself causes a big stimulation.

Coppen and Shaw (1963) and Coppen (1965) have described significant changes in the distribution of sodium between cells and the extracellular space in certain psychiatric patients during episodes of depression and mania. In a more recent study Glen *et al.* (1969) used sodium-sensitive electrodes in the salivary glands of patients and found that the sodium flow during episodes of depression and mania was about one third that of normal. They suggested that the clinical observation of increased sodium transport by the administration of lithium to manic depressive patients may arise from the stimulating effect shown in Fig. 3.18.

Mendels and Frazer (1974) summarized a large body of clinical work with depressed patients and found that those who respond to lithium treatment develop a higher ratio of intracell to plasma lithium than those who do not.

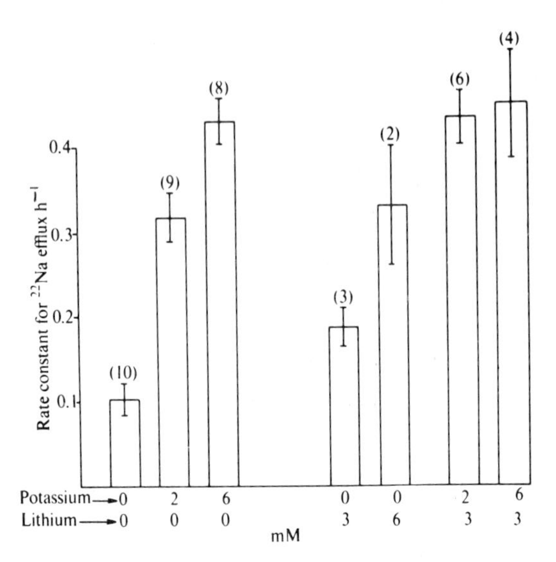

FIG. 3.18 Rate constants for ^{22}Na efflux from normal red blood cells incubated in media containing various concentrations of potassium and lithium. The figures in parentheses above each column indicate the number of experiments in each group. The concentrations of potassium and lithium in the media are shown below each column. [From Glen *et al.* (1972).]

They therefore hypothesize a subgroup of individuals with a genetic difference in the lithium permeability of the cell membrane. In another series of measurements they observed that the absolute concentration of sodium increased in the red blood cells in patients who responded to lithium treatment during such treatment. This suggests that an early clinical test of sodium increase in the red blood cells during lithium treatment will detect those patients who will respond to the treatment.

Also reviewed by these investigators are a series of studies on the genetic factor in manic depressive illness. Patients with families in which there is a history of this illness are more likely to benefit from lithium treatment than those without such a background. To determine whether red blood cell lithium concentration is under genetic regulation, cells obtained from monozygotic and dizygotic twins were incubated in the presence of excess lithium. A significantly higher difference in lithim uptake was found between the pairs of the dizgotic twins than between the monozygotic twins. The calculated heritability index from these data ranged from 0.79 to 0.85, which indicates a substantial genetic factor.

The studies on red blood cell membranes have been shown in animals to correspond to the behavior of nerve membranes, and these intriguing results raise more questions than they answer. However, it appears that the early controversy over lithium treatment was not unreasonable; some patients seem to respond and others don't. It is postulated that it is possible to determine early in the treatment whether response is expected, and that a simple screening of genetic predisposition may now be made.

Although the effect of lithium on sodium efflux shown in Fig. 3.18 is an interesting observation, there are other possible mechanisms currently being discussed in the literature that are related to the reduction of hormonal effect. One of these possibilities will now be described.

The functions of the body are regulated by two major control systems, the nervous system and the hormonal system. Without delving into the biochemistry of the hormonal mechanism we may summarize it briefly in the following way. A hormone attaches to a specific receptor site on a cell. After binding with the receptor this combination of hormone and receptor activates an enzyme in the cell membrane called *adenyl cyclase*. A portion of adenyl cyclase extends through the membrane into the interior or cytoplasm of the cell and, when a hormone is attached, it causes a chemical conversion of *adenosine triphosphate* (ATP) in the cytoplasm to cyclic 3′, 5′-adenosine monophosphate, cAMP. This cAMP initiates a variety of functions, depending upon the particular cell, such as synthesis of chemicals, increase of enzymes, or a changing of the permeability. There is some evidence that lithium inhibits hormones mediated via the adenyl cyclase–cAMP system in the body. In one of these observations an antidiuretic hormone was believed to be inhibited, which affects the water permeability of the renal tubules of the kidney. Thus, someone taking lithium would excrete a greater quantity of water. This would raise the total ion content of the body plasma, thereby increasing the sodium ion concentration of the cells [see Forrest (1975) and Cox and Singer (1975)]. If the body sodium concentration is low or if the sodium pump is weak or inefficient, raising the sodium ion concentration could possibly assure a proper efflux between action potentials. Still another observation and model related to the adenyl cyclase–cAMP system is the effect on the so-called *catecholamine* hypothesis of effective disorders. This hypothesis postulates that mania is characterized by a functional disorder, which results in an excess of noradrenaline at a site in the central nervous system. Inhibition by lithium of the hormonal action would then suppress the effect of noradrenaline. Such inhibition is strikingly shown in Fig. 3.19. In this figure the plasma concentration of cAMP is plotted as a function of time after injection of adrenaline. The upper solid line curve for a mixed group of normal and manic depressive patients shows the effect of adrenaline on humans who had received no lithium. The lower dashed curve is for another mixed group who had received lithium prior to adrenaline injection. Clearly, the cAMP is suppressed. Other comparisons of nonmixed groups, i.e., normal and patients, showed no

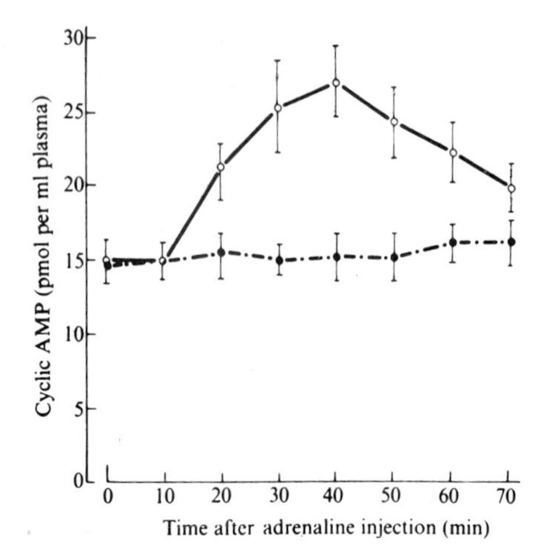

FIG. 3.19 Plasma cAMP concentration versus time after adrenaline injection in humans receiving lithium (dashed curve) and those who have not received lithium (solid curve). [From Ebstein *et al.* (1976).]

differences between their responses. In other words, all humans show the same suppression of cAMP by lithium (see Ebstein *et al.* (1976)).

Although much more clinical data are necessary, the contributions that may be made by biochemists and biophysicists to a fundamental understanding of chemical origins of mental illnesses are becoming more clearly defined.

REFERENCES

Arndt, R. A., and Roper, L. D. (1972). "Simple Membrane Electrodiffusion Theory." Physical Biological Sciences Misc., Blacksburg, Virginia.

Arndt, R. A., Bond, J. D., and Roper, L. D. (1970). An exact constant field solution for a simple membrane, *Biophys. J.* **10**, 1149.

Blaustein, M. P. (1968). Barbiturates block sodium and potassium conductance increases in voltage-clamped lobster axons, *J. Gen. Physiol.* **51**, 293.

Connelly, C. M., and Cranefield, P. F. (1953). The oxygen consumption of the stellar nerve of the squid, *Int. Physiol. Congr., 19th, Montreal.*

Coppen, A. (1965). Mineral metabolism in affective disorders, *Br. J. Psychiatry* **111**, 1133.

Coppen, A., and Shaw, D. M. (1963). Mineral metabolism in melancholia, *Br. Med. J.* **2**, 1439.

Cox, M., and Singer, I. (1975). Lithium and the adenylate cyclase-cAMP system, *N. Engl. J. Med.* **293**, 46

Easton, D. M., and Swenberg, C. E. (1975). Temperature and impulse velocity in giant axon of squid loligo pealei, *Am. J. Physiol.* **229**, 1249.

Ebstein, R., Belmaker, R., Grunhaus, L., and Rimon, R. (1976). Lithium inhibition of adrenaline-stimulated adenylate cyclase in humans, *Nature* (*London*) **259**, 411.

Forrest, J. N., Jr. (1975). Lithium inhibition of cAMP mediated hormones: a caution, *N. Engl. J. Med.* **292**, 423.

Frankenhaeuser, B. (1952). The hypothesis of saltatory conduction, *Cold Spring Harbor Symp. Quant. Biol.* **17**,27.

Frankenhaeuser, B., and Hodgkin, A. L. (1956). The aftereffects of impulses in the giant nerve fibers of *loligo*, *J. Physiol.* (*London*) **131**, 341.

Gardner, D. R., and Kerkut, G. A. (1968). A. Comparison of the effects of sodium and lithium ions on action potentials from *helix aspera* neurones, *Comp. Biochem. Physiol.* **25**, 33.

Glen, A. I. M., Ongley, G. C., and Robinson, K. (1969). Effect of a sensory stimulus on an impaired membrane transport system in man, *Nature* (*London*) **221**, 565.

Glen, A. I. M., Bradbury, M. W. B., and Wilson, J. (1972). Stimulation of the sodium pump in the red blood cell by lithium and potassium, *Nature* (*London*) **239**, 399.

Goldman, D. E. (1943). Potential impedance and rectification in membranes, *J. Gen. Physiol.* **27**, 37

Gordon, H. T., and Welsh, J. H. (1948). The role of ions in axon surface reactions to toxic organic compounds, *J. Cell. Comp. Physiol.* **31**, 395.

Guyton, A. (1971). "Textbook of Medical Physiology. Saunders, Philadelphia, Pennsylvania.

Harris, E. J. (1956). "Transport and Accumulation in Biological Systems." Academic Press, New York.

Hartline, H. K. (1934). Intensity and duration in the excitation of single photo receptor units, *J. Cell. Comp. Physiol.* **5**, 229.

Hodgkin, A. L. (1958). Ionic movements and electrical activity in giant nerve fibers, *Proc. R. Soc. London Ser. B* **148**, 1.

Hodgkin, A. L. (1964). "The Conduction of the Nervous Impulse." Thomas, Springfield, Illinois.

Hodgkin, A. L., and Huxley A. F. (1939). Action potentials recorded from inside a nerve fibre, *Nature* (*London*) **144**, 711.

Hodgkin, A. L., and Huxley, A. F. (1952a). The components of membrane conductance in the giant axon of loligo, *J. Physiol.* (*London*) **116**, 473.

Hodgkin, A. L., and Huxley, A. F. (1952b). Currents carried by sodium and potassium ions through the membrane of the giant axon of *loligo*, *J. Physiol.* (*London*) **116**, 449.

Hodgkin, A. L., and Huxley, A. F. (1952c). The dual effect of membrane potential of sodium conductance in the giant axon of *loligo*, *J. Physiol.* (*London*) **116**, 497.

Hodgkin, A. L., and Huxley, A. F. (1952d). A quantitative description of membrane current and its application to conduction and excitation in nerve, *J. Physiol.* (*London*) **117**, 500.

Hodgkin, A. L., and Katz, B. (1949). The effect of sodium ions on the electrical activity of the giant axon of the squid, *J. Physiol.* (*London*) **108**, 37.

Hodgkin, A. L., and Keynes, R. D. (1955). Active transport of cations in giant axons from *Sepia* and *Loligo*, *J. Physiol.* (*London*) **128**, 28.

Huxley, A. F., and Stämpfli, R. (1951a). Direct determination of membrane resting potential and action potential in single myelinated nerve fibers, *J. Physiol.* (*London*) **112**, 476.

Huxley, A. F., and Stämpfli, R. (1951b). Effect of potassium and sodium on resting and action potentials of single myelinated nerve fibers, *J. Physiol.* (*London*) **112**, 496.

Junge, D. (1976). "Nerve and Muscle Excitation." Sinauer Assoc., Sunderland, Massachusetts.

Katz, B. (1966). "Nerve, Muscle and Synapse." McGraw-Hill, New York.

Keynes, R. D., and Lewis, P. R. (1951). The sodium and potassium content of cephalopod nerve fibers, *J. Physiol.* (*London*) **114**, 151.

Lee, A. G. (1976). Model for action of local anaesthetics, *Nature* (*London*) **262**, 545.

Maizels, M. (1968). Effect of sodium content on sodium efflux from human red cells suspended in sodium-free media containing potassium, rubidium, caesium or lithium chloride, *J. Physiol.* (*London*) **195**, 657.

McConaghey, P. D., and Maizels, M. (1962). Cation exchanges of lactose-treated human red cells, *J. Physiol.* (*London*) **162**, 485.

Mendels, J., and Frazer, A. (1974). Alterations in cell membrane activity in depression, *Am. J. Psychiat.* **131:11**, 1240.

Moore, J. W., Ulbricht, W., and Takata, M. (1964). Effect of ethanol on the sodium and potassium conductances of the squid axon membrane, *J. Gen. Physiol.* **48**, 279.

Narahashi, T. (1971). Neurophysiological basis for drug action: Ionic mechanism, site of action and active form in nerve fibers, *in* "Biophysics and Physiology of Excitable Membranes" (W. J. Adelman, Jr., Ed.). Van Nostrand-Reinhold, Princeton, New Jersey.

Narahashi, T., Anderson, N. C., and Moore, J. W. (1967). Comparison of tetrodotoxin and procaine in internally perfused giant squid axons, *J. Gen. Physiol.* **50**, 1413.

Narahashi, T., Moore, J. W., and Frazier, D. T. (1969). Dependence of tetrodotoxin blockage of nerve membrane conductance on external pH, *J. Pharmacol. Exp. Therap.* **169**, 224.

Narahashi, T., Frazier, D. T., Deguchi, T., Cleaves, C. A., and Ernan, M. C. (1971). The active form of pentobarbital in squid giant axons, *J. Pharmacol. Exp. Therap.* **177**, 25–33.

Rothenberg, M. A. (1950). Studies on permeability in relation to nerve function, *Biochem. Biophys. Acta* **4**, 96.

Seeman, P. (1972). The membrane actions of anesthetics and tranquilizers, *Pharmacolog. Rev.* **24**, 583.

Stein, W. D. (1967). "The Movement in Molecules Across Cell Membranes." Academic Press, New York.

Tasaki, I. (1952). Conduction of impulses in the myelinated nerve fiber, *Cold Spring Harbor Symp. Quant. Biol.* **17**, 37.

CHAPTER 4

Muscle: Energy and Mechanism

INTRODUCTION

Muscle is a highly organized system of organic materials that uses chemical energy to produce mechanical work.

There are three kinds of muscle in the body: (1) smooth muscle, which produces intestinal movement, capillary dilation, etc.; (2) cardiac muscle; and (3) skeletal or striated muscle. Although many of the principles of contraction for all three kinds are the same, there are significant differences, and the literature reporting muscle experiments is vast. Because of the relative ease of performing experiments on skeletal muscle, the understanding of its behavior was achieved first, so that smooth and cardiac muscle behavior are usually compared to that of skeletal. We will therefore restrict our consideration in this chapter to some of the major characteristics of skeletal muscle.

Much is known about the structure and the nature of contraction in skeletal muscle and a reasonably consistent explanation exists for the macroscopic behavior based on microscopic knowledge. However, proof of the mechanism is still elusive because it involves the chemistry of a few molecules that survive for only a few milliseconds. Therefore the biochemists cannot state with certainty what has occurred by analysis of the end products. They can, however, make a rather good inference, and we will examine some of the physical data that lead to this inference and show that it is compatible with macroscopic behavior. We will do this by first considering the macroscopic behavior.

MUSCLE MOTION

Whole muscle is built up of a large number of individual fibers, and mechanical experiments have been performed on both the whole muscle and individual fibers with similar results. The fibers can be excited by an electrical impulse when it is imparted through the motor nerve. A simple electrical impulse can produce a single contractural event known as a *twitch.* Although the duration varies among different muscles, generally a twitch lasts about 0.1 sec. A series of electrical impulses will produce a longer contraction, known as a *tetanus.* When the impulses are stopped the muscle may be returned to its initial length by a very small force; such a condition is known as *relaxation.* Under normal conditions muscles do not shorten more than 35% nor stretch more than 40% of their resting length. The normal range of operation is generally within the range of 85% to 120% of the resting length, and it is within this range that most experiments are performed. A muscle can exert its maximum tension when it has a load large enough to prevent any shortening. This is known as *isometric* contraction. Contraction against a constant load is called *isotonic.*

HEAT OF MOTION

The experiments of Hill (1938) that led to the Hill equation will now be described. A sensitive thermopile was imbedded in the sartorius muscle of a frog. Weights could be added to the muscle for different loads and the contraction measured by a pen attached to the system. The voltage reading of the thermopile, for a given muscle, is proportional to the heat produced in the muscle.

When a load greater than that which the muscle can lift is attached, the muscle contracts isometrically. In Fig. 4.1 curves A and E show the heat production versus time for a tetanized isometric contraction. If a small load is placed on the muscle and a stopping pin is placed so that the muscle can contract only to a preset length, an extra amount of heat is produced. This is shown in Fig. 4.1a, where curves B, C, and D represent increasing lengths, i.e., increasing amounts of shortening, for isotonic contraction under a constant load. The companion experiment was to permit a constant shortening with different loads. The results are the curves F, G, H, and J of Fig. 4.1b, which represent decreasing loads.

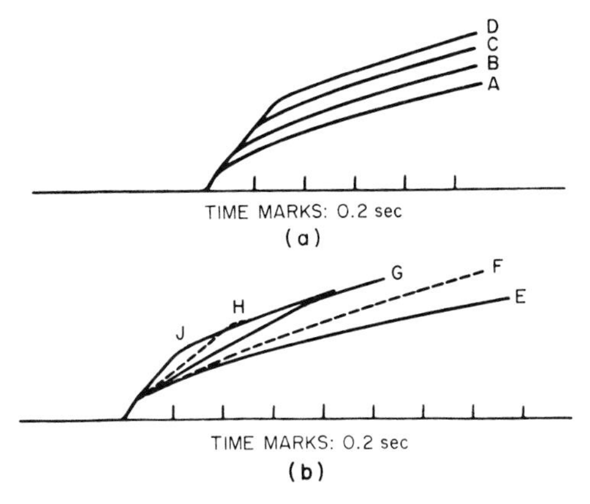

FIG. 4.1 Heat production versus time during isotonic contraction of muscle (curves superimposed). Tetanus at 0°C; muscle 32.5 mm long. (a) Heat produced during shortening different distances under constant load of 1.9 g: A, isometric; B, 3.4 mm; C, 6.5 mm; D, 9.6 mm. (b) Shortening constant distance of 6.5 mm under different loads: E, isometric; F, 31.9 g; G, 23.7 g; H, 12.8 g; J, 1.9 g. [From Hill (1938).]

If the muscle is first stimulated isometrically and then released under the same conditions as the above experiments the results are the superpositioning of the curves of Figs. 4.1a and 4.1b on the isometric curve. Examples of this are shown in Fig. 4.2a, 4.2b. and 4.2c. The sudden extra heat release when the muscle is allowed to shorten is clearly proportional to the velocity of shortening and to the total shortening.

Referring again to Fig. 4.1b it is seen that, except for curve F, which did not reach steady state, the heat liberation rate was greater the lesser the load, but that the total extra heat produced was the same for all at this constant distance. This phenomenon is also evident in the constant distance, variable load experiment shown in Fig. 4.2b. Note that for a constant distance we would expect that the greater the load the more work done by the muscle, even though it shortens more slowly. But the experiment shows that the total heat of shortening a given distance is the same regardless of the load. The conclusion is that when a muscle shortens it liberates extra energy in two independent forms: (1) heat of shortening proportional to the shortening; and (2) mechanical work.

Since the shortening heat is linearly proportional to the shortening distance we may write

$$\text{shortening heat} = ax$$

where x is the shortening distance and a is a constant with dimensions of force. One way of looking at this relation, as suggested by Hill (1938), is

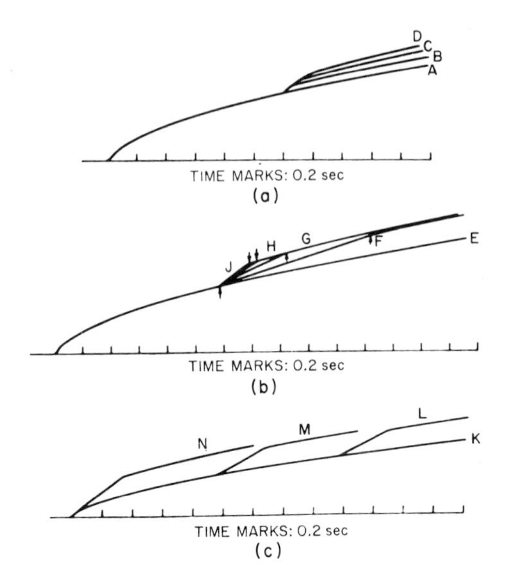

FIG. 4.2 Heat production versus time in isotonic shortening when released during isometric shortening (superimposed curves). Tetanus at 0°C; muscle lengths for (a) and (b) 29.5 mm, for (c) 30 mm. (a) Release at 1.2 sec, shortening different distances at constant load of 3.0 g: A, isometric; B, 1.9 mm; C, 3.6 mm; D, 5.2 mm. (b) Release at 1.2 sec, shortening constant distance 5.2 mm at different loads: E, isometric (45 g); F, 24.9 g; G, 13.9 g; H, 5.7 g; J, 3.0 g (end of shortening shown by arrows). (c) K, isometric; L, release at 1.8 sec; M, release at 0.95 sec; N, release at start (L, M, and N are all for constant distance of 9.1 mm against a constant load of 2.9 g). [From Hill (1938).]

to consider a as a frictional resistance. The heat produced by the muscle in contracting is as if it did work in moving the muscle a distance x against a frictional force a. The quantity a will clearly depend on the cross section of the muscle and in any computation this fact must be considered.

THE HILL EQUATION

In the previous section the experiments of Hill showed that an extra heat production of magnitude ax occurs when a muscle shortens. If P is the load lifted by the muscle in shortening distance x the work done is Px. Thus the total work done in excess of isometric is the sum of these, $(P + a)x$. The rate of energy liberated is the derivative of this with respect to time, $(P + a)(dx/dt)$, or $(P + a)v$, where v is velocity of shortening.

Further experiments were done to determine the velocity dependence on the load. The velocity was read directly from the mechanical record and a was known from the previous experiments. Figure 4.3 shows the quantity

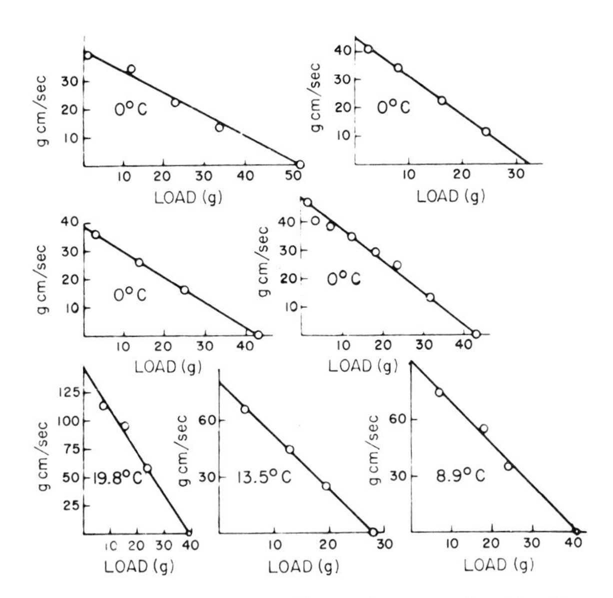

FIG. 4.3 Relation between excess energy liberated per second and load for various muscles at different temperatures. x axis is total load lifted, P. y axis is the quantity $(P + a)v$, the rate of energy liberated, where a was calculated from the heat of shortening ax and $v = dx/dt$ was obtained from the speed of shortening under load P. [From Hill (1938).]

$(P + a)v$ plotted against different loads P for different muscles at different temperatures. In all cases there is a linear relationship between the energy liberation rate and the load. If P_0 is taken as the isometric load of zero motion and b as the linear slope, the relations of Fig. 4.3 can be expressed

$$(P + a)v = b(P_0 - P) \tag{4.1}$$

which is known as the *Hill equation*. The constant b has dimensions of velocity and is also proportional to the size of the muscle.

Equation (4.1) can be rearranged as

$$(P + a)(v + b) = (P_0 + a)b = \text{constant} \tag{4.2}$$

This is the equation of a rectangular hyperbola with asymptotes at $P = -a$ and $v = -b$. The linear relationship of Eq. (4.1) involves one thermal measurement, heat liberated during shortening from which a is derived, and further measurements verified that a is the same for different loads. However, when the resulting Eq. (4.1) is written in the form of Eq. (4.2) it can be verified by mechanical measurements of P and v alone, from which the constants a and b can be derived for verification. An example of this is shown in Fig. 4.4, in which the data points are fitted with the solid curve. The

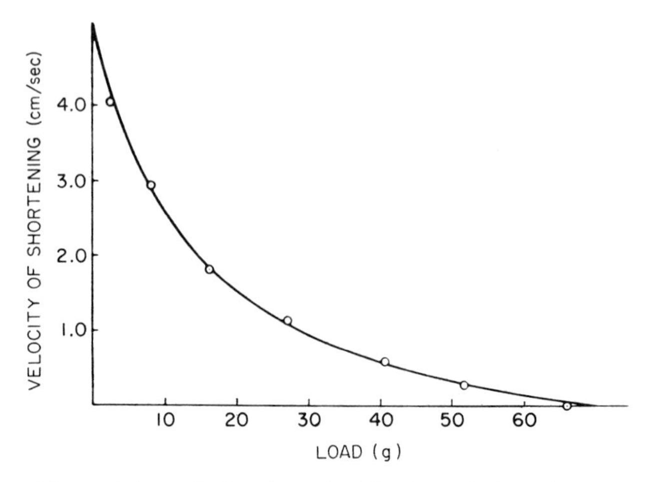

FIG. 4.4 Relation between load and speed of shortening in isotonic contraction. Points are experimental data and solid curve is calculated from Eq. (4.2). [From Hill (1938).]

constants a and b of the hyperbola are within 10% of the values derived from thermal measurements, and thus the simple relation of the Hill equation is substantiated.

STRUCTURE OF SKELETAL MUSCLE FIBER

The skeletal muscles consist of muscle fibers in bundles. Surrounding the muscle is a connective tissue sheath called the *epimysium.* In some muscles the large bundle of fibers is subdivided into smaller bundles and the connective tissue surrounding them is called the *perimysium.* Within this framework the muscle is composed of individual fibers which are more or less tubular in shape and have diameters ranging from 10–100 μm, depending upon the muscle and the animal. These fibers often are the full length of the muscle, although not necessarily. Each muscle fiber is covered with a membrane called the *sarcolemma*, which is about 0.1 μm thick. This sarcolemma follows the fiber during its contraction. Just inside the sarcolemma lie bodies of ovoid form, about 8–10 μm long called *nuclei.* Each fiber contains several hundred to several thousand smaller fibers called *myofibrils* imbedded in an interstitial gel called *sarcoplasm.* An illustration of the sarcolemma, nuclei, and myofibrils of a section of a single fiber is shown in Fig. 4.5. The myofibrils are 1 μm or less in diameter but each one is packed with as many as 2500 smaller elements called *myofilaments.* The filaments are of polymerized protein molecules primarily of two kinds, *myosin* and *actin.*

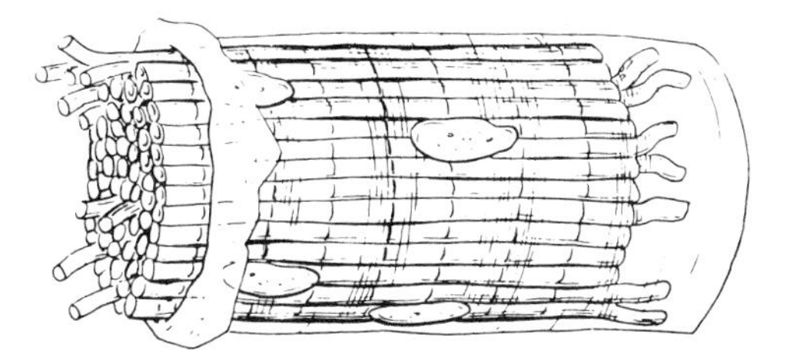

FIG. 4.5 Diagram of a section of a single muscle fiber with the sarcolemma turned back to show the myofibrils; the ovoid bodies are the nuclei. [From Clark (1952).]

THE NEUROMUSCULAR JUNCTION

Before discussing the details of motion in the myofilaments we will briefly describe the transmission of the nerve impulse. The *neuromuscular junction* is the connection between a nerve fiber and the muscle fiber. This connection is a complicated structure called the *end-plate*, which lies outside of the muscle fiber membrane. However, at the end of the nerve there are many little branches called *sole feet*. At the end of the sole feet the membrane has many small folds to increase the surface area. This region of the folds is called the *synaptic gutter* and the space between the sole foot and the gutter is called the *synaptic cleft*, as seen in Fig. 4.6. Within the sole foot are many small *vesicles*, in size about 300–400 Å, which presumably store the chemical

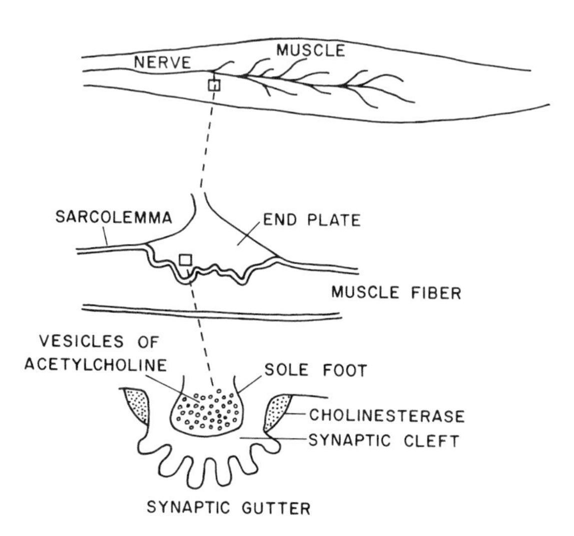

FIG. 4.6 Diagram of the neuromuscular junction showing the attachment of the end-plate to the fiber membrane. Schematic detail of the synaptic gutter.

acetylcholine and the *mitochondria* cells, which manufacture it. Around the rim of the synaptic gutter is stored an enzyme *cholinesterase*, which can destroy acetylcholine.

The chemistry of the neuromuscular junction is quite complex and is discussed in the volumes edited by Bourne (1960). In summary, the model works as follows. When a nerve impulse reaches a sole foot, calcium ions move from the extracellular fluid into the membrane of the sole foot and rupture some of the vesicles that contain acetylcholine. The muscle membrane is similar to the nerve membrane in that it has a negative interior resting potential and an excess of K^+ ions inside and Na^+ ions outside. The acetylcholine changes the permeability of the membrane and it undergoes an action potential similar to that of the nerve. Within about 2 msec after the release of the acetylcholine it diffuses out of the synaptic gutter and is destroyed by the cholinesterase. The impermeability of the membrane is thus restored and after refractory period of a few milliseconds the muscle membrane may be stimulated again.

STRUCTURE OF MYOFIBRILS

A myofibril in isolation has a striated appearance (Fig. 4.7). There are two general repetitive regions, one isotropic to polarized light, and called the I *band*, and one anisotropic to polarized light and called the A *band* (although the isotropies may be reversed upon changing the polarization direction in the light microscope). An I band plus an A band is the linear unit cell for the construction of a myofibril and as such is called a *sarcomere*. One sarcomere at resting length is about 2 μm long in vertebrate animals. As seen in Fig. 4.8, the A band has a lighter center region, called the H *zone*, and the I band has a dark center line, called the Z *line*. In a muscle fiber these dark and light regions are aligned transversely to the fiber axis, giving the fiber a striped or striated appearance under a light microscope.

A combination of x-ray diffraction and electron microscopy has shown that the myofibril is constructed of two types of protein filaments parallel to the long axis and arranged in overlapping arrays. One of the filaments is about twice as thick as the other and they do not overlap completely. Also they interdigitate with one another so that each thick filament is surrounded by six thin filaments in a hexagonal array. A photograph and a schematic of the cross-sectional appearance of these filaments is shown in Fig. 4.9. In Fig. 4.10 the dashed region of the right-hand cross-sectional view of this figure shows that as a result of this hexagonal arrangement each pair of thick filaments has two thin filaments between them. This figure also illustrates the reason for the lighter H zone and I zones. The diameter of the

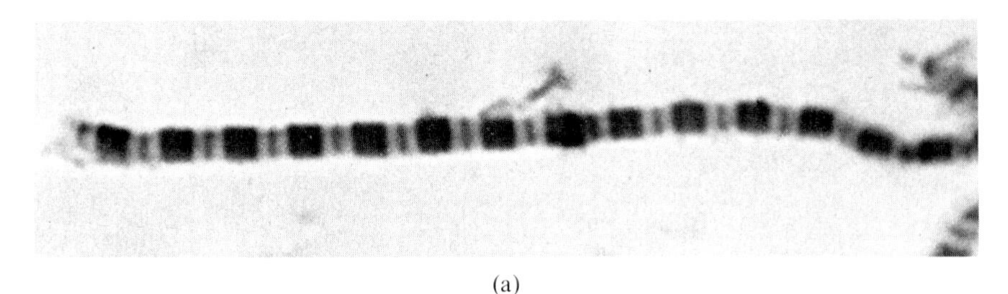

(a)

(b)

FIG. 4.7 (a) Photograph taken with an interference microscope of a single myofibril (magnification 2,480 ×). (b) Electron microphotograph of a longitudinal section through several myofibrils. Between the fibrils are sarcoplasmic elements and the upper part is the surface of the muscle fiber (magnification 53,000 ×). [From Huxley and Hanson (1960).]

thick filaments is about 110 Å, whereas that of the thin ones is about 50 Å. The distance between two thick filaments in vertebrate muscle is about 450 Å.

Myosin can be extracted from muscle fibers with a salt solution. When this is done (Hanson and Huxley (1955)) the A band, except for the H zone, achieves the same optical density as the I band and effectively is indistinguishable. The H zone becomes lighter and the Z lines remain. These results clearly

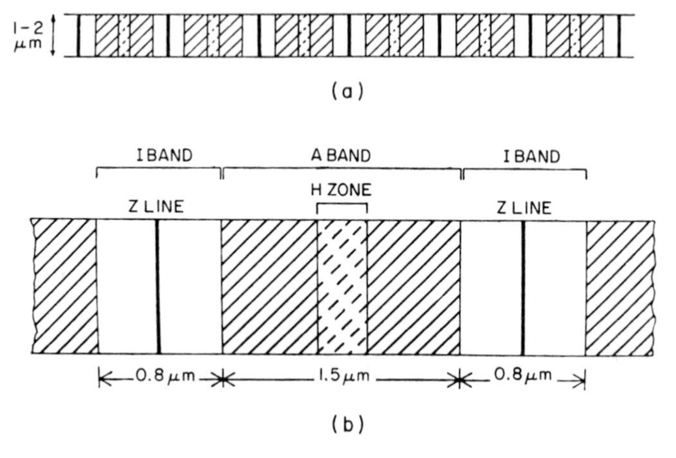

FIG. 4.8 Diagram illustrating the dimensions and nomenclature of the contractile elements of muscle. (a) Isolated myofibril. (b) Myofibril showing band pattern at resting length. [From Huxley and Hanson (1960).]

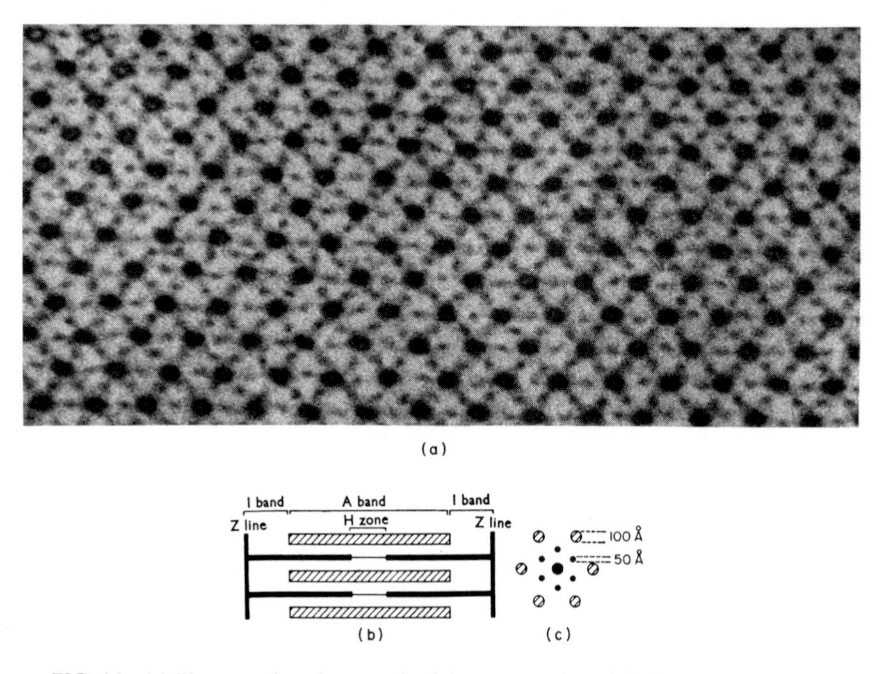

FIG. 4.9 (a) Electron microphotograph of the cross section of the flight muscles of the fly *Calliphora* (magnification 170,000 ×). [From Huxley and Hanson (1960).] (b) Schematic of the filaments in striated muscle showing the longitudinal arrangement and (c) the cross section. [From Huxley (1956).]

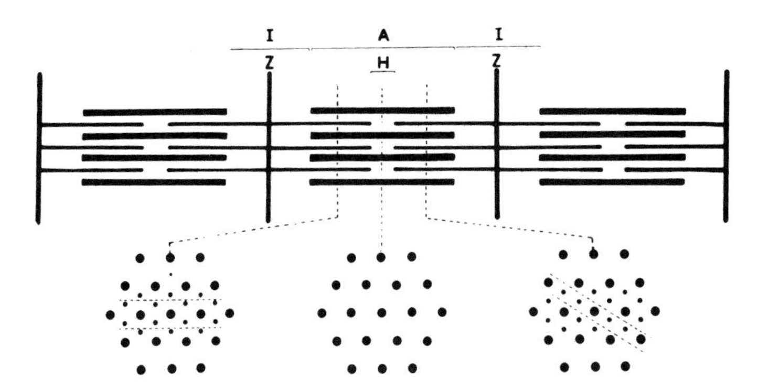

FIG. 4.10 Schematic of longitudinal and cross-sectional segments of the A band. The dashed lines on the cross sections illustrate the differing appearances in an electron microscope of a longitudinal cut with differing angles of the slices. [From Huxley and Hanson (1960).]

indicate that the thick fibers are myosin. Because the thin fibers do not fall apart there must be some threadlike protein that holds them in line. We will discuss this below.

It was found by Szent-Gyorgyi (1960) that potassium iodide could extract actin from muscle fiber by depolymerizing the actin fibers. When this is applied to myosin-free fibrils it is found that most of the I substance leaches out. In a whole fibril such extraction will remove the I substance and the Z line while leaving behind the myosin. Thus, the evidence is that the thin filaments contain actin.

Figure 4.11 shows successive photographs of the contraction of a muscle. These pictures clearly indicate that the two types of fibrils slide longitudinally against each other during contraction and extension of the muscle. This motion is diagrammed in Fig. 4.12.

Along the thick myosin filaments are regularly spaced short lateral projections at essentially right angles to the filament axis (Fig. 4.13). In a region where there are both thick and thin filaments these lateral projections appear to touch projections on the thin filaments. Although in a planar view through a microscope the myosin projections appear to be about 400 Å apart, further detailed study indicates that the arrangement is helical, so that a myosin projection occurs every 60–70 Å. Recall that there are six actin filaments around each myosin and that a longitudinal slice deletes the upper and lower actins. In other words, the x-ray evidence indicates that the myosin molecules lie with their polypeptide chains lengthwise in the muscle and that these chains have a helical configuration. Thus the electron micrographs, which show the protrusions in a planar projection, miss this helical formation. Further experiments have shown that the myosin filaments are

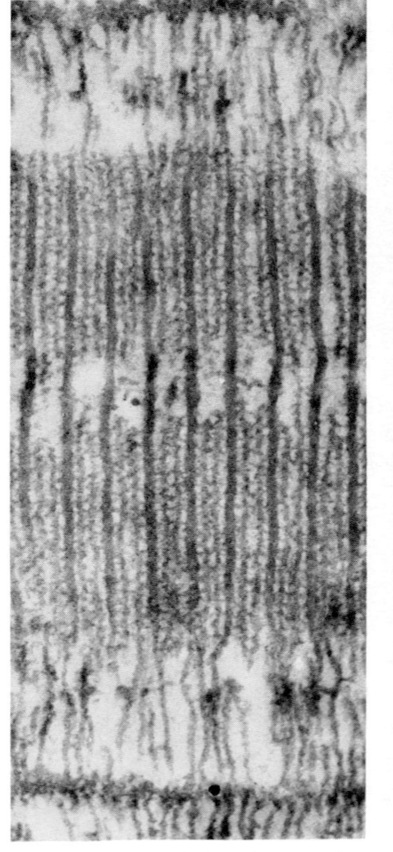
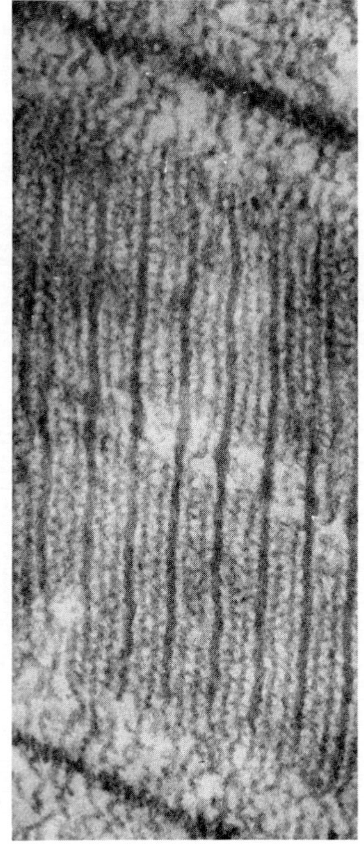
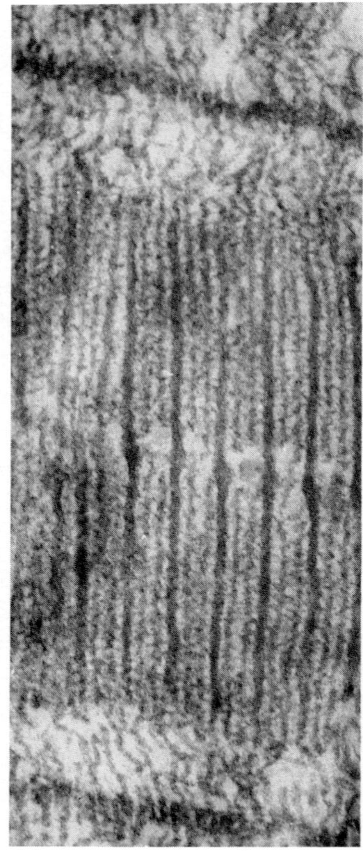

FIG. 4.11 Electron micrographs (magnification 93,000 ×) of a single sarcomere, first in a slightly stretched position (left figure). Subsequent figures show the positions of the thin filaments as the muscle contracts. [From Huxley and Hanson (1960).]

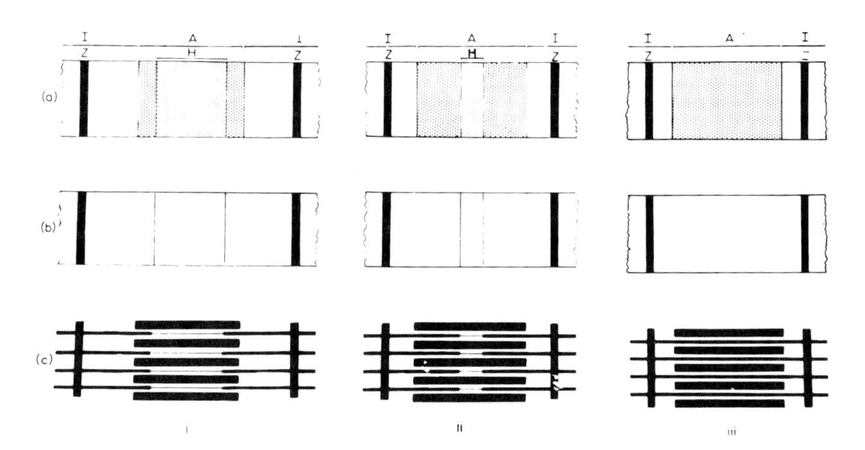

FIG. 4.12 Diagram illustrating the changes in the appearance of a sarcomere (a) with intact fibrils, (b) after the extraction of myosin, and (c) position of the filaments. i, extension; ii, resting length; iii, contraction. [From Huxley and Hanson (1960).]

100 Å

FIG. 4.13 High-magnification (510,000 ×) electron microphotograph of a longitudinal section of the A band, including the H zone, of a rabbit myofibril. [From Huxley and Hanson (1960).]

probably composed of two types of myosin, H-meromysin and L-meromyosin, in a periodic array (H denotes heavy with a molecular weight of 236,000, and L light with a molecular weight of 94,000). Thus, the myosin filament may well be L-meromyosin filaments attached longitudinally with a length that determines the axial periodicity of the projections from the H-meromyosin units. It is these projections which are visible in the electron microscope.

The small filaments are composed of actin (molecular weight 70,000) and tropomyosin (molecular weight 53,000) in a proportion of approximately 1.7 molecules of actin to 1 of tropomyosin for rabbit and chicken breast muscle. X-ray evidence shows that the actin filament is held in a two-chain helical structure by tropomysin with a periodicity of 400 Å (Fig. 4.14). This is significant in that, as already stated, the myosin filament projections also have a periodicity of 400 Å.

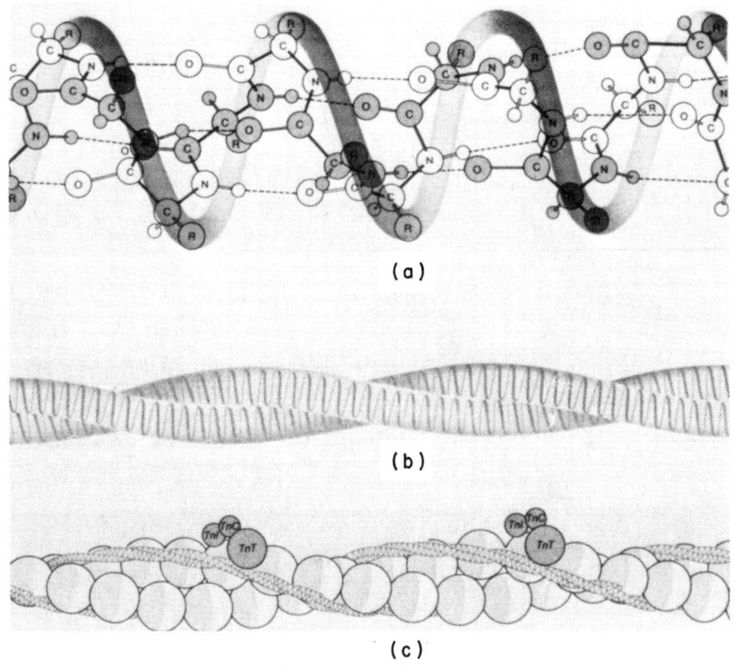

FIG. 4.14 (a) A configuration of many proteins called the *alpha helix*; tropomyosin has this structure. Each subunit consists of a carbon atom C with a side group or radical R, flanked by CO and NH groups. The helix of repeating structures is indicated by the tube. The bracing of the helix is accomplished by hydrogen bonds, indicated by dashed lines. (b) Two alpha helices of (a) are twisted together to form a coil or ropelike structure with side chains interlocking to form a stable structure of tropomyosin. (c) The actin molecules, large circles, held together in a filamentary structure by the rope of tropomyosin. The troponin complex is indicated by the TnT, TnC, and TnI labels. [From C. Cohen, The protein switch of muscle contraction, November. Copyright © 1975 by Scientific American, Inc. All rights reserved.]

Chemical studies have shown that H-meromyosin combines with actin to form an actomyosin complex but that L-meromyosin does not form the complex. This suggests that actin–myosin cross-links are formed, probably by the lateral projections of the myosin, with an active site on the actin, a situation in which the muscle filaments will not slide past one another. It has also been shown that a chemical present in muscle, adenosine triphosphate (ATP), can break the actomyosin complex by the phosphate group breaking away from the ATP and attaching itself to the actin. (A detailed review of the biochemistry involved is given by Needham (1960)). Thus, the model poses that in the resting state there is ATP present and therefore no cross-links. When the muscle is activated something is released that breaks down ATP, and the actin and myosin filaments are drawn to each other and cross-linked. When activation stops ATP separates the cross-links and the muscle can be reextended. Although this chemistry can explain muscle rigor and rest, the mechanism of contraction will be left till the next section.

The mechanism of chemical activation of the cross-linking has been established [see reviews by Cohen (1975) and by Ebashi *et al.* (1969)]. In the bottom drawing of Fig. 4.14 there is shown a cluster of three molecules, labeled TnT, TnI, and TnC, arranged in a periodic fashion along the thin, or actin, filament. In this drawing only a planar slice is shown but this molecular cluster is on different locations along the helix, with a periodicity of about 400 Å.

This cluster of molecules is a complex of the protein troponin, which has three subunits: (1) the largest subunit, designated TnT, binds strongly to the tropomysin chain; (2) a smaller subunit, designated TnI, binds to the actin molecules and can inhibit the actin–myosin interaction; (3) the smallest subunit, designated TnC, binds to calcium.

It has been demonstrated that the bonding of these subunits is dependent upon the calcium ion concentration in the solution surrounding them. Very little TnC is bound to TnT when little calcium is present, but when the calcium concentration is increased above a certain level considerable bonding takes place between these subunits. A second calcium-dependent bonding has been demonstrated in the bonding of the TnI–TnC complex to the actin. In this case the dependence is inverse to the calcium concentration, i.e., the greater the calcium concentration the lesser the interaction of the troponin subunits with the actin molecules.

With these findings a model now exists of the action of the proteins in establishing the cross-linking of the actin and myosin molecules. Consider a cross-sectional slice through Fig. 4.14 at the position of a troponin complex, as shown in Fig. 4.15. Figure 4.15a shows the resting state with low calcium concentration, and Fig. 4.15b the situation in the presence of high calcium concentration. At low calcium concentration the TnI bonds strongly to the

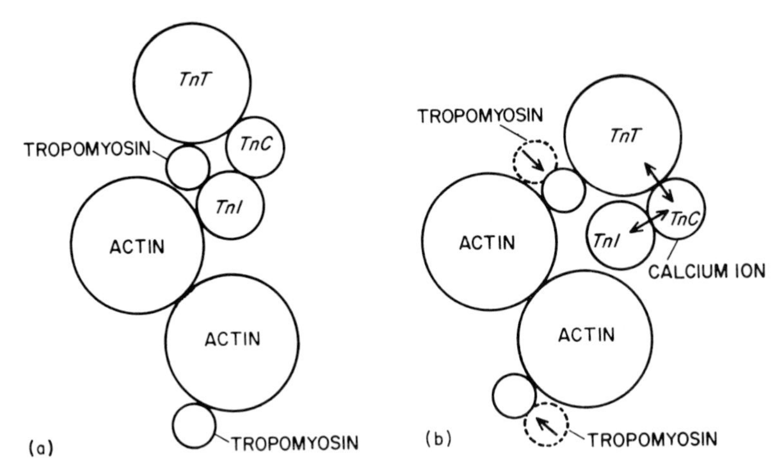

FIG. 4.15 A cross section of the actin fiber at the troponin cluster. (a) The position of the troponin subunits in the absence of calcium ions. (b) The position of the troponin subunits in the presence of calcium ions. Note that the tropomyosin has moved away from a position in which it had blocked the formation of an actin–myosin cross-link. [From C. Cohen, The protein switch of muscle contraction, November. Copyright © 1975 by Scientific American, Inc. All rights reserved.]

actin, which places the tropomyosin in a position to prevent the myosin projection from attaching to the actin molecules; recall that the myosin projections have the same 400 Å periodicity as the troponin complex. If a high concentration of calcium is present the bonding of the TnC to the subunits is increased while the TnI-actin interaction is decreased. The TnI subunit then either moves away from the actin or to a different position. This permits the tropomyosin to move from the blocking position and thereby allows the myosin–actin cross-linking to occur. Another view of the motion of the tropomyosin is shown in Fig. 4.16. This figure is a cross section of the myosin molecule and the actin molecules at the junction of a troponin complex and a myosin projection, the latter called a *myosin head*. The troponin complex is omitted for clarity. The upper position shown in the condition in the presence of calcium, in which the myosin head is bound to the actin molecule. The tropomyosin. which can block the bonding, is out of the way. The lower figure represents the situation in the absence of calcium, in which the tropomyosin has moved to a position which blocks the bonding.

Given the role of changing calcium concentration in switching muscle on and off, where does it come from? The diffusion time of calcium ions from the end plate of the nerve to the individual muscle molecules would be too great to account for the time involved in a muscle twitch. There must be calcium

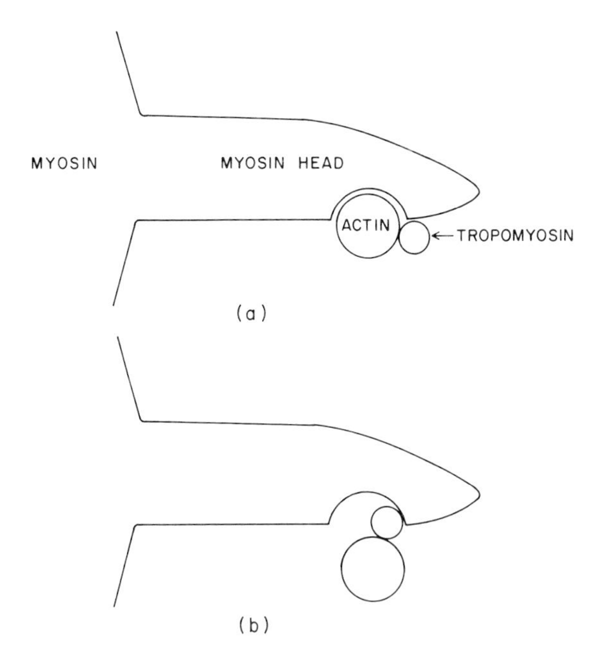

FIG. 4.16 Cross-sectional view of the linkage between the projection from the H-meromyosin, called the head, and the actin. (a) In the presence of calcium the tropomyosin is out of the way and the linkage is formed. (b) In the absence of calcium the tropomyosin moves into a position which blocks the linkage.

stored internally, whose release is signaled electrically. This storage and signal network has been found and is illustrated in Fig. 4.17 (see the review by Hoyle (1970)). Transverse to the muscle fiber is a system of tubules through the sarcolemma. There is also a system of tubules around the fibrils called the *sarcoplasmic reticulum*, which does not open to the exterior. These two systems meet at a number of junctions called *dyads* or *triads*. The sarcoplasmic reticulum terminates in a series of sacs that store calcium when the muscle is turned off. Winegrad (1968) has shown by radioactive tracer techniques that when the muscle is at rest the calcium is concentrated in the I band, which is the outer region of the sarcomere. When the muscle is turned on the calcium becomes concentrated in the region of the A band, where the fibrils interact. The mechanism of turning on a muscle does not require the actual flow of calcium from the sarcolemma but only of an electrical signal, similar to that of nerve conduction, which causes the local release and return of calcium ions within each sarcomere. The return occurs because of the subsequent electrical depolarization of the membrane. This, of course, is the gross picture and the details are currently under study.

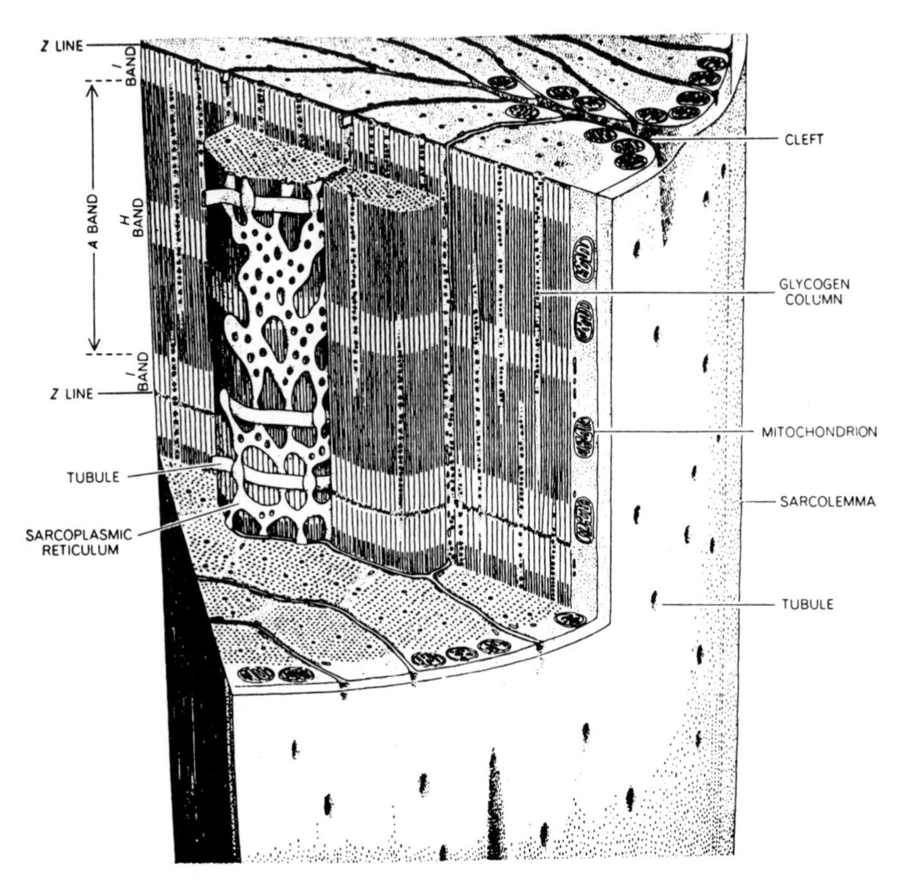

FIG. 4.17 Structure of a muscle fiber showing the tubules through the sarcolemma and perpendicular to the axis of fibers. Also shown is the sarcoplasmic reticulum in the interior, which stores the calcium in sacs not open to the tubules. [From G. Hoyle, How is muscle turned on and off? April. Copyright © 1970 by Scientific American, Inc. All rights reserved.]

THE CONTRACTILE MECHANISM

It should be noted in the above discussion of the biochemistry of the actin–myosin linkage that there is insufficient distance involved to explain the large fraction of resting length to which muscles may contract. It is therefore necessary to postulate that the cross-linkages form, contract the sarcomere by a few angstroms, disconnect, stretch out to attach and form a different cross-linkage, and repeat this process many times. A detailed demonstration of this is currently beyond our present instrumental techniques. However, Huxley (1957) has proposed an empirical model that yields close agreement with the Hill equations and the fiber projection distances. This model will now be described.

For simplicity of writing, let A = actin, M = myosin, AM = the actin–myosin cross-link, XP = a high energy phosphate compound, AXP = this compound attached to actin, and P = the phosphate group. It is assumed that AM forms spontaneously, with the combined state having the lower energy. Thus it is postulated that energy must be provided to break the link. This energy comes from a high energy phosphate compound XP, which has been created at the expense of metabolic energy. This XP compound unites with the site on A after splitting AM. When another chemical event occurs to split the AXP into fragments of XP and A, it is assumed that immediate recombination with the adjacent M cannot occur. This sequence of events can be written

$$\begin{aligned} A + M &\xrightarrow{f} AM \\ AM + XP &\xrightarrow{g} AXP + M \\ AXP &\longrightarrow A + X + PO_4 \end{aligned} \tag{4.3}$$

where f and g are the chemical rate constants (see Section A.7 in Appendix A).

When the muscle is at rest the first reaction does not take place and the AXP form is assumed to dominate under resting conditions. During activation the above sequence of events takes place at a number of sites on each filament in an A band. These sites are probably staggered by the helical configuration so that they come into action asynchronously as the muscle shortens. If the overall tension of the muscle arises from only a fraction of cross-links at any one time then an average constant tension exists during the shortening period.

The hypothesis for the model is that one of the filaments, presumably the myosin, has side pieces M that can move some distance along the main backbone of the filament by thermal vibration and form cross-links at A sites on the actin filaments, these links being broken by the XP compound. A schematic of this is shown in Fig. 4.18. In this figure it is seen that a spring with spring constant k governs the motion of the moving side piece M within its limited range. The M piece vibrates with thermal motion but when it attaches itself to the A site it contracts to the equilibrium O position and the two filaments have made a relative sliding motion of distance x. Note that the nearest Z band to which the A filament is anchored is out of the picture on the right-hand side.

The rate constants f and g of the reactions are assumed to have the form shown in Fig. 4.19. This figure has the same x coordinate system as Fig. 4.18, in that M is to the right of O. It is assumed that f, the rate constant of forming the AM linkage, increases with distance to the right of O until point h, where

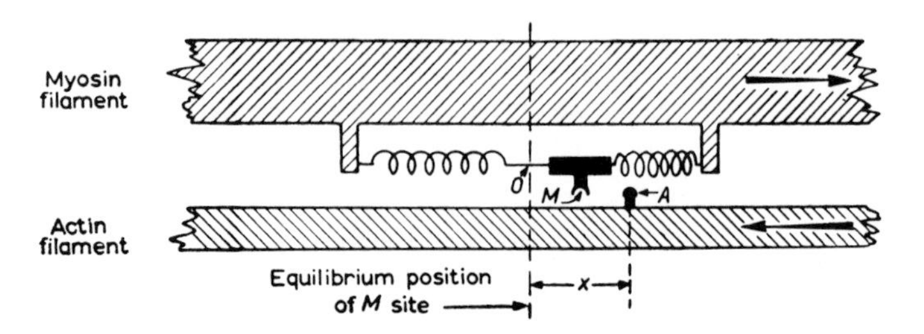

FIG. 4.18 Schematic of an assumed model of the right-hand half of the A band of a single site of muscular contraction. The myosin head is labelled M, an actin bonding site is labelled A, and O is the equilibrium position of the myosin head. The arrows show the directions of relative motions of the two filaments when the muscle shortens. [From Huxley (1957).]

the linkage is formed. g, the rate constant for breaking the linkage, is smaller than f on the right but increases abruptly to a large finite value to the left of the zero position. The spring tension pulls the linked AM to point O, where the rate constant g for the breakage suddenly increases and the formation rate f goes to zero. This sudden increase in the link breakage rate prevents the spring from opposing further contraction by other links. Note that g to the left of point O remains finite. This means that at high rates of shortening there will still remain some unbroken links and they will create a resistance to muscle contraction. Thus the limit to the speed of shortening will occur when the resistance of the unbroken links to the left of O equals the force of contraction of the links to the right of O. Note also that although the rate constants have been made linear and given the values required to match the quantitative values produced by Hill's experiments, Huxley showed that the functional form of the Hill equation could also be generated with rate constants that vary exponentially with distance, although the resulting parameters do not agree as well with the experimental ones.

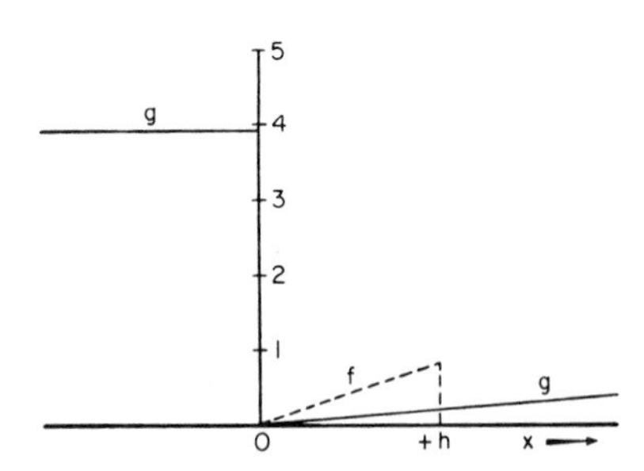

FIG. 4.19 Dependence of rate constants f and g of (Eq. (4.3)). ---, rate constant f for formation of actin–myosin bond; ———, rate constant g for breaking bond. The unit of the y axis is the value of $f + g$ at $x = h$. [From Huxley (1957).]

Let there be a large number of contractile sites, all having the same value of x, and let n be the fraction of these in which the AM linkage exists. We may write the rate of change of n with time (Section A.7) as

$$dn/dt = (1 - n)f - ng \tag{4.4}$$

where the rate of formation is $(1 - n)f$ and the rate of breakage is ng.

If the velocity of motion of the A and M sites sliding past each other by the force of the spring is v we may write $v = -dx/dt$ or $dt = -dx/v$, and therefore write Eq. (4.4) as

$$-v(dn/dx) = f - (f + g)n \tag{4.5}$$

If the sarcomere has length s and V is the rate of shortening of the whole muscle then $v = sV/2$. The factor 1/2 is necessary because in Fig. 4.18 it is seen that in sliding only half the distance is traversed by the actin, and hence it has only half the relative velocity. Thus, in terms of muscle length per second, Eq. (4.5) may be written

$$-(sV/2)(dn/dx) = f - (f + g)n \tag{4.6}$$

Referring now to the reaction scheme of Eqs. (4.3), one phosphate group is split off, liberating energy e for each site in one cycle of the reaction. If l is the distance between the reaction sites on the A filament and v the sliding velocity, the frequency of A sites coming into interaction distance with the M sites is v/l. The number of reactions is then the product of this frequency times the number available (rate of formation $(1 - n)f$) times a time increment dt, i.e.,

$$(v/l)\int_{x=\infty}^{x=-\infty} f(1 - n)\,dt, \quad \text{or} \quad (1/l)\int_{-\infty}^{\infty} f(1 - n)\,dx \tag{4.7}$$

It should be noted here that f is a function of x (see Fig. 4.19), and although the limits of the integral are nominally from $x = -\infty$ to $x = \infty$, in actuality it has value only in the region where f is nonzero. If m is the number of M sites per cubic centimeter of muscle, the rate of energy liberation per cubic centimeter E is therefore me times Eq. (4.7), or

$$E = (me/l)\int_{-\infty}^{\infty} f(1 - n)\,dx \tag{4.8}$$

The tension P in the muscle can be found by introducing the spring constant k of the moving M site. Since the force required to extend a spring is kx and work is Fdx, the work done at one myosin site while one actin site is carried past it is

$$\int_{-\infty}^{\infty} kx\,dx \tag{4.9}$$

and for n sites it is n times this. Again, although the nominal integration limits of Eq. (4.9) are $x = -\infty$ to $x = \infty$ the actual length involved is l, the distance between the A sites. We may then write that the average force exerted by the combined sites as the total work divided by the distance, or

$$\text{average force} = (1/l) \int_{-\infty}^{\infty} nkx \, dx \tag{4.10}$$

Using the notation of Hill, where P = tension = average force, we note that this tension is the sum of the tensions generated by all of the contraction sites in half a sarcomere (assuming that an average of half the A sites are on one side of the M sites). The number of these sites for a muscle of 1-cm^2 cross-sectional area is $ms/2$, where m is the number of sites per cubic centimeter and s is the sarcomere length. Therefore, the tension P per cm^2 of muscle is

$$P = (msk/2l) \int_{-\infty}^{\infty} nx \, dx \tag{4.11}$$

In order to evaluate the integrals for E and P the functional form of the rate constants shown in Fig. 4.19 are written as

$$x < 0 \qquad (f = 0 \text{ and } g = g_2)$$

$$0 < x < h \qquad (f = f_1(x/h) \text{ and } g = g_1(x/h))$$

$$x > h \qquad (f = 0 \text{ and } g = g_1(x/h))$$

With f and g now expressed functionally, Eq. (4.6) can be solved for n, which is then substituted into Eqs. (4.8) and (4.11) to obtain E and P. The solution of Eq. (4.6) with V being positive as the muscle is shortened is

$$x < 0 \qquad n = \frac{f_1}{f_1 + g_1}\left[1 - \exp\left(-\frac{\phi}{V}\right)\right]\exp\left(\frac{2xg_2}{sV}\right) \qquad \text{where}$$

$$\phi = (f_1 + g_1)h/s$$

$$0 < x < h \qquad n = \frac{f_1}{f_1 + g_1}\left\{1 - \exp\left[\left(\frac{x^2}{h^2} - 1\right)\frac{\phi}{V}\right]\right\}$$

$$x > h \qquad n = 0$$

When these relations for n in the different regions are inserted into Eq. (4.8) and the integral evaluated, the energy E is the sum of two terms, or

$$E = me\frac{h}{2l}\frac{f_1}{f_1 + g_1}\left\{g_1 + f_1\frac{V}{\phi}\left[1 - \exp\left(-\frac{\phi}{V}\right)\right]\right\} \tag{4.12}$$

When $V = 0$ this equation gives the isometric heat liberation rate of

$$E = me\frac{h}{2l}\frac{f_1 g_1}{f_1 + g_1} \tag{4.13}$$

which is a quantity determined experimentally by Hill (1938) that can be used in the evaluation of the unknown constants.

The forms of n and f can be used in a similar fashion to evaluate the integral of Eq. (4.11). The result is

$$P = \frac{msk}{2l}\frac{f_1}{f_1 + g_1}\frac{h^2}{2}\left\{1 - \frac{V}{\phi}\left[1 - \exp\left(-\frac{\phi}{V}\right)\right]\left[1 + \frac{1}{2}\left(\frac{f_1 + g_1}{g_2}\right)^2 \frac{V}{\phi}\right]\right\} \tag{4.14}$$

The rate of doing mechanical work is the product PV and the rate of heat liberation is $E - PV$.

Hill's data (Figs 4.3 and 4.4 and Eq. (4.1) and (4.2)) were used by Huxley to verify Eqs. (4.12) and (4.14). The adjustable parameters

$$\frac{kh^2}{2e}\frac{f_1 + g_1}{f_1} \quad \text{and} \quad \frac{f_1 + g_1}{g_2}$$

were varied to obtain the best fit. The results are shown in Figs. 4.20 and 4.21. Hill had found from experiments on a variety of muscles that $a/P_0 = \frac{1}{4}$

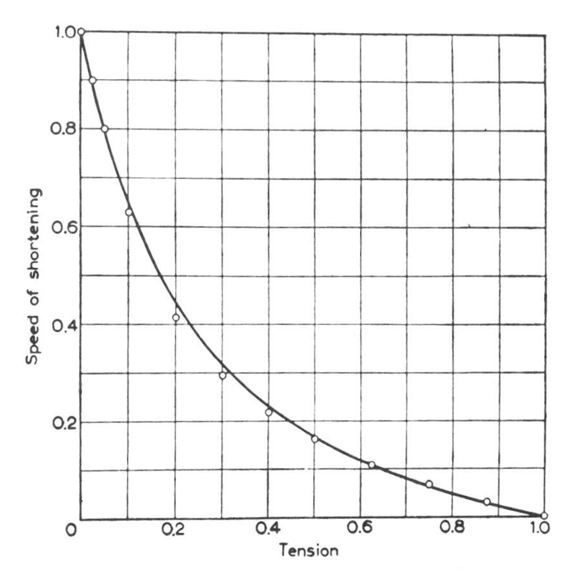

FIG. 4.20 Comparison of the Hill experiment with the A. F. Huxley theory. ——, Hill's equation obtained from experimental data; O, points obtained from Eq. (4.14). [From Huxley (1957).]

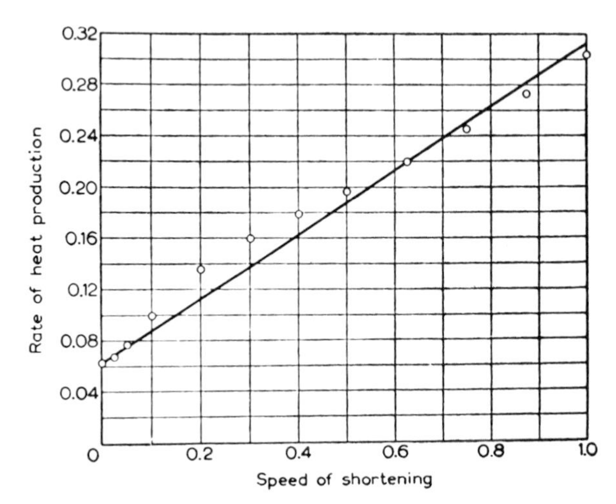

FIG. 4.21 Relationship between rate of heat production and speed of shortening. ———, Hill's data; ○ Eq. (4.12). [From Huxley (1957).]

was a fairly constant value. The solid line of Figs. 4.20 and 4.21 were calculated from the Hill equations on this basis. The open circles are calculated points from Huxley's Eq. (4.12) for Figs. 4.21 and from Eq. (4.14) for Fig. 4.20. The agreement is seen to be quite good and the total rate of heat liberation seen in Fig. 4.22 also agrees favorably with the data of Fig. 4.3.

The time constant following muscle stimulus has been found to be about 150 msec. Since the time constant is of the order of $1/g_1$, g_1 can be estimated to be about 10 sec^{-1}. The sarcomere length s is found by electron microscopy, and the other constants are obtained from Hill's data. When these values are used with the adjustable constants in Huxley's equations, $l = 153$ Å and $h = 156$ Å. It is disturbing that l and h have nearly the same value since an implicit assumption in the model was that l is considerably greater than h. This discrepancy has not been resolved, although with so many parameters involved and some measurements taken on dead fibers complete agreement is not really expected. The l distance does agree favorably with the x-ray electron microscopy measurements of Huxley (1953a, b) and of Hanson and Huxley (1955). Estimates of the spacings of the actin and myosin molecules along their respective filaments can be made from their molecular weights and lattice spacings. If the myosin filament has six molecules abreast, one facing each of the six surrounding actin filaments, and longitudinal spacing comes out to be 400 Å. Referring to Fig. 4.10, it is seen that each actin filament is surrounded by three myosin filaments. Thus the spacing of sites on the actin filaments is expected to be 400/3 or 135 Å, in rather good agreement with the calculated l distance of 153 Å.

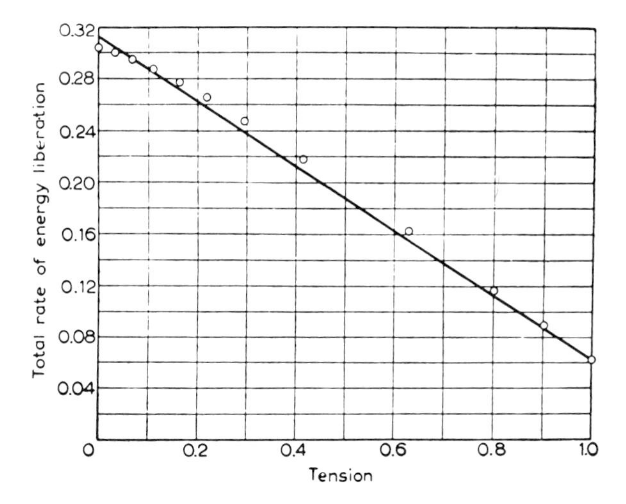

FIG. 4.22 Total rate of heat liberation versus tension. ———, Hill's data; ○, Huxley's theory. [From Huxley (1957).]

The model presented here is a physical one. A chemical model of the implicit mechanisms has been proposed by Davies (1963). Even the physical model at this point is not complete, however, because it does not explain the tension and energy data for muscle when it is in a stretched or contracted position. The implications of this will be discussed in Chapter 6 under Starling's law. However, the present approach is a satisfactory interpretation of existing data and it is expected that new models or theories must use this basic one as a starting point.

REFERENCES

Bourne, G. H. (ed.) (1960). "Structure and Function of Muscle," Vols. I–III. Academic Press, New York (see also the new edition, 1973).

Clark, W. E. LeG. (1952). "The Tissues of the Body." Oxford Univ. Press, London and New York.

Cohen, C. (1975). The protein switch of muscle contraction, *Sci. Amer.*, November, p. 36.

Cold Spring Harbor Laboratory (1973). The mechanism of muscle contraction, *Cold Spring Harbor Symp. Quant. Biol.* **37**.

Davies, R. E. (1963). A molecular theory of muscle contraction: Calcium dependent contractions with hydrogen bond formation plus ATP dependent extensions of part of the myosin-acton cross-bridges, *Nature (London)* **199**, 1068.

Ebashi, S., Endo, M., and Ohtsuki I. (1969). Control of muscle contraction, *Quart. Rev. Biophys.* **2**, 321.

Gordon, A. M., Huxley, A. F., and Julian F. J. (1966). The variation in isometric tension with sarcomere length in vertebrate muscle fibers, *J. Physiol.* **184**, 170.

Hanson, J., and Huxley, H. E. (1955). The structural basis of contraction in striated muscle, *Symp. Soc. Exp. Biol.* **9**, 228.

Hill, A. V. (1938). Heat of shortening and the dynamic constants of muscle, *Proc. R. Soc. London* **126B**, 136.

Hill, A. V. (1970). "First and Last Experiments in Muscle Mechanics." Cambridge Univ. Press, London and New York.

Hoyle, G. (1970). How is muscle turned on and off? *Sci. Amer.*, April, p. 85.

Huxley, A. F. (1957). Muscle structure and theories of contraction, *Progr. Biophys. Biophys. Chem.* **7**.

Huxley, A. F., and Simmons, R. M. (1971). Proposed mechanism of force generation in striated muscle, *Nature* (*London*) **233**, 533.

Huxley, H. E. (1953a). Electron microscope studies of the organization of the filaments in striated muscle, *Biochim. Biophys. Acta* **12**, 387.

Huxley, H. E. (1953b). X-ray analysis and the problem of muscle, *Proc. R. Soc. London* **B141**, 59.

Huxley, H. E. (1956). Muscular contraction, *Endeavor*, October.

Huxley, H. E., and Hanson J. (1954). Changes in the cross-striations of muscle during contraction and stretch and their structural interpretation, *Nature* (*London*)**173**, 973.

Huxley, H. E., and Hanson, J. (1960). The molecular basis of contraction, *in* "Structure and Function of Muscle" (G. H. Bourne, ed.), Vol. I. Academic Press, New York.

Morales, M. F. (1959). Mechanisms of muscle contraction, *Rev. Mod. Phys.* **31**, 426.

Needham, D. M. (1960). Biochemistry of muscular action, *in* "Structure and Function of Muscle" (G. H. Bourne, ed.), Vol. II. Academic Press, New York.

Ridgway, E. B., and Gordon, A. M. (1975). Muscle activation: Effects of small length changes on calcium release in single fibers, *Science* **189**, 881.

Szent-Gyorgyi, A. G. (1960). Proteins of the myofibril, *in* "Structure and Function of Muscle" (G. H. Bourne, Ed.), Vol. II. Academic Press, New York.

Walls, E. W. (1960). The microanatomy of muscle, *in* "Structure and Function of Muscle" (G. H. Bourne, ed.), Vol. I. Academic Press, New York.

Wilkie, D. R. (1968). "Muscle" (*Stud. Biol. No. 11*). Arnold, London.

Winegrad, S. (1968). Intracellular calcium movements of frog skeletal muscle during recovery from tetanus, *J. Gen. Physiol.* **51**, 65.

CHAPTER 5

Bone: Mechanical and Electrical Properties

INTRODUCTION

Bone is a living tissue of the body, composed of organic and inorganic material in dynamic equilibrium with the body fluids. Although the proportions vary between parts of the skeleton and with age, roughly one third is organic and two thirds inorganic.

Cells in bone are continually being dissolved and reformed so that the bone is being remodeled during life. The remodeling rate in man is quite noticeable during the first three decades of life. It is a widely held belief among students of bone that such remodeling is not only necessary for growth but also to keep the mineral part of the bone available for exchange with body fluids.

It is a well-known phenomenon that bone develops in regions of stress and dissolves where there is no stress. This is quite observable in x rays of fractures some months after they are healed; the bulky growth that surrounds the fracture site when it is immobilized will readjust by dissolution and growth to the original shape of the bone. It has also been observed that a patient during bed rest can lose up to 0.5 g of calcium per day. In the first space orbital flights in gravity-free conditions the astronauts lost up to 3 g of calcium per day. It was at first feared that space flight for man was not possible, but a controlled exercise program was developed that greatly reduced this loss.

A living system develops in a way that minimizes the energy required to exist. Sufficient bone develops to support the body adequately with some reasonable safety factors. Wolff (1892) stated that *bone responds to mechanical stress by differential growth so as to resist the applied stress.* This is now known as Wolff's law. There is clearly a fundamental mechanism at work that controls this growth and dissolution, and much study has been directed at explaining this phenomenon. Within the past decade it has been demonstrated that the mechanism is the piezoelectric effect, which characterizes the connection between stress and electric potential in solids. Based on this knowledge it has also recently been shown that the proper use of electric potential can accelerate the healing of fractures and this will be discussed later. It is the aim of this chapter to review some of these mechanical properties of bone and to clarify the growth phenomenon.

MECHANICAL STRUCTURES

If a long axially symmetric wooden beam is supported at the ends it will sag in a curve as in Fig. 5.1. Because of this sag the fibers of the lower surface will be stretched and those of the upper surface will be compressed. This is obvious because the radii of curvature of the upper and lower surfaces differ and therefore the arc lengths differ. If one surface is longer and the other shorter than the original beam length then clearly somewhere between the two is an arc that has the length of the unbent beam. This is called the neutral axis and in this region the fibers are not stressed either way. Any supporting structure such as a column in a building or the shaft of a thigh bone that is unevenly loaded will have a tendency to bend and therefore have a compressional stress on the concave side and tensile stress on the convex side.

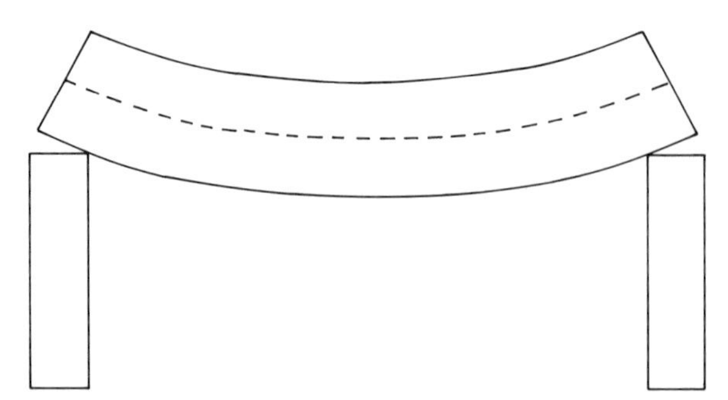

FIG. 5.1 Sketch of a wooden beam, symmetric about the x axis, supported at the ends (exaggerated). The dashed line is the only concentric arc that has the original length.

An engineer, in designing a supporting structure, has to locate the lines of tension and compression and use materials most suitable to resist the two types of stress. Thus a suspension bridge has steel cables of high tensile strength to support the road bed and concrete pillars of high compressive strength to support the entire bridge structure. The situation is more complex in designing a dynamic rather than a static structure because other mechanical properties must be considered, such as torsional strength, impact strength, fatigue life, etc.

Returning now to our wooden beam model, if the neutral zone is under neither tension nor compression there is no longitudinal stress on it and it is making no contribution to the structure except to hold together the parts under stress. Since such a holding operation requires less material than the stress parts, to save weight and material much of it can be whittled away. The result is the well-known I-beam, which contributes almost the same support as a solid bar with a square cross section of the same dimensions. If the material of the I-beam were stronger in compression than in tension one side would have to be made thicker than the other. Then all I-beams could be used in only one direction unless both sides were made thicker. Thus there is economy if a material is selected that has comparable strength both in tension and compression. Still the I-beam is designed to resist bending in one direction. It could be made equally strong in all directions by designing an I-beam that has been rotated about its long axis. The result is a cylinder and there is no need for a substance in the center to hold the sides together. If the walls of the cylinder are made too thin for the load to be borne, the cylinder can buckle. At this point an engineering study must be made for material economy. Material must be used either to thicken the walls or to construct struts inside the cylinder. Material economy is still possible at this point because in the bending of a cylinder the stress is greatest at the middle (a cylinder of large length to width ratio never buckles near the ends). Thus the thickening of the walls or the internal supports must be greatest near the middle and can be reduced towards the ends.

BONE STRUCTURE

Every physics student has seen the way in which iron filings line up along the flux lines in a magnetic field. If the deposition of bone is along the electrical fields associated with the stress lines one expects a similar alignment, and indeed this is the case.

In a bird, reduction in bone weight is highly desirable. The metacarpal wing bone, which supports the long primary flight feathers, must resist bending. Nature's economy was therefore to use cross members instead of

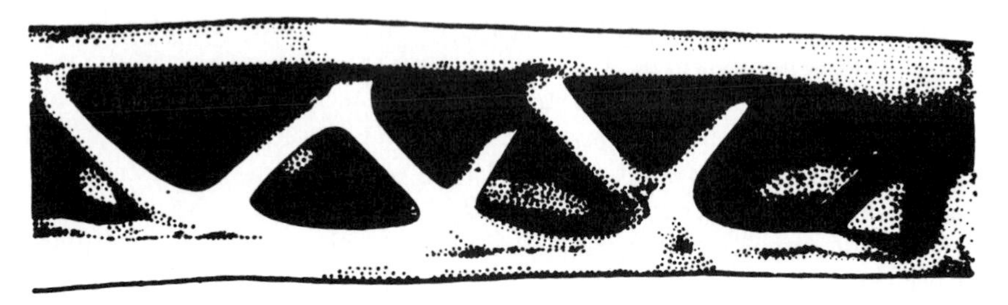

FIG. 5.2 Metacarpal bone from a vulture's wing. [From O. Prochnow, "Formenkunst der Natur," (1934). Reproduced in Thompson (1971).]

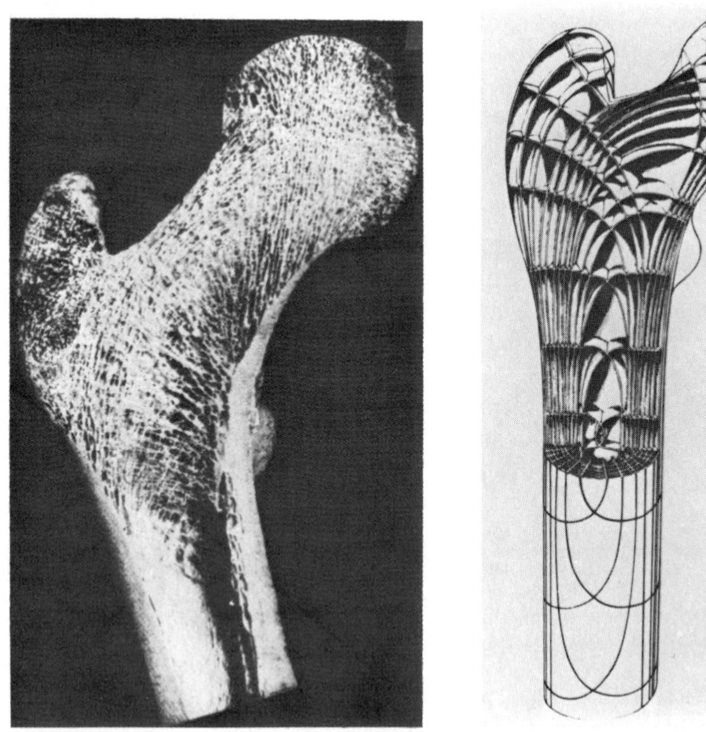

FIG. 5.3 FIG. 5.4

FIG. 5.3 Cross section of the head of a human femur showing some of the trabeculae. [From Spalteholz (1943).]

FIG. 5.4 Theoretical construction of three-dimensional trajectorial system of stress in a femur model. [From Kummer (1966).]

thickening the cylinder. Figure 5.2 shows the metacarpal bone of a vulture's wing, which looks identical to the truss bridge designed in steel by engineers.

The human thigh bone is an example of both kinds of strengthening. The long straight portion is a hollow cylinder with walls thicker at the center and becoming thinner towards the ends. This long region supports the body weight in a vertical direction in the same manner as a column may support a building, and the strengthening is to prevent buckling. The end is curved as it approaches the joint and the bending stresses are appreciable. Very thin spidery formations of bone, called *trabeculae*, fill the interior here, as shown in Fig. 5.3. A computer solution of the three-dimensional equal stress lines is shown in Fig. 5.4 and it is seen that the trabeculae follow these. The right-angle crossing of stress lines is particularly visible in Fig. 5.5. Trabeculae have been referred to by bioengineers as crystallized stress lines.

The moduli of elasticity and breaking strengths in both compression and tension of bone are shown in Fig. 5.6 for the human femur. It is seen, however, that it is not designed as strongly in the transverse direction. Table 5.1 gives some typical measured values of the properties indicated in this figure. These data were taken with static loads.

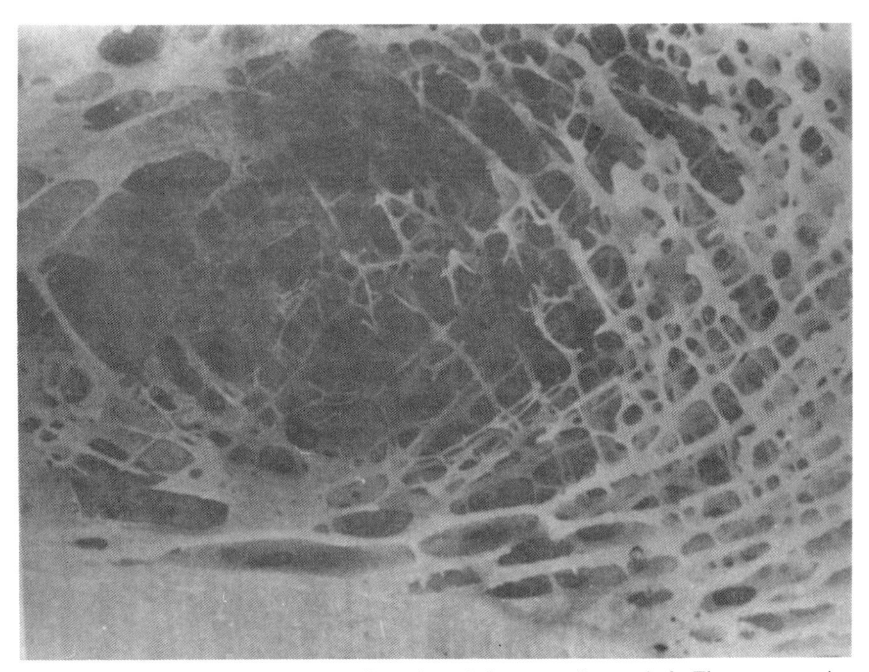

FIG. 5.5 The trabeculae in a small section of the upper femur shaft. The empty region surrounds a zone of zero stress. [From Kummer (1966).]

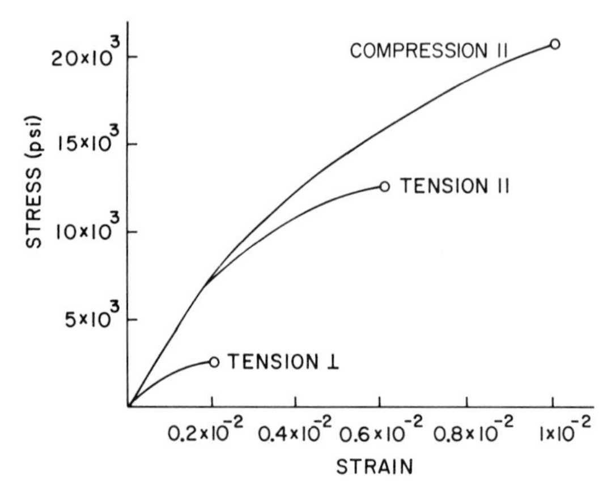

FIG. 5.6 Schematic of typical response of compact mammalian bone in stress–strain experiment. ○, failure point; ||, parallel to bone axis; ⊥, perpendicular to axis.

TABLE 5.1[a]

Mode direction	Modulus ($\times 10^{-6}_{psi}$)	Fracture Stress ($\times 10^{-3}_{psi}$)
Tension parallel	2.3	12
Tension perpendicular	—	1.4
Compression parallel	1.2	20
Synthetic hydroxyapatite	15	
Collagen	0.2	

[a] Original data tabulated and averaged by Welch (1970).

MICROSCOPIC STRUCTURE

If one examines a cross section of bone under a light microscope the appearance is something like the sketch in Fig. 5.7. The basic structure is of tubular elements known as *osteons*. It has been shown by Hulbert *et al.* (1971) that, when using a calcium aluminate ceramic for implantation, the bone will fully impregnate only those implants with porosity diameters of 100 μm or greater. This indirect measurement gives the viable growth size of the osteon.

The osteon tubes are oriented roughly in line with the axis of the bone, although sometimes they spiral around the axis. Since circular columns do not pack tightly, there are filled regions between the tubes called *interstitial lamellae*. Both the osteons and the interstitial lamellae have a laminated structure and between these laminates are disk-shaped cavities called *lacunae*. Within these lacunae are fine canals called *canaliculi*. These canaliculi

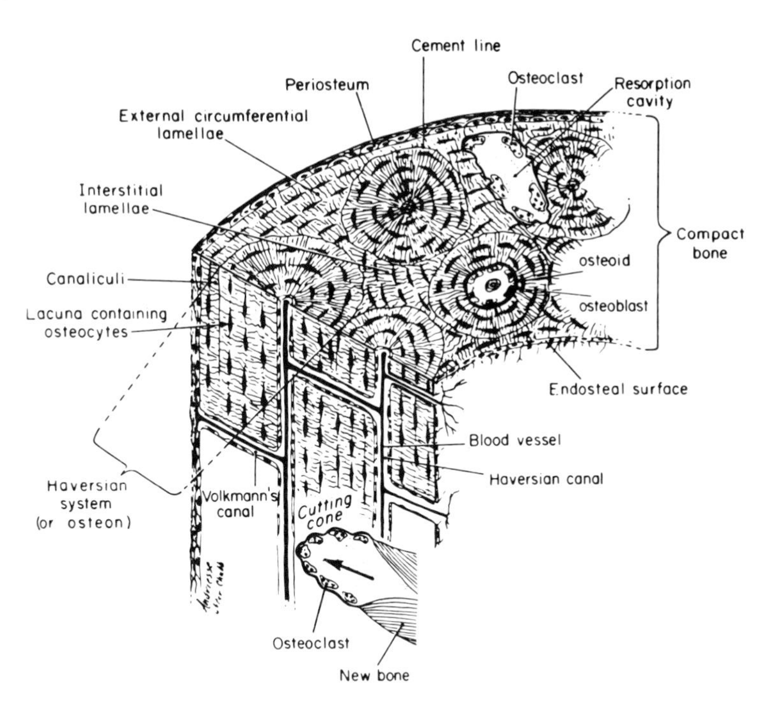

FIG. 5.7 Three-dimensional diagram of shaft of long bone showing both a longitudinal section and a cross section. [Reprinted by permission from W. H. Harris and R. P. Heaney, *New England J. Med.* **280**, 193 (1969).]

feed the bone cells, the *osteocytes*. Since this feeding is through such a fine system, which probably does not operate well over long distances, the concentric arrangement allows the osteocytes to be located near a capillary that runs through the center of the osteon. This central tube that contains the capillary is called the *Haversian canal*. Within the walls of this canal are cells called *osteoblasts*, which form bone cells. In addition, if bone resorption is taking place, a special cell called the *osteoclast* appears from an unknown source, devours large portions of bone and then disappears, all within a 48-hr period. In addition to the other canals, a larger type, called *Volkmann's canals*, carry larger blood vessels which supply the capillaries in the Haversian canals. They pass at right angles to the long axis of the bone and open on either the inside (*endosteal*) or outside (*periosteal*) surface of the bone.

The osteoblasts apparently secrete an organic substance which is polymerized into a *collagen* fiber. These fibers form the organic matrix of bone. Within the extracellular fluid of the body are both calcium ions Ca^{++} and monohydrogen phosphate ions HPO_4^{--} in concentrations of 0.0012 mole/liter and 0.0011 mole/liter, respectively. If the product of the concentrations of two ions that form a salt rises above a certain critical value, called the *solubility product*, the salt will precipitate. The solubility product for the precipitation of $CaHPO_4$ at body temperature is 3.4×10^{-6}. It is seen from the above numbers that the product of the ion concentration is only 1.32×10^{-6}, which means that calcium phosphate would not be expected to precipitate. However, the collagen fibers formed by the osteoblasts seem to provide a special condition that permits the precipitation of $CaHPO_4$. This initial precipitate forms the nucleus for further precipitation of Ca^{++} and HPO_4^{--} ions, and the crystal not only grows but transforms into a more stable type of bone mineral, $[Ca_3(PO_4)_2] \cdot Ca(OH)_2$, called *hydroxyapatite.* Note that since this is a chemical system, a change in body fluid pH by a prolonged condition of either acidosis or alkalosis can alter the mineralization of the bones.

ULTRASTRUCTURE

The collagen causes the precipitation of calcium phosphate on itself epitaxially, as shown by Marino and Becker (1970), that is, the oriented overgrowth of bone mineral on certain comparably crystallized areas of the organic matrix. Such epitaxial growth apparently also causes conversion of the mineral into hydroxyapatite. The role of collagen is not finished at this point. It controls the shape and the size of the crystals, roughly 250-Å by 50-Å diameter rods, although they may have flat surfaces. The collagen bonds itself to the crystal and distributes the crystals in a uniform array on itself somewhat like the diagram in Fig. 5.8. Bundles of these fibers with their attached crystals form the osteon lamellae with the fiber axes in the plane of the lamellae.

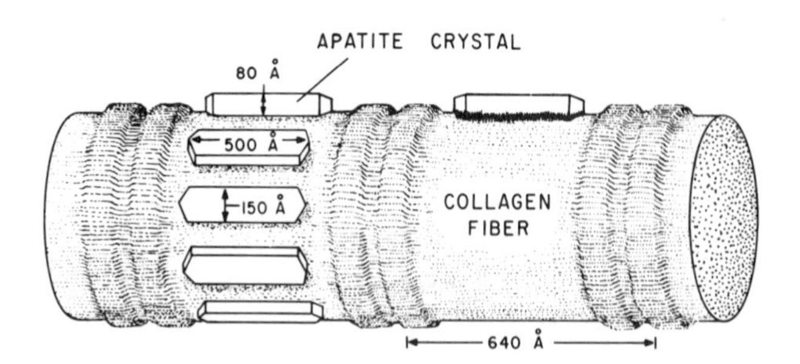

FIG. 5.8 Schematic of the hydroxyapatite arrangement on a collagen fiber. [From Becker (1967).]

This arrangement is an ideal form of a simple composite material, and Currey (1964) has reasonably argued that it behaves more like a two-phase material than as a prestressed material such as prestressed concrete. The mechanical properties of two-phase composite materials are derived in engineering texts (e.g., Lowenstein, 1966), and it is instructive to see how good the approximation is. Young's modulus of a composite material can be written as the sum of the volume fractions of the two materials times their respective moduli with an angular function, called the *efficiency of reinforcement*, that depends on the relative angle between the stress axis and the fiber angle. The modulus of the bone G_B would be written as

$$G_B = G_{HA} V_{HA} \eta + G_C(1 - V_{HA}) \tag{5.1}$$

where G_{HA} is the modulus of hydroxyapatite, G_C the modulus of collagen, V_{HA} the volume fraction of hydroxyapatite, and η the efficiency of reinforcement equal to $\cos^4 \phi$, where ϕ is the angle. For $G_{\parallel}$ parallel to the stress axis, which will be assumed parallel to the fiber axis, $\phi = 0°$ and $\cos^4 \phi = 1$. For $G_{\perp}$ perpendicular to the stress, $\phi = 90°$ and $\cos^4 \phi = 0$. Therefore we may write the ratio of $G_{\parallel}$ to $G_{\perp}$ from Eq. (5.1) as

$$\frac{G_{\parallel}}{G_{\perp}} = \frac{G_{HA} V_{HA} + G_C(1 - V_{HA})}{G_C(1 - V_{HA})} = 1 + \left(\frac{V_{HA}}{1 - V_{HA}}\right)\frac{G_{HA}}{G_C} \tag{5.2}$$

Since the mineral content of bone is about twice that of organic material, $V_{HA} = \frac{2}{3}$. The elastic modulus of synthetic hydroxyapatite has been measured to be 15×10^6 psi, and that of collagen as 0.2×10^6. Inserting these into Eq. (5.2) the result is that $G_{\parallel} \equiv 150 G_{\perp}$. The data of Table 5.1 suggest a ratio of about 2; thus the calculation is in error by about a factor of 75. This discrepancy is analyzed by Welch (1970) in terms of the applicability of Eq. (5.1). The major error may arise from using the moduli of a large crystal of hydroxyapatite for that of a crystal of about 100 Å in size.

PIEZOELECTRICITY

Piezoelectricity means *electricity caused by pressure*. We will consider its origin in an ionic crystal. Figure 5.9 shows two types of crystals in both normal and stressed condition. These crystals illustrate two types of symmetry. Figure 5.9a is called *centrosymmetric* because the figure can be rotated about a vertical axis by a multiple of 90° and reproduce the figure. (The multiple need not be 90°—any integral divisor of 360° will do.) In contrast, Fig. 5.9b cannot be reproduced by any angular rotation other than 360°. When stress is applied to crystal in Fig. 5.9a as indicated, the crystal expands in the horizontal direction. But because of its original symmetry the ions expand uniformly and there remains a charge balance on the surfaces.

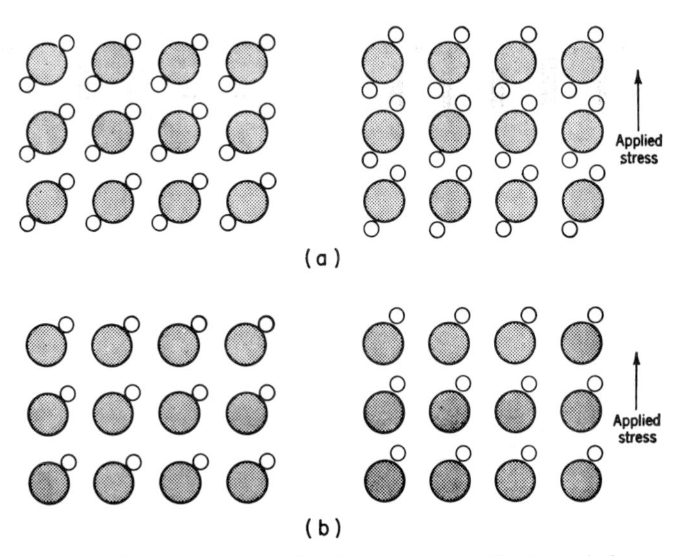

FIG. 5.9 Schematic of piezoelectric effect in ionic crystals. The large circles represent ions of charge opposite to that of the small circles. (a) When a stress is applied to a crystal with a center of symmetry no net change in charge on the surfaces will appear. (b) In contrast, when there is no center of symmetry an applied stress will cause a change in relative charge on the surfaces. [From "Introduction to Solids" by L. V. Azaroff. Copyright 1960 McGraw-Hill Book Company. Used with permission of the McGraw-Hill Book Company.]

The ions of the basic molecule of the crystal in Fig. 5.9b form small dipoles because the charge on a molecule that is not symmetric is never symmetrically distributed. Thus when the noncentrosymmetric crystal is subjected to linear stress the molecules tend to rotate slightly, resulting in more of one type of ion appearing on one surface than the other. Thus there is a potential difference between the surfaces.

If the applied stress is σ and the resultant strain $\Delta l/l$ is ε, Hooke's law relates them by the elastic modulus G as

$$\sigma = G\varepsilon \tag{5.3}$$

In the case of a piezoelectric charge, a polarization density P (which could be measured with a voltmeter) is also proportional to the stress, or

$$P = \delta\sigma \tag{5.4}$$

where δ is called the *piezoelectric constant*. It is evident that just as the elastic modulus would vary with direction of applied stress in an anisotropic crystal, so would the piezoelectric constant. Therefore, several of these constants are required to completely describe each type of crystal. Although we have used only a two-dimensional crystal to describe a piezoelectric system, a noncentrosymmetric polymer may exhibit the same properties. One such polymer, poly(vinylidenefluoride), is currently used as a microphone transducer.

PIEZOELECTRIC PROPERTIES OF BONE

It was first reported by Fukada and Yusada (1957) that if a fresh dog femur was clamped at one end and weights added at the other end then an electric potential difference between the top and bottom of the bone was measurable; the underside, which was in compression, was negative with respect to the upper side, which was in tension. Subsequent measurements indicated that the piezoelectricity originated in the collagen rather than in the hydroxyapatite.

More detailed measurements showed that the piezoelectric potential does not remain constant with a constant applied stress. The actual situation in whole bone is illustrated by the experiments of Steinberg *et al.* (1973). They placed fresh femurs from rats in a device that produced a four-point bending mode, as shown in Fig. 5.10. They placed a fixed load on the bone and measured both the deformation and the electric potential as a function of time. Their data are shown in Fig. 5.11. The load was put on the bone at a deformation rate of 0.5 in./min and immediately removed. The lower part of the Fig. 5.11 shows the deformation and its duration for different loads from 1 to 5 lb. The upper part of Fig. 5.11 shows the resulting potential—an immediate increase and a subsequent decrease, both coincident with the deformation. However, the potential reverses with complete unloading and takes about 20 sec to recover to 0. The second type of measurement was loading at the same deformation rate, maintaining constant deformation for 30 sec, and then measuring the potential and load for constant deformation as a function of time. These results are shown in Fig. 5.12 (loadings from 1 to

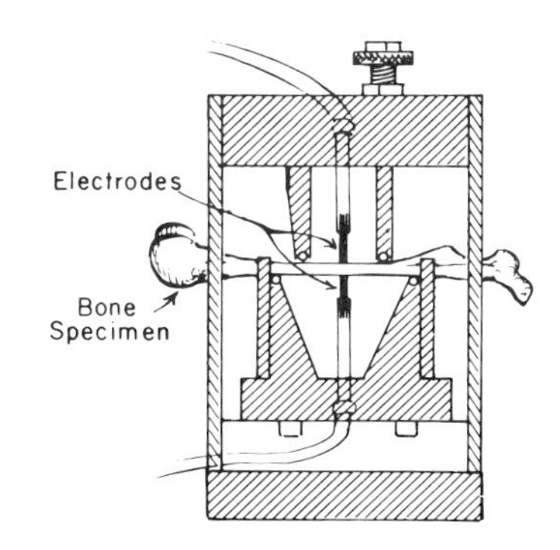

FIG. 5.10 Schematic of bone specimen in a four-point bending apparatus for simultaneous measurement of elastic and piezoelectric response of bone. [From Steinberg *et al.* (1973).]

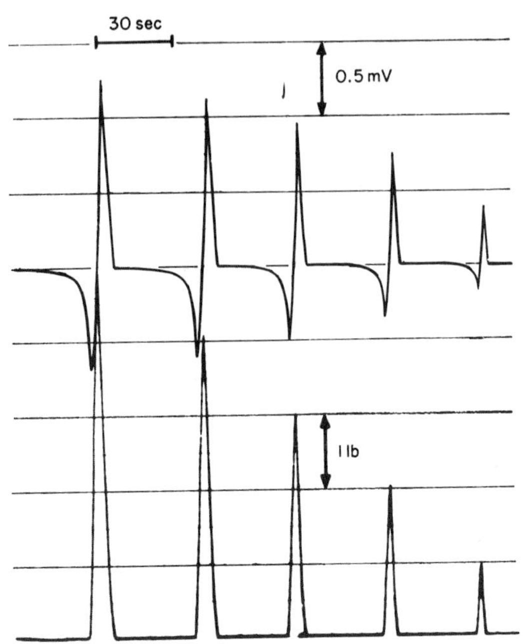

FIG. 5.11 Recording of electric potential (top) and load/deformation curve (bottom) for normal rat femur subjected to sequential loading of 1, 2, 3, 4, and 5 lb at a deformation rate of 0.5 in./min with immediate release. Time reads from right to left. [From Steinberg *et al.* (1973).]

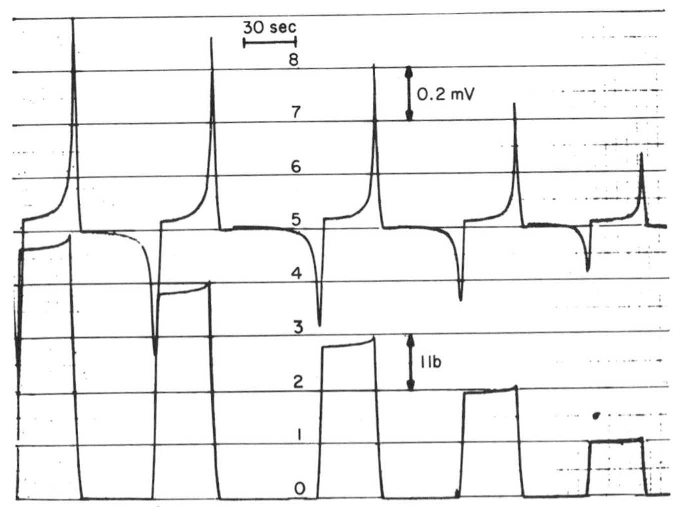

FIG. 5.12 Recording of electric potential (top) and load/deformation (bottom) from normal rat femur subjected to sequential loading of 1, 2, 3, 4, and 5 lb at a deformation rate of 0.5 in./min with each deformation maintained for 30 sec prior to release. Time reads from right to left. [From Steinberg *et al.* (1973).]

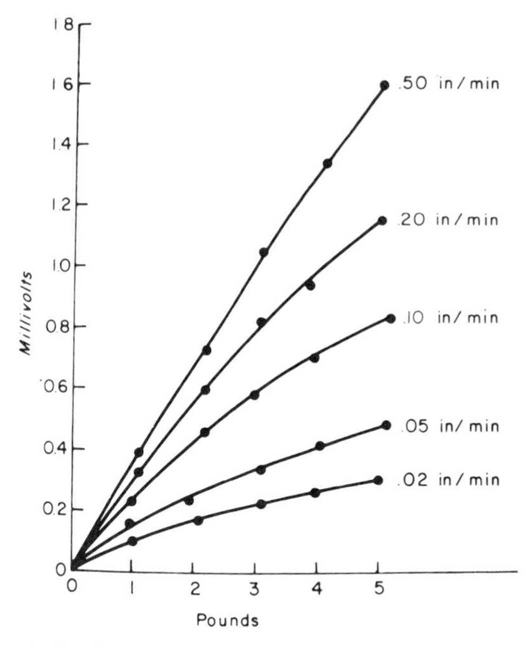

FIG. 5.13 Graph of peak voltage versus load at different rates of deformation of rat femur. [From Steinberg *et al.* (1973).]

5 lb). The lower part of Fig. 5.12 shows that the load does not remain constant but decreases with time. This indicates stress relaxation and demonstrates that bone is viscoelastic. Other measurements of viscoelasticity will be described in the next section. In the upper part of Fig. 5.12 the potential is seen to decay in a few seconds to a small constant value, which remains while the load is applied. Upon removal of the load it reverses itself as in Fig. 5.11 and under no load it returns to zero value in about the same time as it did under the experimental conditions of Fig. 5.11.

These experiments indicate that the unraveling of the piezoelectric behavior of whole bone is not so simple as in crystals. From further data, similar to that in Figs. 5.11 and 5.12, the curves of Figs. 5.13 and 5.14 can be plotted. These figures show that the maximum voltage increases with increasing load, as expected from Eq. (5.4), but that time, expressed in the rate of loading, plays a significant role. This time dependence indicates that the voltage decreases either by charge leakage, through the bulk or on the surface of the bone, or by release of strain by viscoelastic relaxation. Possibly both occur, but the time constant of the decay of potential is comparable to that of stress relaxation, as seen in the curves of Fig. 5.12, which indicates that viscoelastic relaxation is the dominant factor in potential decrease.

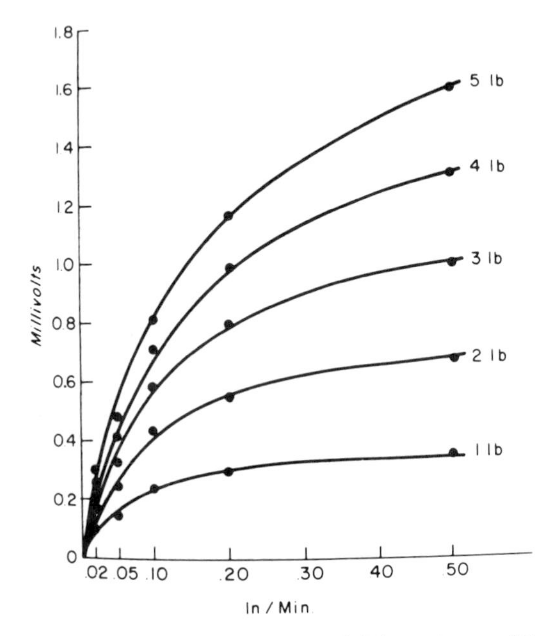

FIG. 5.14 Graph of peak voltage versus rate of deformation at different loads for rat femur. [From Steinberg *et al.* (1973).]

VISCOELASTIC PROPERTIES

The stress relaxation of sections of bovine bone, cut in different orientations, have been measured by Korostoff and his associates. Curves similar to the loading curves of Fig. 5.12 were studied in detail, as shown in Fig. 5.15, for specimens called, L-longitudinal, O-oblique, and T-transverse (0°, 45° and 90° to the long axis of the bone).

Following the treatment of viscoelasticity in Chapter 2 we wrote the stress as a function of time as an exponential decay or

$$\sigma = \sigma_0 e^{-Gt/\eta} \tag{2.4}$$

where σ is stress, G the elastic modulus, and η the viscosity. A logarithmic plot of such simple behavior would be linear in time with a slope of $-G/\eta$. However, the data do not lend themselves to such a simple analysis, which implies that there is a distribution of relaxation times rather than a single one. The curves of Fig. 5.15 are suggestive of at least two major processes, one with a short relaxation time of about 50 min, and another much longer one. The short-time relaxation of Fig. 5.12 is not seen in Fig. 5.15. However, Korostoff (1971) also observed a piezoelectric voltage with a corresponding reversal for 40-sec loadings.

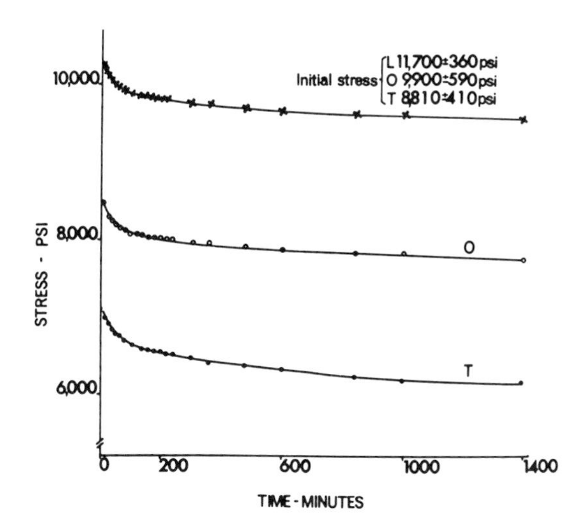

FIG. 5.15 Stress relaxation of bovine femur as a function of time. L, longitudinal to bone axis; T, transverse; O, oblique (45°). [From Korostoff (1971) *in* "Biomaterials" (A. L. Bement, Jr., ed.), University of Washington Press, Seattle, Washington. Copyright 1971 by University of Washington Press.]

The voltage data of Figs. 5.11 and 5.12 (as well as Korostoff's, which are not shown) indicate that there is a voltage reversal, as if after a partial relaxation the element somehow becomes stressed in the opposite direction. These figures suggest that this electric relaxation time is about 10 sec, a time much too short to be seen on the stress relaxation curves of Fig. 5.15, but clearly evident on the lower curves of Fig. 5.12.

In the initial model of viscoelastic elements proposed by Korostoff, he considered the short-time electric and stress relaxations, similar to those of Fig. 5.12. He proposed a model of random Voigt elements in parallel with springs, as in Fig. 5.16. In this situation the overall elastic response of bone is governed by an elastic constant G, the stress relief of Fig. 5.15 is controlled by a slow-moving and high-viscosity dashpot of viscosity η_1 and elastic constant G_1, and the fast piezoelectric response is controlled by a low viscosity element η_2 and low spring constant G_2. The upper part of this figure represents the tension aspect. In the lower part the same scheme is used for the compression aspect, in which the relaxation governing the piezoelectric response is for compression. Upon release of the applied stress this element is in tension and can give a reverse voltage response, which decays with the same time constant, as seen in the experimental behavior.

Although this is a preliminary approach to the explanation of the rather unusual viscoelastic and piezoelectric behavior of bone, a more sophisticated interpretation will certainly be along these lines.

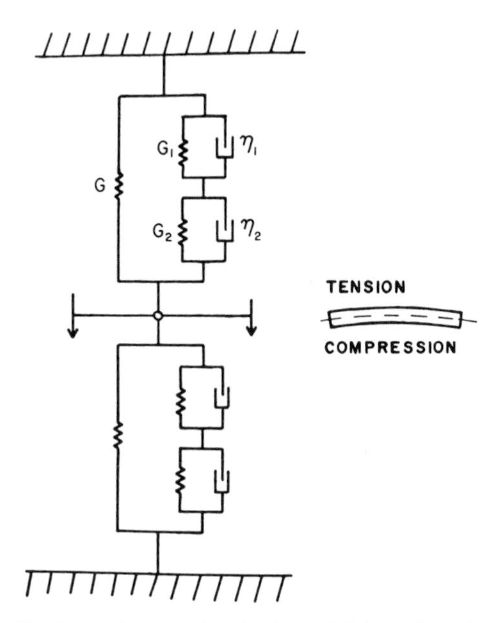

FIG. 5.16 Simplified two-element viscoelastic model for a bent bone specimen, where G_1, G_2, and η_1, η_2 are (respectively) the elastic moduli and viscosities of the two elements and G is a parallel elastic modulus. The upper part of this figure represents tension and the lower part compression. [From Korostoff (1971) *in* "Biomaterials" (A. L. Bement, Jr., ed.), University of Washington Press, Seattle, Washington. Copyright 1971 by University of Washington Press.]

These studies are reasonably fundamental, and a question may well be asked about the potential use of these results in medical physics. Two possible applications are immediately obvious. If one wished to accelerate bone growth by applying stress, or, alternatively, retard dissolution by exercise, the optimum frequency of stress application to maintain the highest average electric potential is obtainable from these data. Secondly, bone implantation is a useful surgical procedure. In designing implants, is it preferable that the material be piezoelectric, and if so what would be the optimum piezoelectric constant and how should the implant be oriented with respect to the stress axis? Also, if the implant is metallic, will the resulting charge leakage have a deleterious effect on the piezoelectric potentials produced by the bone and thereby retard healing or bone growth around the implant? These are important questions, which remain to be answered.

ACCELERATED FRACTURE HEALING BY ELECTRIC FIELDS

In the foregoing sections it has been shown that collagen fibrils develop first and thereby orient the growth of the hydroxyapatite crystals. How do

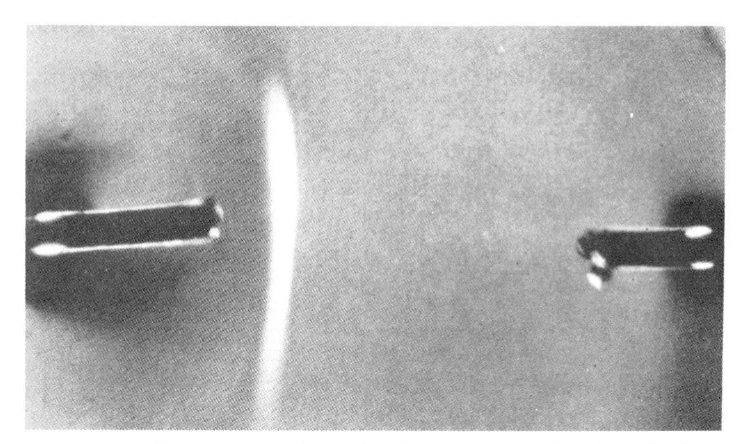

FIG. 5.17 Band of collagen in solution forming near a negative electrode. [From R. O. Becker, C. A. Bassett and C. H. Bachman, *in* "Bone Biodynamics" (H. M. Frost, ed.), Little, Brown and Co., Boston, Massachusetts, 1964. Copyright 1964 by Little, Brown and Co.]

the collagen fibrils know how to orient themselves? A simple experiment was performed by Becker *et al.* (1964), which answers this. They took a solution of acid-soluble collagen and placed two electrodes in it. In about 5 min a long curved band of collagen appeared, as in Fig. 5.17, closer to the cathode than to the anode and with the concave side toward the cathode. The weaker the current the closer was the band formation to the cathode. If the field was turned off, the band disappeared, and upon reversing the field the effect appeared around the other electrode. The currents involved, about 1 μA, are about the same as that developed in stressed bone. If salt was added to the solution when the band was formed, the band did not disappear when the current was turned off or reversed. Clearly, the piezoelectric fields of stressed bone orient collagen fibers. Recall that when a bone was stressed the side in compression was negative. From the skeletal structure of man the femur tends to bend outward, i.e., compressional stress is on the inward side. This becomes the negative side where the collagen fibers would preferentially align themselves and extra bone growth would be expected. These experiments offer a reasonable explanation for Wolff's law.

The finding that electric fields cause bone growth and dissolution led to a period in the last decade of implantation of small electrodes and a battery. This technique was shown to accelerate fracture healing and dissolution of cancerous bone growth. However, there was considerable controversy as to its true effectiveness, and coupled with the problems associated with such an invasive technique, it has not been generally accepted. A recent experiment in which electric fields were applied by a noninvasive technique will now be described.

There are two approaches to the production of an electric field in the region of a fracture. The first consideration would involve two capacitor plates, one on either side of the fracture of, for example, a femur. Experiments by Bassett *et al.* (1974a) on rabbits showed that fields of 100 V/cm increased the fracture repair rate. However, scaling such a system to a man's leg would require voltages of a hazardous level. A variation of this has been developed by Becker *et al.*(1977) who insert an insulated silver wire (negative) to the fracture site. This is connected to an external d.c. generator and to a large carbon electrode (positive) attached to the skin. In tests on thirteen difficult fractures in humans they report healing of nine, the failures caused by technical difficulties.

The second possibility is the production of electric fields inductively by pulsing external magnetic fields. Such inductive coils, about 2.5 by 3 cm, were placed about 8-cm apart next to dogs' legs where the leg bones had been cut (Bassett *et al.*, 1974b). Two different pulses were used in two different sets of dogs. In one set the pulse was 1-msec long at 1 Hz (walking frequency)

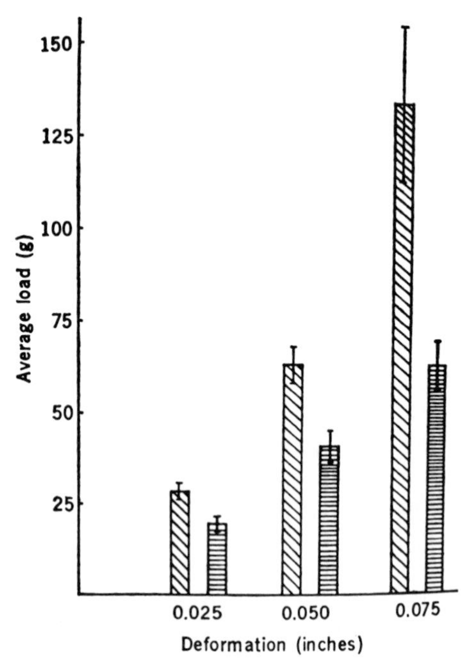

FIG. 5.18 Average load in grams versus deformation of excised dog femurs after 28 days of fracture healing. The data with horizontal slashes are from a control group, and those with oblique slashes are those treated with pulsed field at 65 Hz. [From C. A. L. Bassett, R. J. Pawluk, and A. A. Pilla, *Science* **184**, 575 (1974a). Copyright 1974 by the American Association for the Advancement of Science.]

and produced a peak voltage in the bone of 2 mV/cm. The second group had a pulse of 150-μsec duration at 65 Hz and produced a field of 20 mV/cm in the bone.

After 28 days the bones were excised and stress–strain data taken. No significant differences were found by comparing the bones of the dogs exposed to the 1 Hz 2 mV/cm with those of a control group. However, the bones of the dogs exposed to the combination 65 Hz 20 mV/cm showed a remarkable effect. The averaged data for the group with the inductive fields are shown in Fig. 5.18 by the bars with oblique slashes; for comparison the averages for the control group are shown by bars with horizontal slashes. These constitute data points on a stress–strain curve and the modulus of elasticity of such a curve is given by the slope. Inspection of Fig. 5.18 indicates that the modulus of the data for electromagnetic field healing is about twice that of the control group. Since most of the formation of the bone is expected to occur at its weakest point, these data clearly indicate an accelerated rate of healing in the electric field.

Many natural body rhythms (heart beat, walking, etc.) have a frequency of 1 Hz that would be expected to play some role. However, this cannot be ascertained from these experiments because in going to the successful higher-frequency experiments the investigators simultaneously changed two other variables, pulse duration and field strength in the bone. Clearly, however, this is a fruitful area and further investigation is continuing with systematic control of the variables.

BIOCERAMICS AS BONE IMPLANTS

The interior of the body is hostile to anything but the body's own tissues: various electrolytic exchanges take place; it has a relatively high NaCl content; and oxygen partial pressures may vary between its adjacent areas. The environment, composition, and pH within body fluids change from time to time. There is a pH change in damaged tissue, with an initial fall to about 5.4, followed by a gradual rise and the development of a pH of 7.35 after about 10 days of the healing process.

Metals are used surgically for fixation devices for bones in such forms as screws, plates, and rods for internal support of long bones. Other than gold and platinum, which lack strength, metals are extracted from the lower potential form in the ores in which they are found. Thus there is a tendency for them to revert to the lower energy form by oxidation. Such reversion is called *corrosion*. The resistance of a metal to corrosion by simulated body fluids often has little relation to its resistance within the body. Stainless steels, for example, depend on their resistance to oxidation by a thin layer of oxide which forms on the surface and which, under static conditions, is virtually impenetrable to further oxygen as long as the oxygen partial pressure remains high. Within the body however, there are local differences in oxygen pressure and the equilibrium that favors the oxide layer will shift toward its dissolution in a low oxygen environment. The resulting exposed metal acts as an anode to the surrounding areas, which is cathodic, and electrolytic pitting of the metal can result, with the chloride salts acting as electrolytes. The protective oxide layer can also be punctured by abrasion due to relative motion with bone while in place. If two different metals, or slightly different compositions of the same metal, are near each other, such as screws holding a plate, galvanic action can cause corrosion. There is still another form of attack on metals called *stress corrosion.* This occurs when the metal is subjected to low-level alternating stresses. The oxide layer is insufficiently flexible and microscopic cracks occur. These permit entrance of further oxygen, particularly between the grains of the metal, and leads eventually to fatigue failure.

Although screws, plates, pins, etc., of highly resistant metals have been used as fixation devices in bone with a high percentage of success for many years, the philosophy of the procedure has been questioned. Why use something that can oxidize; why not use a material already oxidized instead? Increasing attention has therefore been directed toward such materials in the past decade and they have been given the generic name of *bioceramics.*

Bioceramics are usually made of various oxides and this class of materials is quite broad. Clearly, hydroxyapatite is an oxide ceramic, one manufactured by the body and rapidly resorbable by the body. An example of an artificial resorbable ceramic is plaster of Paris ($CaSO_4 \cdot \frac{1}{2}H_2O$). Bone implants of this in dogs have shown that it is physiologically compatible and is completely resorbed. It lacks strength, however. One can make a general grouping of the known bioceramics by their mean lifetime in the body before resorption. Figure 5.19 shows a schematic plot of the relative reactivity

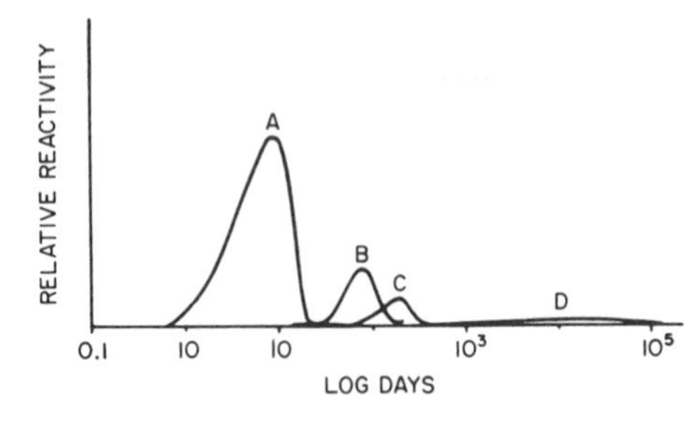

FIG. 5.19 Bioceramic reactivity spectrum. A, resorbable bioceramics, e.g., $Ca_3(PO_4)_2$; B, moderate surface reactive bioglass, e.g., Na_2O–CaO–P_2O_5–SiO_2; C, low surface reactive bioceramic, e.g., Na_2O–CaO–CaF_2–P_2O_5–SiO_2; D, nearly inert bioceramic, e.g., Al_2O_3. [From L. L. Hench, Ceramics, glasses, and composites in medicine, *Med. Instrum.* 7, 136 (1973).]

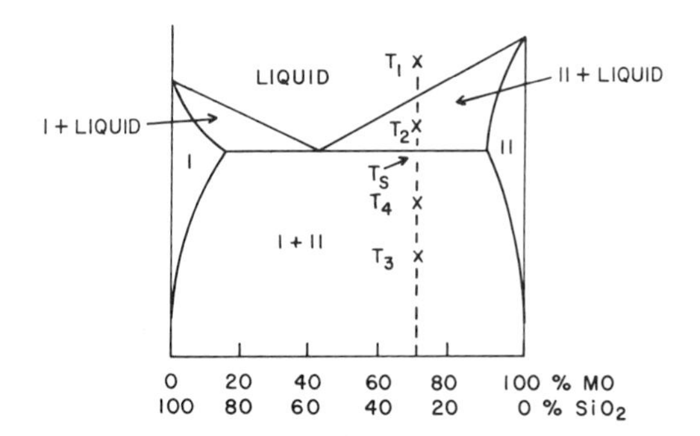

FIG. 5.20 Hypothetical binary phase diagram for a SiO_2–MO (metal oxide) system.

within the body of four types of bioceramics versus their mean lifetime in the body. Plaster of Paris, mentioned above, is an example of group A.

It will be instructive at this point to give a brief review of ceramics and glasses and how they are made. These materials are held together by a mixture of ionic and covalent bonds. In a pure single crystal perfect order of the ions is present. In a glass the ions are in a random array. Often a crystalline material is formed by many very small crystals whose presence are sometimes only detectable by x-ray diffraction techniques.

Consider a hypothetical phase diagram of SiO_2 and a metal oxide MO illustrated in Fig. 5.20. The left-hand side represents the SiO_2 at 100% concentration decreasing to 0% at the right-hand side, while the MO concentration decreases from right to left. The ordinate is temperature. The region labeled I is pure solid SiO_2, and that labelled II is pure solid MO, and the region between is a solid solution of the two. As the temperature increases above T_s, the solidification temperature of the system, three regions are evident: solid I plus liquid, solid II plus liquid, and the region in which the mixture is entirely liquid. If a mixture of a composition along the indicated dashed line in Fig. 5.20 is held in the liquid region at T_1 for a sufficient length of time the liquid will become homogeneous. If it is rapidly quenched and if the mixture contains sufficient SiO_2, which forms random networks, a glass may result. If a mixture of MO and SiO_2 powders is fired at temperature T_2 a liquid phase sintered structure results. Before firing, the structure will be quite porous. Upon heating, phase I starts to melt and the liquid will penetrate between the grains and fill the pores. When this preparation is cooled the liquid will either crystallize in a fire-grained matrix surrounding the grains of phase II or, if the cooling is rapid enough, a glassy matrix will surround the phase II matrix. This process is called liquid phase sintering. It is possible to form a powder compact by a process called solid state sintering. A compacted powder of 100% MO, or a mixture of several different metal oxides, is heated to a high temperature, at which surface diffusion of atoms can occur in a reasonable period of time. Free energy differences, arising from different radii of curvature of surfaces, tend to favor those with larger radii of curvature and a sintering of the grains takes place. This process is discussed in Chapter 11 in the section leading to the development of Eq. (11.33). The resulting sintered material has a decreasing porosity, which depends on the time and temperature of heating.

Another class of microstructures that may be made by the powder mixture of Fig. 5.20 are glass ceramics. The process starts by first heating the mixture of powders to T_1 and forming a glass. Thus, no pores are present in the material. The glass is then cooled to temperature T_3 and held there for a period of time. This heat treatment permits the formation of a high concentration of nuclei about which crystals can grow. When these nuclei have formed the temperature is raised to T_4 and maintained for a period of time. This temperature is not high enough to dissolve the nuclei that have formed but is high enough to permit rapid growth of crystals about these nuclei. The composition of these crystals that result may or may not correspond to the separation of phases indicated by Fig. 5.20. The growth of the crystals is allowed to continue until they impinge upon one another and complete crystallization of the material has occurred. The resulting structure is that of very fine-grained randomly-oriented crystals and it often exhibits considerable improvement in physical and chemical properties over those of the starting materials.

There are three basic concepts of structural strength of bone implants: (1) mechanical attachment to existing bone; (2) mechanical attachment to expected bone growth; and (3) chemical bonding to existing bone. The first concept is the one in present general surgical use. For example, a hole is drilled in a bone and a metal plate can be attached by a screw inserted in the hole. Sometimes gaps between metal and bone are filled by a nonbiodegradable plastic. Even if the metal is not attacked by the body there seems to be a significant failure rate of such attachments after a decade or so, particularly in high mechanical stress regions such as hip joint replacement. Some high-density, i.e., very small pore, bioceramics are also mechanically attached by notching or threading. Sufficient life tests have not been made to know whether these are an improvement over metals.

The second concept is the creation of high porosity ceramics or metals (foam or spongelike casts). The research of Hulbert *et al.* (1971) have shown that osteogenesis (bone growth) will take place in a porous ceramic. The ceramic mixture selected was calcium aluminate, which is formed by heating together a mixture of fine powder of calcium carbonate and aluminum oxide. A series of experiments showed that in order for osteogenesis to occur the pore size had to be 100 μm or larger. Optimization of this pore size with structural strength was accomplished by varying the powder size and heating times. It was learned that implants in bone of such porous ceramics were chemically stable in the body and that the bone would grow through the pores and lock the implant into its structure. This technique is currently under evaluation.

The third concept is to create a bioceramic that is only partly chemically reactive with bone. In this way bone could bond chemically to the ceramic without completely absorbing it in time or without damaging its structural strength. Toward this goal Hench and his associates developed a glass with rather unique biological properties. The glass is composed of Na_2O–CaO–P_2O_5–SiO_2. The sodium oxide initiates a series of chemical reactions between the glass and the bone. The sodium ions on the surface of the glass exchange with hydrogen ions in the body fluids with a resulting increase in interfacial pH. The sodium loss also creates a surface layer with excess SiO_2. Calcium and phosphate ions migrate through this layer into the glass and initially form an amorphous calcium phosphate layer between the bone and the bioglass. Within a two-week period of time the amorphous calcium phosphate begins to crystallize as an apatite mineral very similar in structure to the hydroxyapatite of bone mineral. In this way a firm, chemically bonded attachment of the bone to the bioglass is formed.

The procedure is to flame spray or immersion coat a metal or aluminum oxide implant with the bioglass using a technique that leaves no cracks in the glass or a significant number of pores in the glass–implant interface. The glass makes a firm bond to the implant. Experiments with such implants, usually as femoral head replacements in monkeys or rats, show great strength of attachment to the bone after only 10 days. The animal is sacrificed and the bone–glass interface is examined with a scanning electron microscope and a determination of the elements across the interface made by energy dispersive x-ray analysis (EDXA) and Auger electron spectroscopy. Figure 5.21a shows one such interface, which had been implanted in a rat for 10 days. The regions labelled SS, BG, and B are stainless steel, bioglass, and bone, respectively. The vertical line which passes through c is the interface and it can be seen to be joined in 40–60% of the interface. Were the implant left in longer, complete joining would have occurred. The EDXA scans for the three regions b, c, and d are shown in the right part of the figure. Energy on the abscissa identifies the element, and the ordinate the relative concentration. The EDXA examination shows a decrease in the Si concentration and a variation in the Ca/P ratio across the interface from glass to bone. These concentration gradients were taken through the joined parts, which are structurally contiguous and morphologically uniform.

These and similar results are impressive and convincing. There is no question but that Hench and his associates have developed a glass to which bone will chemically bond. Longer term studies *in vivo* have shown no sign

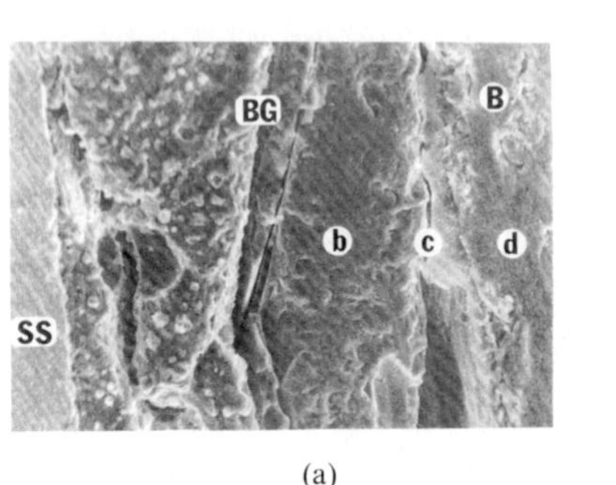

(a)

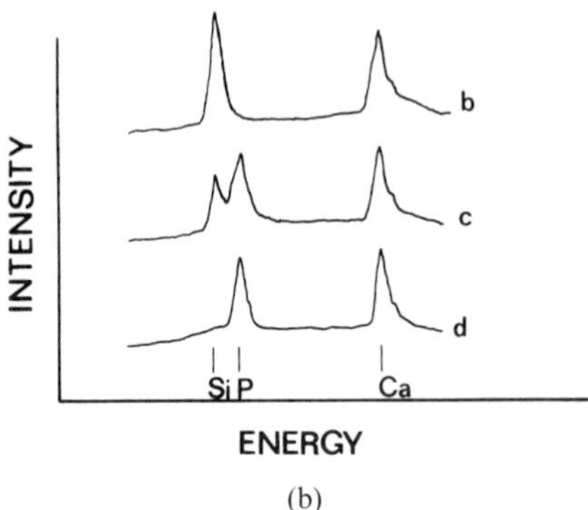

(b)

FIG. 5.21 (a) Bioglas coated stainless steel rat tibia implant (magnification 100×). SS, stainless steel; BG, bioglass; B, bone. (b) Composition of regions b, c, and d identified by EDXA spectra. [From Hench *et al.*, (1977).]

of deterioration. The remaining physical experiments seem to be the choice of variations of composition for greatest structural strength and bonding strength to the implant and the development of reliable implant coating techniques. The remaining physiological tests are further reliability studies in animals and, finally, tests in humans.

What works for bone implants may well work for tooth implants and this group of investigators is also studying bioglass-coated tooth implants (Stanley, 1976).

REFERENCES

Azaroff, L. V. (1960). "Introduction to Solids." McGraw-Hill, New York.

Bassett, C. A. L. (1965). Electrical effects in bone, *Sci. Amer.* **213**, 18.

Bassett, C. A. L. (1968). Biological significance of piezoelectricity, *Calcif. Tissue Res.* **1**, 252.

Bassett, C. A. L., Pawluk, R. J., and Pilla, A. A. (1974a). Augmentation of bone repair by inductively coupled electromagnetic fields, *Science* **184**, 575.

Bassett, C. A. L., Pawluk, R. J., and Pilla, A. A. (1974b). Acceleration of fracture repair by electromagnetic fields. A surgically noninvasive method. *Ann. New York Acad. Sci.* **238**, 242.

Becker, R. O. (1967). The electrical control of growth processes, *Med. Times* **95**, 657.

Becker, R. O. (1974). The significance of bioelectric potentials, *Bioelectricity Bioenergetics* **1**, 187.

Becker, R. O., Bassett, C. A., and Bachman, C. H. (1964). Bioelectrical factors controlling bone structure, *in* "Bone Biodynamics" (H. M. Frost, ed.). Little, Brown, Boston, Massachusetts.

Becker, R. O., Spadaro, J. A., and Marino, A. A. (1977). Clinical experiences with low intensity direct current stimulation of bone growth, *Clin. Orthopaed.* **124**, 75.

Blencke, B. A., Bromer, H., and Pfeil, E. (1974). Rasterelektronenmikroskopische Untersuchungen der Reaktion des Knochens auf glaskeramische Implantate, *Z. Orthop. Ihre Grenzgeb.* **112**, 978.

Bourne, G. H. (ed.) (1956). "The Biochemistry and Physiology of Bone." Academic Press, New York.

Carter, D. R., and Hayes, W. C. (1976). Bone compressive strength: the influence of density and strain rate, *Science* **194**, 1174.

Currey, J. D. (1962). Stress concentration in bone, *Quart. J. Microsc. Sci.* **103**, 111.

Currey, J. D. (1964). Three analogies to explain the mechanical properties of bone, *Biorheology* **2**, 1.

Dempster, W. T., and Liddicoat, R. T. (1952). Compact bone as a non-isotropic material, *Am. J. Anat.* **91**, 331.

Evans, F. G., and Bang, S. (1967). Differences and relationships between the physical properties and the microscopic structure of human femoral, tibial and fibular cortical bone, *Am. J. Anat.* **120**, 79.

Evans, F. G., and Lebow, M. (1951). Regional differences in some physical properties of the human femur, *J. Appl. Physiol.* **3**, 563.

Fukada, E., and Yusada, I. (1957). On the piezoelectric effect of bone, *J. Phys. Soc. Jpn.* **12**, 1158.

Guyton, A. C. (1971). "Textbook of Medical Physiology." 4th ed. Saunders, Philadelphia, Pennsylvania.

Harris, W. H., and Heaney, R. P. (1969). Skeletal renewal and metabolic bone disease, *New England J. Med.* **280**, 193.

Hench, L. L. (1973). Ceramics, glasses, and composites in medicine, *Med. Instrum.* **7**, 136.

Hench, L. L., Pantano, C. G., Jr., Buscemi, P. J., and Greenspan, D. C. (1977). Analysis of bioglass fixation of hip prosthesis, *J. Biomed. Res.* **11**, 267.

Hench, L. L., Piotrowski, G., Paschall, H. A., Miller, G., Buscemi, P., Greenspan, D., and Carr, T. (in press). Prosthetic joint fixation by bioglass-bone bonding, *Clin. Orthopaed.*

Hulbert, S. F., Talbert, C. D., and Klawitter, J. J. (1971). Investigation into the potential of ceramic materials as permanently implantable skeletal prosthesis, *in* "Biomaterials" (A. Bement, Jr., ed.). Univ. of Washington Press, Seattle, Washington.

Korostoff, E. (1971). Properties of bone and dentin: Criteria for specifications, *in* "Biomaterials" (A. L. Bement, Jr., ed.). Univ. of Washinton Press, Seattle, Washington.

Kummer, B. (1966). Photoelastic studies on the functional structure of bone, *Fol. Biotheoret.* **6**, 31.

Kummer, B. K. F. (1972). Biomechanics of bone: Mechanical properties, functional structure, functional adaptation, *in* "Biomechanics, Its Foundations and Objectives" (Y. C. Fung, N. Peroone, and M. Anliker, eds.). Prentice-Hall, Englewood Cliffs, New Jersey.

Lowenstein, K. L. (1966) *in* "Composite Materials" (L. Holliday, ed.). Elsevier, New York.

Lugassy, A. A., and Korostoff, E. (1969). Viscoelastic behavior of bovine femoral cortical bone and sperm whale dentin, *in* "Research in Dental and Medical Materials" (E. Korostoff, ed.). Plenum Press, New York.

McCutchen, C. W. (1975). Do mineral crystals stiffen bone by straitjacketing its collagen, *J. Theor. Biol.* **51**, 51.

Marino, A. A., and Becker, R. O. (1970). Evidence for epitaxy in the formation of collagen and apatite, *Nature* **226**, 652.

Posner, A. G., and Betts, F. (1975). Synthetic amorphous calcium phosphate and its relation to bone mineral structure, *Accounts Chem. Res.* **8**, 273.

Prochnow, O. (1934). "Formenkunst der Natur." E. Wasmuth, Berlin.

Shamos, M. H., and Lavine, L. S. (1964). Physical basis for bioelectric effects in mineralized tissues, *Clin. Orthop.* **35**, 177.

Shamos, M. H., and Lavine, L. S. (1967). Piezoelectricity as a fundamental property of biological tissues, *Nature (London)* **213**, 267.

Shamos, M. H., and Shamos, M. I. (1963). Piezoelectric effect in bone, *Nature (London)* **197**, 81.

Spalteholz, W. (1943). "Hand Atlas of Human Anatomy," 7th ed., Vol. 1. Lippincott, Philadelphia, Pennsylvania.

Stanley, H. R., Hench, L., Going, R., Bennett, G., Chellemi, S. J., King, C., Ingersoll, N., Ethridge, E., and Krentziger, K. (1976). The implantation of natural tooth form bioglasses in baboons, *Oral Surg. Oral Med. Oral Pathol.* **42**, 339.

Steinberg, M. E., Wert, R. E., Korostoff, E., and Black, J. (1973). Deformation potentials in whole bone, *J. Surg. Res.* **14**, 254.

Thompson, D'Arcy (1971). "On Growth and Form." Cambridge Univ. Press, London and New York (Reprint).

Welch, D. O. (1970). The composite structure of bone and its response to mechanical stress, *Recent Adv. Eng. Sci.* **5**, 245.

Wolff, J. (1892). "Das Gesetz der Transformation der Knochen." Hirschwald, Berlin.

CHAPTER 6

Heart Motion: Electrocardiography and Starling's Law

INTRODUCTION

The heart is a pump that receives low pressure input of blood from the venous system and ejects high pressure blood into the arterial system. The gross physical principles of a pump are mathematically trivial but the details of flow rate through valves in a moving system require hydrodynamic calculations beyond the scope of this book. We shall consider some fluid motion in the next chapter.

The heart is composed of three types of muscle; *atrial* (upper chambers), *ventricular* (lower chambers), and specialized excitatory and conductive fibers. The atrial and ventricular fibers are quite similar to the skeletal or striated muscle fibers discussed in detail in Chapter 4. However, the specialized excitatory and conductive fibers contract by only small amounts and their primary function is to provide a rapid transmission system for the electrical impulses of the heart.

Figure 6.1 shows a schematic of a planar section of cardiac muscle. Several sarcomeres are shown horizontally and the actin and myosin filaments shown are similar to those in Fig. 4.10 (although the drawing of Fig. 6.1 is not as detailed). The large horizontal ovoid bodies represent mitochondria. A

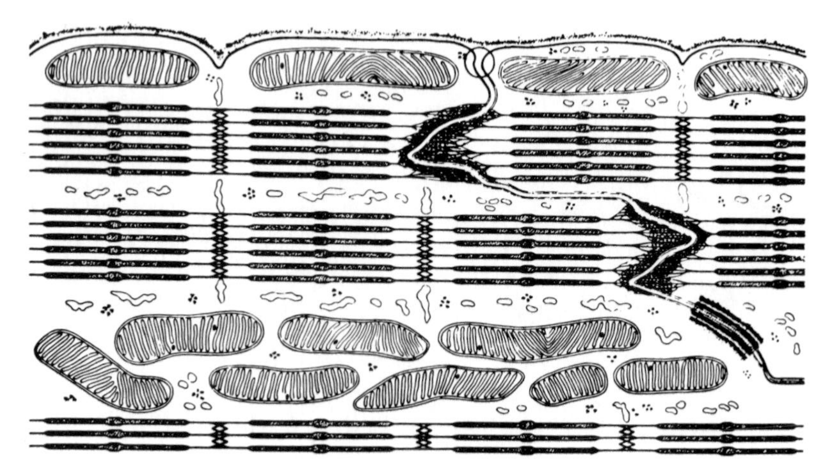

FIG. 6.1 Schematic of mouse cardiac muscle. [From Sjörstrand and Anderson-Cedergren (1960).]

significant feature observed in cardiac but not in skeletal muscle is the body separating the middle sarcomeres in the upper fiber of Fig. 6.1 and the right-hand sarcomere from the left-hand ones in the middle fiber. These bodies, called *intercalated discs* (intercalated = inserted), are actually cell membranes that separate individual cardiac muscle cells from each other. However, these membranes have an electrical conductivity about 400 times greater than that of the outside membrane of the cardiac muscle fiber. Thus, an electrical signal from one excited cell can pass with considerable ease in the fiber direction. The cardiac muscle fibers in turn divide and recombine in a manner that gives direction to the spreading of an exciting stimulus. Such a group of cells combined in a functional mass is called a *functional syncytium.* There are two separate *functional syncytia* in the heart, the *atrial syncytium* and the *ventricular syncytium.* These two syncytia are connected electrically by a system of nerves called the AVB, which will be described later in this chapter.

The time sequence of the action potential in cardiac muscle is different from that of skeletal muscle. The action potential in cardiac muscle has the spike followed by a plateau, which represents depolarization time of about 0.15 sec in the atria and about 0.3 sec in the ventricles. Thus, during these times the cardiac muscle remains contracted rather than responding with the shorter twitch of a skeletal muscle. Furthermore, the membrane of heart muscle cells is more permeable to sodium than is the skeletal muscle membrane, and because of an intrinsic leakage of sodium it is self-excitable.

In addition to the propagation of the action potential through the muscle fibers of the two syncytia, the action potential is also propagated through nerve fibers, with time delay sections so that portions of the heart contract sequentially to give a proper pumping action. The propagation of this action potential involves the motion of ions, which can be detected with a voltmeter either on the heart muscle or on the skin. Such measurements are called *electrocardiography* and a print-out of voltage versus time is called an *electrocardiogram* (EKG or ECG).

In this chapter we shall describe the major parts of the heart and the meaning of the electrical signals. In presenting this we shall follow primarily Schaefer and Haas (1962), Burger (1968), and Guyton (1955, 1971), listed in the references at the end of this chapter.

In addition, we shall introduce some of the current thoughts on explaining why the heart is able to pump out all of the blood that comes into it, a quantity that varies with stress and activity. This behavior is known as Starling's law (1918) and the most recent conference on the subject (Ciba Foundation, 1974) indicates that the reason is not yet understood. However, in view of the rather glib ending of Chapter 4, it will be healthy to recognize the limits of knowledge.

HEART MOTION

In order to interpret the electrical signals that are picked up on the external surface of the body we must understand first the mechanics of the heart motion.

The heart has four chambers, the *right atrium* and *ventricle* and the *left atrium* and *ventricle*. The right side services the pulmonary blood system and the left side services the rest of the body.

In Fig. 6.2, RA and LA are right atrium and left atrium, respectively, and RV and LV are the right and left ventricles, respectively. Venous blood goes into the right atrium from the *superior* and *inferior vena cavae* and drains into the right ventricle through the *tricuspid valve*. The atrium then begins to contact, which causes about 20–30% more filling of the ventricle. As the ventricle begins to contract the tricuspid valve closes. This is a completely passive valve and behaves as a check valve. By its closing, pressure in the ventricle builds up sufficiently to open the *semilunar valve*, which leads to the pulmonary artery. As the ventricular volume, and hence pressure, decreases this valve snaps shut so that there is no back flow. After circulation through the pulmonary system the blood then drains into the left atrium, then through the *mitral valve*. The left atrium contracts, forcing more blood into the left ventricle. The left ventricle then contracts, closing the mitral valve and forcing open the *aortic valve*. It in turn snaps shut to prevent back flow when the pressure in the left ventricle begins to decrease .This sequence does not imply that the right side pumps and the left side waits until that blood has circulated

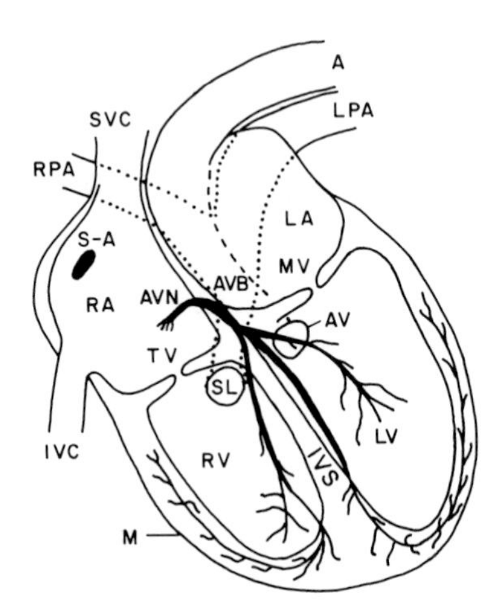

FIG. 6.2 Schematic cross section of a heart. RA and LA are right and left atria; TV, tricuspid valve; MV, mitral valve; IVC and SVC, inferior vena cava and superior vena cava; M, myocardium. AV, the aortic valve, joins the aorta A (shown by dashed lines), which is behind and outside of the heart. SL, the semilunar valve, leads to the pulmonary artery (dotted lines), which branches into the right pulmonary artery RPA and the left pulmonary artery LPA, also outside and behind the heart. IVS is the interventricular septum; S–A, sino-atrial node; AVN, atrial-ventricular nerve fibers; AVB, atrial-ventricular bundle. The black lines represent the Purkinje system. RV and LV are the right and left ventricles.

through the pulmonary system before it pumps. Instead, the flow is continuous and the contraction of the left ventricle occurs almost at the same time as the right.

The tricuspid and mitral valves close with a snap that vibrates the surrounding fluid. This is the "thump" of the first sound of the heartbeat. Then the semilunar and aortic valves snap shut which gives rise to the second "thump." Some home and apartment water systems make a similar sound when a faucet is turned off rapidly.

The time period when the heart is contracting, i.e., pumping, is called *systole* and the period when it is quiet, while being refilled, is called *diastole*.

A summary graph of pressure versus time for the left atrium and ventricle is shown in Fig. 6.3.

In the atrial pressure curve, reading from the left, the first pressure rise is due to the contraction of the atrial muscle. The pressure drops and then rises again as the ventricle begins to contract. This drop is caused by some reflux of blood back into the atrium, the bulging of the mitral valve, and the pulling of the atrial muscles during ventricular contraction. The ventricular pressure curve has a small pressure rise during diastole, caused by the contracting atrium, followed by the major rise during its own contraction.

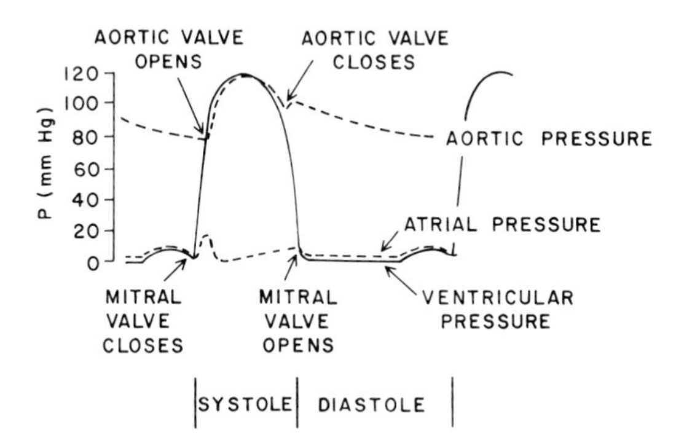

FIG. 6.3 The pressure changes in the left atrium and ventricle of a heart during one cycle. [Redrawn from Guyton (1971).]

Note that the arterial system at the aorta does not drop below 80 mm-Hg because of the action of the aortic valve. In a diagnostic measurement of blood pressure the two important quantities are the lowest and the highest pressures attained near the aorta.

DEPOLARIZATION SEQUENCE

The origin of nerve action potentials and their role in initiating muscular contraction have been discussed in Chapters 3 and 4. For our present purposes, however, it is sufficient to recall that a nerve, in its resting condition, has a deficiency of sodium ions inside compared to the outside. This causes a potential difference between the inside and outside of about 80 to 85 mV in the cardiac muscle. Upon a suitable signal, rapid ion diffusion takes place through the membrane in a fraction of a second, which not only depolarizes the membrane but briefly reverse the potential. Shortly afterward the sodium ions leave the interior and the membrane is repolarized. The time for repolarization is called the *refractory period*. During this period the membrane usually cannot be depolarized again.

If depolarization is initiated at a point the depolarization will be propagated along the membrane, thereby causing muscular contraction, with a velocity of about 0.3 to 0.4 m/sec for atrial and ventricular muscle fibers to about 100 m/sec in large or myelinated nerve fibers. The refractory period during which ventricular cardiac muscle cannot be reexcited is about 0.25 sec and is called the *functional refractory period*. There is an additional *relative refractory period* of about 0.05 sec during which it is difficult but not impossible to reexcite the muscle. The functional refractory period for atrial muscle is about 0.15 sec and the relative refractory period is about 0.03 sec.

In the right atrium there is a small section of specialized muscle called the S–A or *sino-atrial node* which has a low resting membrane potential of about 55 to 60 mV and a constant ion leakage; this causes a periodic self-excitation. Thus, each time the resting membrane potential is established, ion leakage causes a decay of the potential unit until it reaches the threshold for excitation. At this point it depolarizes. This depolarization is then transmitted through both sides of the atrial muscle at a velocity of 0.3 m/sec. As this impulse travels through the atrial muscle the muscle contracts.

Such a high velocity of transmission would cause the ventricles to contract before the atria have filled them, so a delay mechanism exists. The atrial signal reaches a node of AVN (*atrial-ventricular nerve fibers*) within the right atrium, called the A–V node, in about 0.04 sec after its origin at the S–A node. However, the S–A impulse must then travel through the A–V node and its associated fibers (these are, in order, *junctional fibers*, *nodal fibers*, and *transitional fibers*) until it finally reaches the AVB (*atrial-ventricular bundle*), a group of nerve fibers leaving the A–V node. From the AVB the signal goes through high velocity nerve fibers, at about 2 m/sec, called the *Purkinje* (pronounced poor keen′ yeh) *system*, into the muscle of both ventricles. In the ventricular muscle the speed of transmission is about 0.3 m/sec. Figure 6.4 shows the times in seconds of the signal arrival at different parts of the heart.

It should be noted that the initial signal can arise from any membrane that depolarizes. All cardiac membranes have some leakage even under normal conditions, e.g., the rhythmic discharge rate of the S–A node is 70 to 80 times

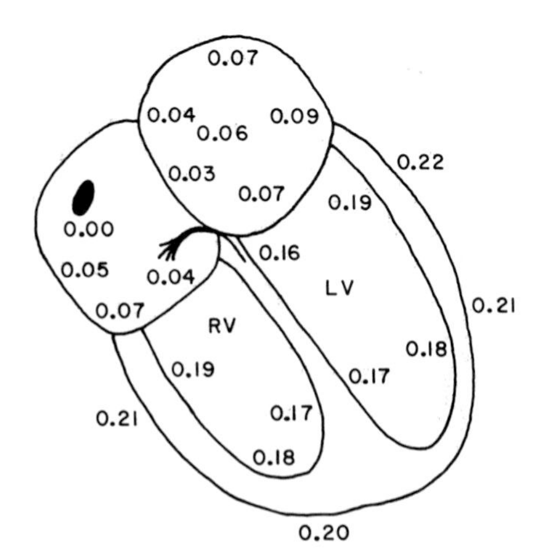

FIG. 6.4 Time in seconds of heart activation. Note that in the ventricular region the upper part of the septum depolarizes first; the depolarization wave then descends the septum and goes up the outer muscle (the myocardium).

per minute, the A–V nodal fibers discharge at 40 to 60 times per minute, and the Purkinje fibers at 15 to 40 times per minute. The S–A node is the pacemaker because it discharges faster than the others and its impulse is conducted to make the others discharge. Sometimes a part of the heart will develop a discharge rate faster than that of the S–A node, and it then becomes the pacemaker. This is called an *ectopic* pacemaker.

If some of the transmission network becomes damaged or blocked, the delay fibers at the A–V node may be by-passed. This can give rise to a rapid circular movement of excitation through the heart muscle, called *fibrillation*, in which there are many rapid beats but no rhythmic pumping of blood. This condition can sometimes be stopped if all of the fibers are thrown into a refractory period simultaneously. The S–A node or some other part of the heart will then become the pacemaker for rhythmic depolarization. This is done by passing an intense current through the heart muscle itself. This current passes through almost the whole heart muscle and depolarizes all of the nerves, causing a simultaneous refractory state in them.

ELECTRICAL SIGNALS

The action potential of the ventricular cardiac muscle is very rapid and behaves like the nerve action potential. However, the repolarization time is slower. This is shown in Fig. 6.5a. Recall that a polarized membrane is like a charged capacitor andthe flow of ions in the action potential is equivalent to a shorting of the capacitor's plates, as indicated in Fig. 6.5b. If a group of membranes is sequentially shorted, as in Fig. 6.5c, there is a continuous flow of a pulse of positive charge, which is equivalent to a current, as seen in Fig. 6.5d.

Consider now a single active cardiac muscle fiber, as in Fig. 6.6a. If the fiber is closed at both ends and is completely in the resting state, as illustrated in the right side of Fig. 6.6a, the potential difference across the membrane is equal at any point along the fiber, and no potential field exists in the surrounding medium because of the homogeneous double layer. This situation also obtains in a completely active state when the inside of the membrane is uniformly positive, as indicated in the left side of Fig. 6.6a. Note that the outside is an infinite source of positive ions.

If, however, an excitation wave has entered a resting fiber from the left, as in Fig. 6.6a, so that the left side is depolarized while the right side is still polarized, a dipole is momentarily formed. Its origin will now be discussed. Both ends of the fiber in Fig. 6.6a may be reduced to electrical ineffectiveness by introducing disks 1 and 4, which have the same potential difference across their surfaces as do the respective ends of the membranes. These closed ends now do not produce an electric field. The introduction of disk 1 must be

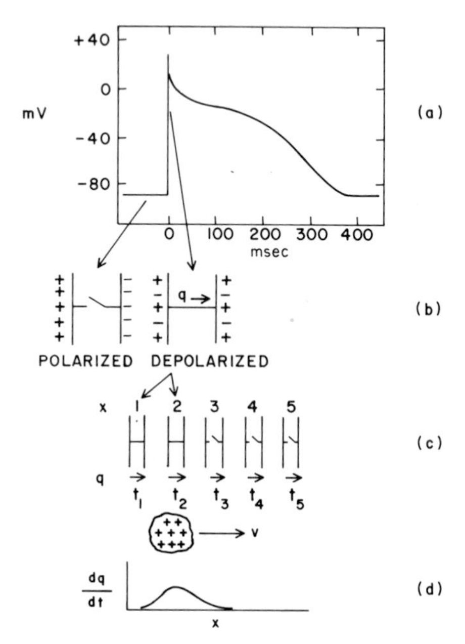

FIG. 6.5 (a) Action potential and repolarization of cardiac muscle. (b) Schematic of the membrane as a capacitor in the polarized and depolarized conditions. (c) A series of membranes depolarizing sequentially from 1 to 5 and the equivalent packet of plus charges moving with velocity v. (d) The current (dq/dt) of this charge packet moving in the x direction.

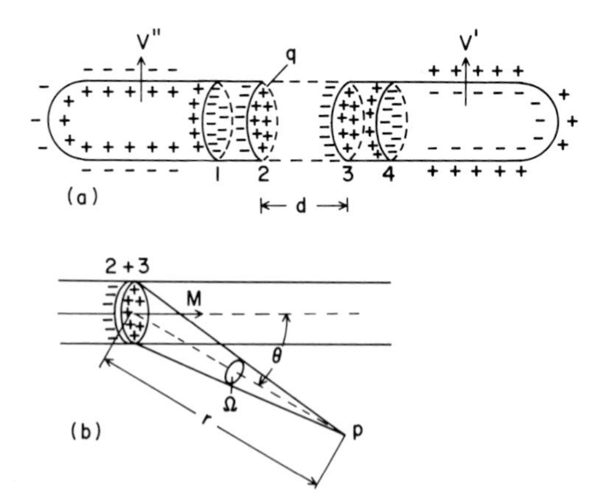

FIG. 6.6 (a) Model of a depolarizing myocardial fiber; the left side is depolarized and the right is still polarized. Disks 1 and 4 are added to these two parts to neutralize them. Two more disks, 2 and 3, of opposite polarity are added to compensate for disks 1 and 4. The electric field produced by disks 2 and 3 forms an electric dipole and produces the same electric field as the action potential. (b) The amount of potential at point p from the dipole M due to disks 2 and 3. [Redrawn from Schaefer and Haas (1962).]

compensated for by the introduction of disk 2, which has an equal area but an opposite charge. Thus disk 2 is the only charge acting to produce an electric field. The same procedure must be done on the resting part of the fiber by introducing disk 3 to compensate electrically for disk 4. The result is that the electrical disturbance of the region of activation can be represented by two disks of charges, 2 and 3. These form a dipole with a dipole moment proportional to the surface area and to the dipole moment per unit surface area. Referring to disks 2 and 3, it is seen that the action potential V has the value $V = -V'' + V'$ and that the absolute amount of voltage change is $|V| = |V''| + |V'|$, which is the magnitude of the action potential.

A dipole of this discoid type is characterized by the magnitude of the cross-sectional area, assuming the disks are close together, and the dipole moment per unit area m. When V is the potential difference between the two surfaces of each disk, the following relation holds:

$$m = V/4\pi.$$

If we consider the potential V_p at some point p at a considerable distance from the discoid dipole it can be expressed as

$$V_p = m \cdot \Omega$$

where Ω is the solid angle under which the disk appears from point p. This is illustrated in Fig. 6.6b. The solid angle can be eliminated by the introduction of a vector dipole moment M. This vector M has a direction normal to the surface of the disk and a magnitude equal to the product of the dipole moment per unit area and the area A:

$$M = m \cdot A$$

where A is the cross-sectional area of the dipole or muscle fiber. Using this vector dipole M, the potential V_p at a point p can be given in terms of M, θ, and r of Fig. 6.6b as

$$V_p = M \cos \theta / r^2 \tag{6.1}$$

This result could have been anticipated by simply recognizing that this relationship is the potential at large distances from a dipole. Note that this equation is valid only in infinite, homogeneous, linear resistive media. If the medium has a boundary the flow lines of the potential are forced to run parallel to the boundary and are therefore compressed. Thus higher potentials are found near the boundaries. The solutions to such problems are quite complicated and are not pertinent to our discussion here.

The meaning of V_p needs clarification. If we use a voltmeter to measure the potential difference between two points, 1 and 2, then the meaning of V_{p_1} minus V_{p_2} is clear. However, in our later discussion of a central terminal (CT)

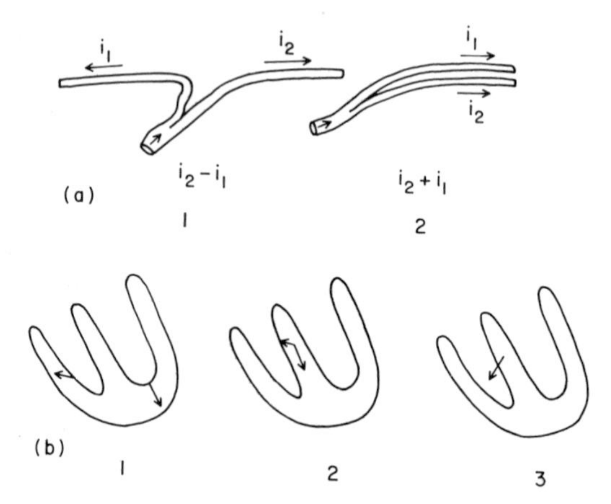

FIG. 6.7 (a) Vector addition of two fiber currents: (1) antiparallel, and (2) parallel. (b) Ventricles of the heart showing dipole vector addition: (1) two dipole vectors of depolarizing myocardium, (2) vectors transposed to a common origin, (3) resultant dipole vector.

probe, V_p can only mean the voltage difference between the CT probe and a zero potential probe. Such a probe is difficult to obtain in electrocardiography because the line of zero potential of a dipole is perpendicular to the dipole and, as we shall see, the effective vector dipole of the heart changes as it depolarizes. In practice, when a central terminal is used, large intervening resistances are required between it and the reference terminal.

Although Fig. 6.4 shows a time sequence of the depolarization of the heart, it is not expected that individual fibers behave this way. Two fibers in parallel will double the dipole moment in the direction of current flow, while two antiparallel fibers will cancel one another's fields as in Fig. 6.7a. Thus, what is seen in electrical signals is an average of fiber behavior. Figure 6.7b illustrates the vector addition of two fiber currents, hence dipoles, in different parts of the heart. Their vector sum can be represented by a single vector at the center of the heart. Thus dipoles of all the fibers can be represented by a single heart dipole vector. This sum is zero when the heart is completely polarized or completely depolarized.

THE EINTHOVEN TRIANGLE

We have seen that if the heart is changing its state of polarization then at each instant in time the sum of all of the currents can be represented by a single dipole vector through the center of the heart. This center is usually taken as the septum, but this is arbitrary and, because of the many other assumptions in the model to be discussed, this assumption doesn't matter.

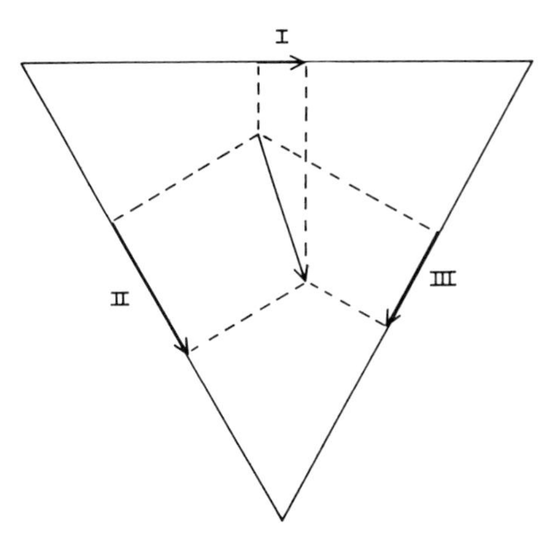

FIG. 6.8 Einthoven triangle showing the projection of an instantaneous heart dipole vector on the three legs of an equilateral triangle.

Consider an instant in time in which the heart is depolarizing, i.e., the muscle fibers are contracting. In Fig. 6.8 such a dipole is represented by the arrow. If a triangle is constructed about this dipole, its projections on each of the legs of the triangle are shown. These are induced electric fields due to the dipole, and a voltmeter connected between the head and tail of any of the projections on the triangle legs will indicate a potential difference.

Einthoven, one of the pioneers of electrocardiography, chose three leads as shown in Fig. 6.9: (I) between the right and left arms, (II) between the right arm and left leg, and (III) between the left leg and left arm. He placed leads at the extremities, where the impedance from the lead to the heart is large. Thus the circle of Fig. 6.9 is very roughly a common potential region and

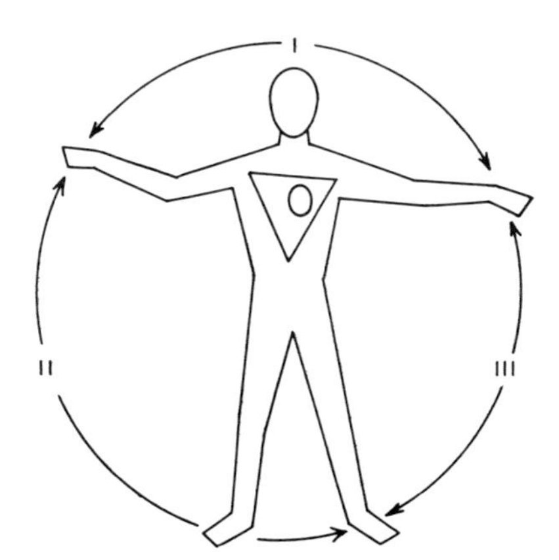

FIG. 6.9 The three leads selected by Einthoven. (I) Between left and right arms, (II) between right arm and left leg, and (III) between left leg and left arm. The Einthoven triangle is drawn on the chest.

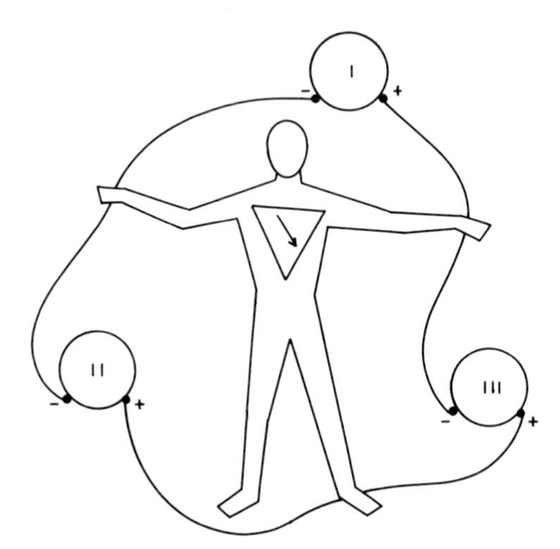

FIG. 6.10 Voltmeters inserted into the leads of Fig. 6.9.

brief changes in potential in the leads will be induced by changes in the heart's vector dipole magnitude and direction.

If voltmeters are inserted in the loops labeled I, II, and III the motion of the net dipole of the heart can be determined by the changing potentials. The arrangement of the voltmeters is shown in Fig. 6.10. Note that while the voltmeters in sections I and III are connected with negative and positive terminals together, the connections of voltmeter II are reversed, with the negative terminal of II connected to the negative terminal of I and the positive terminal of II connected to the positive terminal of III. In this arrangement, if the right arm is negative with respect to the left leg, voltmeter II will show a positive voltage. If the left arm is positive with respect to the left leg, voltmeter III will show a positive voltage.

The reason for this arrangement is that a reading of only two of the voltmeters is required, and from these the third may be calculated by addition. This is seen in the following way. The reading on voltmeter I will be the potential difference between the left arm V_L and right arm V_R, or $I = V_L - V_R$. Voltmeter II will read the potential difference between the right arm V_R and the foot V_F, or $II = V_R - V_F$. Voltmeter III will read the potential difference between the foot V_F and the left arm V_L. Since I, II, and III connect points on an equipotential surface their sum must be zero,

$$I + II + III = (V_L - V_R) + (V_R - V_F) + (V_F - V_L) = 0 \tag{6.2}$$

Note that because of the reversal of the leads of voltmeter II there is a sign change and

$$I + III = II$$

Therefore, if voltmeters I and III are recorded, II is simply their sum.

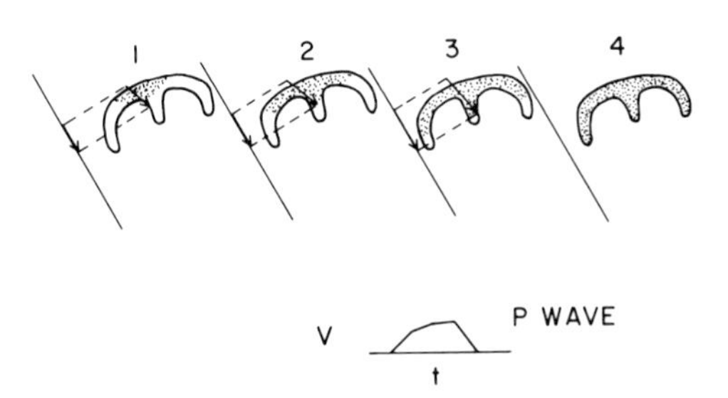

FIG. 6.11 Depolarization of the atria as projected on leg II of the Einthoven triangle (1–4). The shaded area represents the part depolarized and the central arrows are the resultant vector dipole. The time sequence is from left to right. A voltmeter in lead II will have readings proportional to the projection of the dipole. Voltage as a function of time is shown in the lower curve and is called the *P wave*. [After Guyton (1971).]

Why have three voltmeters instead of one? The heart is not an ideal dipole and the dipole does not move in the plane of the Einthoven triangle. Thus Einthoven's arrangement gives three different electrical "views" of the heart and can assist in locating a malfunctioning part by triangulation.

In the time sequence of depolarization shown in Fig. 6.4 it is seen that the atria depolarize first, then there is a considerable delay at the A–V junction followed by the depolarization of the ventricles. If we follow the dipole changes with a voltmeter in, for example, lead II, we can separate the events into an atrial stage followed by a ventricular stage. Figure 6.11 represents the depolarization of the atria. It is initiated at the S–A node and spreads evenly and rapidly throughout the muscle. The resultant vector dipole is shown by the central arrow and its projection on leg II of the Einthoven triangle is indicated. The time course is from 1 to 4. If a voltmeter is inserted in lead II the result is shown in the insert. The voltage versus time curve measured for the *depolarization of the atria is known as the P wave*.

The time sequence of depolarization of the ventricles is shown in Fig. 6.12. It begins in the upper part of the septum, which first receives the signal from the AVB. The time sequence is from 1 to 6 and a voltmeter in lead II will indicate the voltage versus time curve at the bottom of the figure. The *ventricular depolarization is known as the QRS wave*. Note that no curve is shown at Q. In this particular sequence the voltage in lead II increases continuously from 1 to 3. However, it is seen in Fig. 6.12(1) that a projection on lead I will be negative, followed by a positive voltage from Figs. 6.12(2) and 6.12(3). The negative part is the *Q wave*.

After depolarization the atria repolarize, starting from the part that depolarized first. Although this produces a dipole and a corresponding voltage, it occurs at the time of the QRS wave and it is therefore not seen separately.

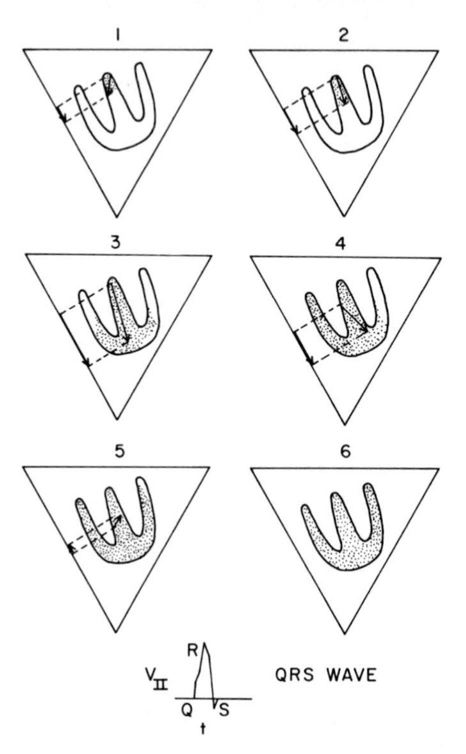

FIG. 6.12 Depolarization of the ventricles with the projection of the resultant vector dipole on leg II of the Einthoven triangle (1–6). The resulting voltage versus time curve is known as the *QRS wave*. [After Guyton (1971).]

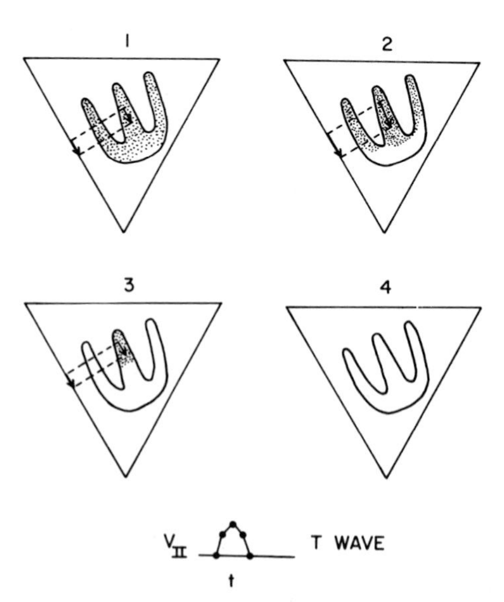

FIG. 6.13 Repolarization of the ventricles with the resultant dipole vector projected on leg II of the Einthoven triangle (1–4). The resulting voltage versus time curve is known as the *T wave*. [After Guyton (1971).]

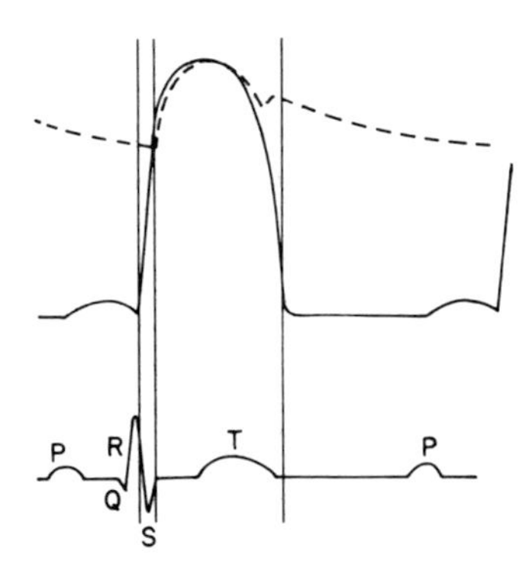

FIG. 6.14 Figure 6.3 of aortic and ventricular pressure versus time of the heart with the PQRST electrocardiograph curve drawn below in the same time sequence. Voltage scale is arbitrary. [After Guyton (1971).]

The ventricles repolarize in the opposite way. That is, the apex depolarizes first and the septum depolarizes last. Although there is much speculation, the reason for this is not known for certain. The repolarization of the ventricles is indicated in Fig. 6.13 and the resulting voltage versus time curve is known as the *T wave*. The sequence of the PQRST waves is shown in Fig. 6.14.

Note that although much of the interpretation of EKG signals seems to be inferred, measurements made with electrodes attached to the heart itself substantiate the model.

THE CENTRAL TERMINAL

It is desired to measure a "potential" at various points near the heart, and Wilson, another pioneer in the field, introduced the *central terminal*, CT. If the electrodes of the extremities are connected together through three equal resistances R, greater than 5000 Ω, then a point known as the central terminal is obtained that can be used as a reference point on the body surface, as shown in Figs. 6.15a and 6.15b. Note that R must be large compared with skin resistance but small compared with the input impedance of the voltmeter. With the condition of no current flow from CT to the voltmeter, Kirchhoff's law holds,

$$\frac{V_R - V_{CT}}{R} + \frac{V_L - V_{CT}}{R} + \frac{V_F - V_{CT}}{R} = 0$$

from which

$$V_{CT} = (V_R + V_L + V_F)/3 \tag{6.3}$$

The connections are usually made so that the CT is placed on the chest and connected to the positive terminal of a voltmeter, while the negative

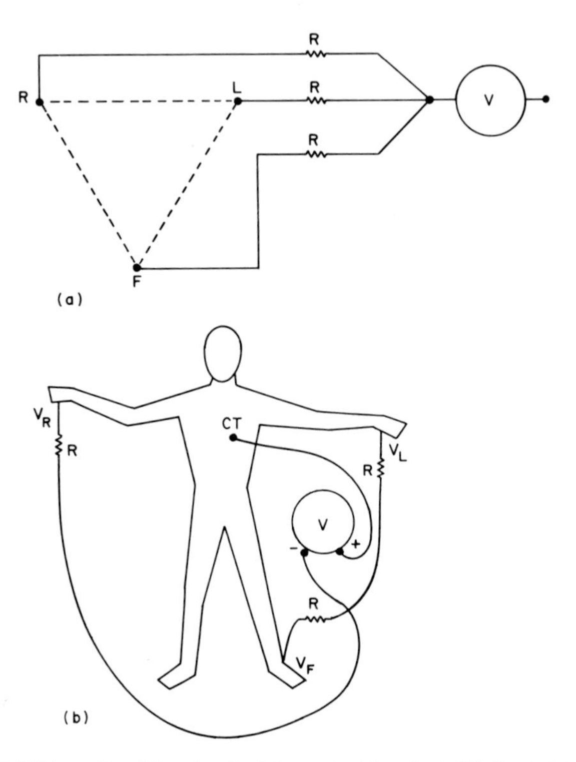

FIG. 6.15 (a) Schematic of the circuit of the central terminal. (b) Central terminal voltage probe as arranged for measurement.

terminal is connected to the junction of the three limb terminals. The CT also records the PQRST wave, but it can be placed in different positions so that it is close to various point of the heart. Following Wilson's studies, six positions are generally used, as indicated in Figs. 6.16a and 6.16b. Figure 6.16b indicates that each lead position sees a certain solid angle of the heart at a different location. Thus the heart can be divided into sections, such as the right and left atria, the right and left ventricle, and the septum. Then by geometric means the relative percentage of the dipole potential of each part of the heart on each of the lead positions can be determined. By this means not only the total dipole behavior as read from leads I, II, and III can be determined, but also through the CT readings various portions of the heart can be examined for their performance.

As an example, the percentage contribution to the potential at V_5 is the product of the solid angle and thickness. For the situation illustrated in Fig. 6.16b Schaefer gives the following values in Table 6.1. The numbers

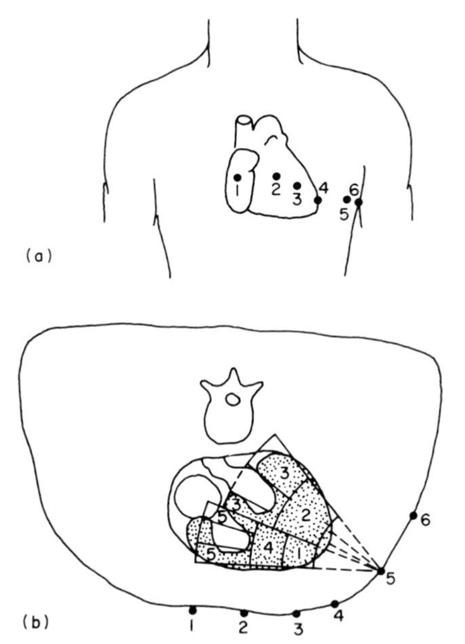

FIG. 6.16 (a) Conventional positions of the central terminal, front view. (b) Conventional arrangements of the central terminal, top view. Numbered areas indicate amounts of different sections of the heart "seen" by terminal 5. [Redrawn from Schaefer and Haas (1962).]

shown on the right are the percentage of the whole heart, e.g., the total of the products is 163 and the total of compartments 1 + 2 + 3 is 118, and thus 118/163 = 0.72, Since compartments 4 and 5 represent the right ventricle it is seen that even with the electrode in position 5, very close to the left, the contribution from the right is still greater than one fourth. Thus the task of locating small areas of heart malfunction by this means alone is still difficult.

TABLE 6.1

Compartment	Thickness (cm)	Relative solid angle	Product	
1	9	2.3	21	
2	13.5	3.5	47	72%
3	15	3.3	50	
4	8.5	2.3	20	28%
5	15	1.7	25	

VECTORCARDIOGRAPHY

The Einthoven triangle assumes that the motion of the vector dipole of the heart is in a single frontal plane, and the improvement of the central terminal was the gain of information out of the plane of the triangle. The

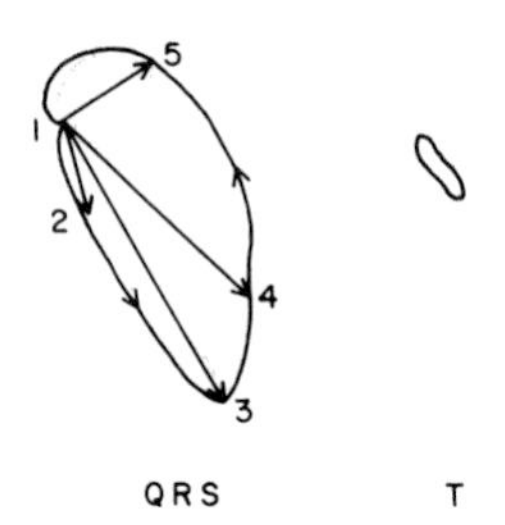

FIG. 6.17 Trace on an oscilloscope when two probes are placed on the chest and connected to the x and y axes of an oscilloscope. Arrows 1 to 5 represent time changes of vector dipole.

next logical development was to try to measure the three-dimensional components directly. This is called a *vectorocardiograph.*

For example, instead of measuring the voltage between the central terminal and an assumed zero potential of an extremity, a vectorcardiograph is a measure of the potential difference between two electrodes in contact with the chest. If two electrodes are attached to the chest and one is connected to the y axis of an oscilloscope and the other to the x axis, the resulting scope trace might appear as in Fig. 6.17. In this figure the vector dipoles are drawn by arrows as they change sequentially from 1 to 5, and a trace of the QRS wave is indicated by the continuous loop. A short time later a small loop in the same position would appear, which corresponds to the T wave.

Consider now a three-dimensional vectorcardiograph. Figure 6.18a shows the placing of the probes and the axes. Figure 6.18b shows in the center three-dimensional motion of the vector dipole. On the three planes, frontal, sagittal and horizontal, are the projections of this motion. It is these projections that would be seen on the oscilloscope. Example two-probe traces could appear as illustrated in Fig. 6.19.

The difficulty with vectorcardiography is in the leads themselves. All tissue between the probe and the dipole becomes part of the lead. Thus there must be corrections for the different conductivities of different tissues of the body, the differences in individuals, and the changing center of electrical activity of the heart. Several systems have been developed but none without defects. The current attitude is that one system should be chosen and clinical experience, i.e., empirical studies, should be used to obtain knowledge of the figures generated by normal and abnormal heart activity.

This brief coverage of vectorcardiography does not do justice to the enormous amount of careful work done by physicists and physiologists on this technique. Dummies have been built filled with conducting solutions, to imitate the body fluids, and mechanically rotated electric dipoles have been inserted. The field equations have been solved and a variety of electrical lead schemes have been proposed to give the best average of the changing lead resistance as the heart dipole undergoes its three-dimensional motion. It is

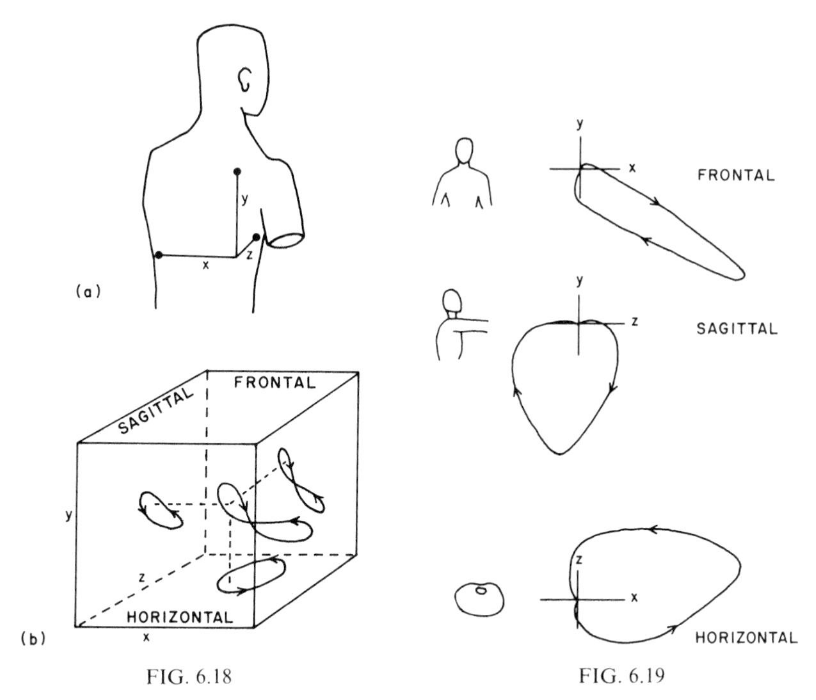

FIG. 6.18

FIG. 6.19

FIG. 6.18 (a) Position of probes in vector cardiography. (b) Central loop is actual motion of the vector dipole of the heart. The projection of this motion is shown on the three planes. [Redrawn from Grishman and Scherlis (1952).]

FIG. 6.19 Example of three planar projections of a heart dipole as seen on an oscilloscope trace.

an interesting problem and an interesting study, but the details are beyond our immediate range of interest. The reader is referred to books on the subject in the references.

COMPUTER DIAGNOSTICS OF ELECTROCARDIOGRAPHS

When increased health care for a population is planned, the decreased use of medical personnel will, in principle, increase the speed and decrease the unit cost of routine tests. Toward this end considerable work has been going on in the utilization of computers in electrocardiograph screening. There are several distinct steps in this process which must be solved independently. Four of these will be discussed as examples. First, because of the large cost of a computer, one that is regionally located could handle the input from many medical centers, and therefore there is the problem of accuracy of transmission of the signals. Second, a storage and retrieval system must exist so that digital transmission can be used and the computer can compare changes in an individual. Third, the computer must be programmed for pattern recognition. Fourth, the computer should be taught to diagnose by comparing a given electrocardiograph with a memory bank of abnormalities.

The energy content of the body surface is predominantly below 100 Hz, although Geselowitz *et al.* (1962) have shown that a frequency range up to 500 Hz is necessary to recover the detail of small notches and slurs during ventricular depolarization. However, most direct-writing electrocardiographs have an upper frequency limit of 100 Hz and it has been shown that with such a limit amplitude errors of the main deflections do not exceed 0.1 mV. The lower limit of the frequency band must extend to 0.05 Hz to minimize recording errors in the slow-moving portions such as the ST segment.

For satisfactory data reproduction and analysis a signal-to-noise ratio of about 40 dB (a power ratio of 10,000) is required. Transmission of the signals by telephone lines can be unsatisfactory on occasion due to local equipment, weather conditions, or length of transmission line. An example of noise superimposed on an electrocardiograph transmission is illustrated in Fig. 6.20a. Such a noisy record could be erroneously interpreted by a computer and is therefore unacceptable. It has been found that the use of a low pass filter that eliminates all signals of frequency greater than 40 Hz would reduce the noise of Fig. 6.20a to that of Fig. 6.20b. Although some noise is still visible the ECG is now usable. However, in the elimination of heart signals in the frequency range of 40–100 Hz by this filtering process the overall percentage accuracy of computer diagnosis decreases. A different approach to this problem is under investigation. Instead of transmitting the

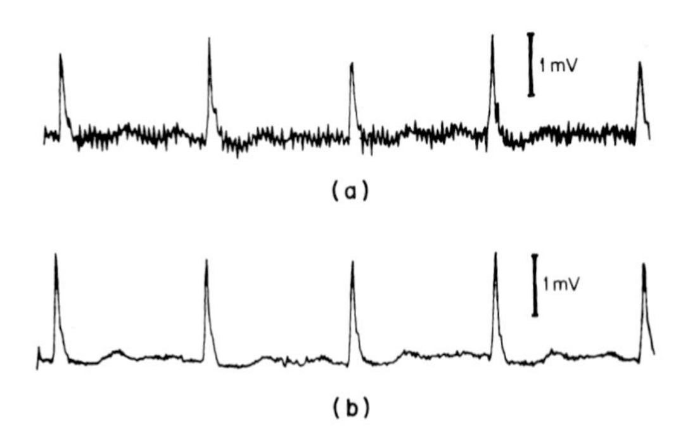

FIG. 6.20 (a) Noise superimposed on an electrocardiograph signal transmitted by telephone over a distance of approximately 500 miles. (b) The same signal after applying a digital filter with a cutoff at 40 Hz. [From Pipberger *et al.* (1975). Reproduced with permission from the *Annual Review of Biophysics and Bioengineering*, Volume 4. © 1975 by Annual Reviews Inc.]

analog (voltage) data directly, the analog data can be converted to digital data for transmission and a digital filter used. However, voice-grade transmission lines have limitations on the number of bits (of digital information) per second that can be transmitted. Therefore real-time transmission cannot have sufficient accuracy and the information must be stored and transmitted more slowly.

If real-time on-line data transmission from the patient to the computer is not used, then an intermediate storage system must be employed. This is not necessarily bad since once storage is accomplished the data may be placed in a memory bank for later recall and comparison. The most commonly used type of storage is FM (frequency modulation) tape. FM is necessary to record the low frequencies with sufficient fidelity. Tape reels and tape cassettes are being used. For long-term storage FM tape strips are bonded to standard punch cards.

A number of computer programs have been developed for analysis of electrocardiographs. Such programs are not trivial because they must deal with both normal and abnormal records. The first step is to have the computer identify the beginning and end of the waveform. Usually smoothed voltage derivatives with respect to time are used with the goals of determining P wave onset and end, QRS onset and end, and T end. These are accomplished by comparing with a standard curve, which has been created by the average of a number of visually identified waveforms. The program then minimizes the variance around the standard curve. The use of signals from several leads obviously increases the accuracy of this process. Once the beginning and end of a waveform has been identified the computer can quickly calculate time intervals, wave durations, and the amplitudes of various deflections. There are differences between two waveforms, called "beat-to-beat variations" because of respiration, muscle movement, line voltage fluctuations, etc. If the record is long enough to contain several beats the computer can calculate the mean measurements for all beats. However, cost factors must be considered. The more information available the better the computer average but the longer the transmission time and computer time. Work in this area seems to be leading to a compromise time somewhere between 2.5 and 10 seconds. This is not sufficient time for analysis of arrhythmias, however. A cost compromise for this situation is to have the computer print out an alert to the physician and the analysis can be done by hand.

An important philosophical question is whether or not to program the computer to diagnose a condition from its analysis of the data. One problem encountered in such programming is the amount of information given to a computer memory bank. If precise deviations from normal are stored then the computer can make a comparison of a given electrocardiograph with its memory to determine a best fit. However, if a sufficiently large sample of deviations is taken, then pathological conditions overlap and the computer can be "confused" in its comparisons. Since programs must be designed, individuals must make an arbitrary selection of diagnostic decision rules. How well do individuals agree with the decision rules? Simonson *et al.* (1966) asked ten expert electrocardiographers to interpret 105 unknown electrocardiographs from patients whose conditions were known by independent procedures such as cardiac catheterization, surgery, or autopsy. Correct diagnoses by the ten experts averaged 54%. Such a gross averaging is not a fair summary. For some conditions the specialists achieved nearly 100% accuracy while for some obscure ones the precision was low. Also, difficulties were encountered in the diagnoses of compound cases in which more than one type of defect was present.

It is clear that because of these results and the above-mentioned confusion in the programming of a broad spectrum of overlapping conditions the optimum approach for the present seems to be the introduction of clearly defined patterns into the computer memory bank. The computer would then alert the physician of any electrocardiograph that does not fit a clearly recognizable waveform.

STARLING'S LAW OF THE HEART

The writings of Harvey in 1628, Hales in 1740, and many physiologists since then have noted that when there is an increased flow of blood to the heart from the vena cava, the heart pumps it out through the arterial system. There are many ways of expressing this physiological behavior, one being that *within physiological limits, the heart pumps all the blood that comes to it without allowing excessive damming of blood in the veins.*

Starling was a great physiologist and gave a famous Linacre lecture in 1918 in which he summarized the findings and attempted to explain the reason. Although he was by no means the discoverer, this behavior is now known as Starling's law of the heart. Surprisingly, there is still not a satisfactory understanding of this law. We shall consider the difficulties that arise from a muscle fiber model alone and indicate some recent attempts to introduce pump efficiency into the model. First, however, we shall describe the type of experiment that demonstrates Starling's law.

The tissues of the body are able to regulate the supply of blood to themselves by means of dilation or constriction of the small blood vessels. A sudden dilation would increase the flow of blood into the veins and consequently into the right atrium of the heart. This increased tissue demand is signaled to the heart both through the nervous system and through the hormonal system, and the heart responds by beating faster. There is still another mechanism by which the heart accommodates the increased volume of blood, but in order to study it alone the heart must be separated from the nervous and hormonal systems.

One arrangement frequently used is called the heart–lung preparation. An animal's chest is opened and a tube is placed in the aorta. This tube is led through heaters, flow meters, reservoirs, etc., and back into the heart at the right atrium, all other blood vessels being tied off. Thus, the only parts of the animal's body that remain alive are the heart and lungs, and the behavior of the heart as a function of filling pressure or volume is separated from the nervous and hormonal systems. The venous pressure can be regulated by raising or lowering the reservoir connected to the right atrium while the venous flow rate can be controlled by a screw clamp on the tube between the reservoir and the right atrium. The arterial pressure is measured by a manometer attached to the tube from the aorta, and the pressure is controlled by a screw clamp on the tube.

Consider first the effect of altering the arterial pressure while maintaining a constant venous inflow. If the aortic screw clamp is suddenly tightened so that the arterial pressure goes from 100 to 130 mm-Hg then during the first few beats after this the ventricles do not pump out as much blood as they receive. As a consequence the systolic and diastolic volumes of the ventricles increase, and remain increased until the arterial pressure is reduced to the original value. The detailed measurements indicate the following. Suppose that before the screw clamp is applied the heart pumps 10 cm^3 of blood into the aorta against a mean arterial pressure of 100 mm-Hg, 90 mm-Hg diastolic and 110 mm-Hg systolic. (At each ventricular systole the aortic valve opens at 90 mm-Hg, and blood flows into the aorta until the pressure rises to 110 mm-Hg. The pressure then falls to 90 mm-Hg, at which point the aortic valve closes.) Now the arterial resistance is increased to give a mean pressure of 130 mm-Hg. The aortic diastolic pressure does not fall as far as before, so that when the next ventricular beat takes place it is perhaps 100 mm-Hg. Because of this increased arterial resistance and aortic pressure the ventricle has not ejected as much blood and some remains in the ventricle. For example, the ventricle on its first beat has forced out only 3 cm^3 of its total of 10 cm^3, leaving a 7-cm^3 residue. The venous flow remains constant, so that by the end of the next diastole the ventricular volume is 17 cm^3. This enlarged ventricle contracts much more forcibly and the maximum pressure in the aorta rises to 140 mm-Hg and the ventricle sends out 8 cm^3 of blood. At the end of this beat the ventricle will still be fuller by an excess of 9 cm^3 and the next beat will be still more forceful; the pressure rises to a maximum of 145 mm-Hg and the ventricle pumps out 10 cm^3 of blood. After this the heart goes on beating regularly, pumping out the amount that comes in. However, the heart remains dilated. If the heart were connected to the central nervous

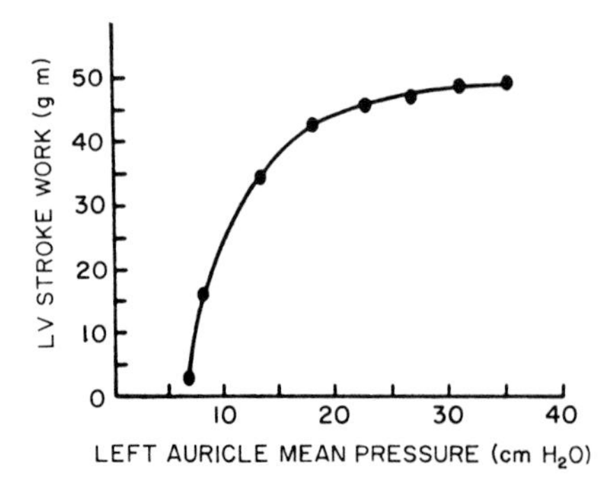

FIG. 6.21 Left ventricular (LV) stroke work versus the mean atrial (auricle) pressure of a dog's heart. [From Sarnoff (1955).]

system it would beat faster to expel the excess blood, thereby returning to normal size.

Consider now the effects of increasing the venous flow while maintaining a constant arterial pressure. For a few beats the ventricles eject less blood than they receive so that their diastolic volume increases. This is followed by an increasing output per beat until the output equals the input and no further increase in heart volume occurs. [See Starling and Evans (1962) for a more detailed description of these experiments.] These latter experiments have been performed carefully by Sarnoff (1955) and Fig. 6.21 shows a plot of his data of left ventricle mean stroke work as a function of left atrial mean pressure. As will be seen in the next chapter, the major part of the work of the heart is in increasing the pressure of the blood. Thus, the stroke work is proportional to the product of the volume per stroke and the pressure increase. Figure 6.21 shows that the stroke work of the heart can increase rapidly and approaches a saturation value that is the physiological limit of the given heart.

The Hill experiments described in Chapter 4 showed that the greater the distance of muscle movement, the greater the heat output and, as a conclusion, the greater the work done by the muscle. Thus Starling's law may be expressed in another form: *the energy of contraction is a function of the length of the muscle fibers prior to contraction.*

MUSCLE TENSION AND SARCOMERE LENGTH

In Chapter 4 we considered the mechanism of skeletal muscle motion with the implicit assumption that there were an adequate number of actin–myosin cross-links to give the muscle sufficient strength at all points on the curves considered. However, one would not expect a constant strength for all muscle lengths because the amount of overlap of the myofilaments would change, particularly at long sarcomere lengths there would be a small amount of overlap.

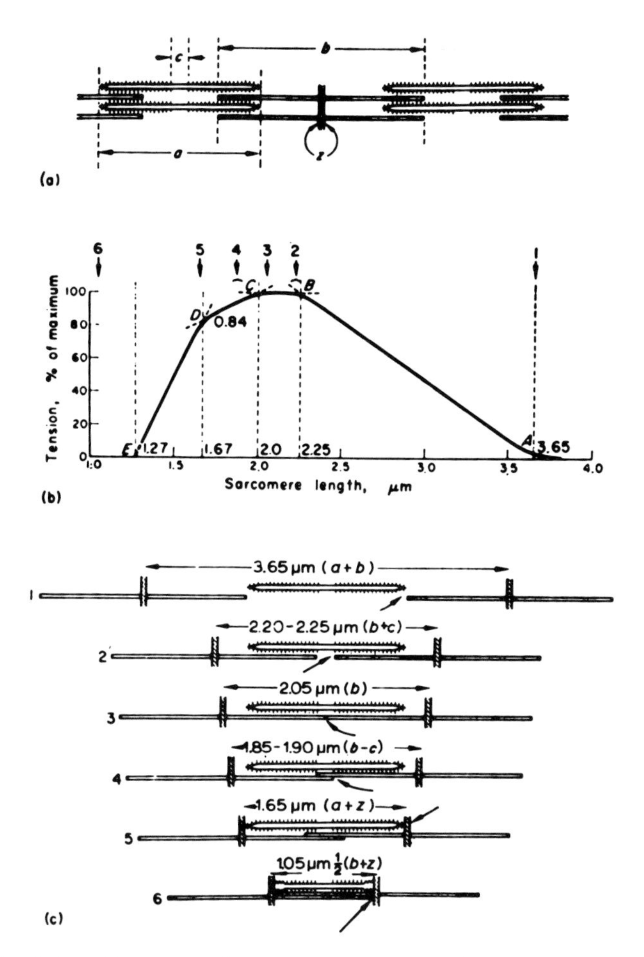

FIG. 6.22 Sliding filament lengths and the isometric length–tension curve of electrically stimulated frog muscle fibers. (a) Arrangement of filaments with assumed lengths: *a*, myosin, 1.6 μm; *b*, actin and tropomyosin + troponin, 2.05 μm; *c*, region without projections in the middle of myosin filaments, 0.20–0.25 μm. (b) Length–tension curve for a single fiber of a frog skeletal muscle. (c) The sarcomere lengths corresponding to the numbered arrows in (b). Note that in position 6 there must be some folding of the filaments. [From Gordon *et al.* (1966).]

Gordon *et al.* (1966) measured the isometric tension versus length of single fibers of striated frog muscle fibers when subjected to electrical stimulation. Their results are shown by Fig. 6.22b. In this same figure are sketches of the relative positions of the actin and myosin filaments for different sarcomere lengths. The descending slope of the curve in Fig. 6.22b at large lengths is believed to arise from the fact that less cross-linkages are formed. The decline of tension on the ascending part (sarcomere lengths below 2.0 μm) is believed to arise from interference of the mechanism of tension generation. This interference is first (1.67–2.0 μm) caused by a double overlap of the thin filaments in the A bands, and then by compression of the ends of the thick filaments against the Z lines at sarcomere lengths shorter than 1.65 μm.

In view of the above results of the Starling's (1918) law experiments, the normal or resting length of the myofilaments in cardiac muscle is not known. That is, do cardiac muscle sarcomere lengths range over the ascending or descending part of the curve of Fig. 6.22b? Arguments for the ascending part are based on the belief that this would allow reserve strength for large cardiac distensions. Arguments for the descending part are based on the fact that during contraction maximum pressure and near-maximum ejection rate are maintained until as much as two thirds of the stroke volume has been ejected. Although the pressure can be maintained by a smaller tension in the wall of the heart as it empties (see Eq. (1.14)) the same rate of ejection requires a faster shortening of sarcomeres. Because of this, the demands on cardiac muscle remain high even after significant shortening. By starting with relatively long sarcomeres and returning to the plateau of Fig. 6.22 the disadvantageous filament folding as the sarcomeres shorten is avoided. It may well be that the cardiac muscle sarcomeres are at different lengths with some average as a resting configuration. This would assure an adequate reserve of cross-links at all positions.

THE ROLE OF VENTRICULAR GEOMETRY

It is not sufficient to decide on a sarcomere length, to measure a length–tension curve, and thereby to conclude that Starling's law is understood. Gabe (1974) has shown that the way in which the fibers are arranged in a ventricle plays a significant role in the flow rate that is developed. It is worthwhile to examine his elementary one- and two-dimensional models to see the importance of fiber arrangement.

Figure 6.23a shows three models of a pump. The one-dimensional (1D) pump is a cylindrical surface of muscle fibers all parallel to the cylindrical axis. The cylinder is flat on top and bottom and nonporous. As the fibers contract ejection is through the upper tube, which corresponds to the aorta. In the two-dimensional (2D) configuration all the fibers are around the circumference of the cylinder and when they contract the cylinder is reduced in radius but not in length. The three-dimensional (3D) model is a combination of the other two and has not been analyzed.

Figure 6.23b shows the 1D and 2D pumps in greater detail. Assume that both have the same relaxed radii and lengths and that therefore their volumes are the same; V_1 and V_2 will designate the 1D and 2D volumes, respectively.

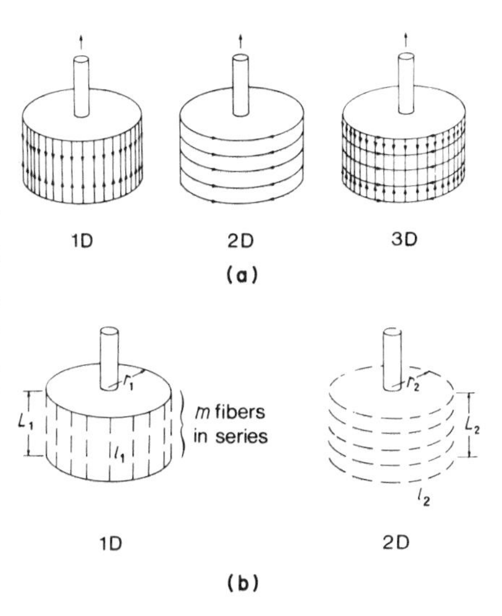

FIG. 6.23 (a) One-, two-, and three-dimensional (resp., 1D, 2D, 3D) pumps. In the 1D pump the fibers are in the surface of the cylinder parallel to the long axis. In the 2D pump the fibers are circumferential. The 3D pump combines the fibers of the 1D and 2D pumps. (b) Details of the 1D and 2D fibers. [From Gabe (1974).]

Although not a necessary requirement, we can further assume that they are constructed of the same number of muscle fibers simply by adjusting the density of fibers on the surface. Thus if there are n fibers each of length l the 1D pump is formed of $m_1 l = L$ fibers end to end, making a length L, and the cylinder is formed of a parallel array of these compound fibers. The fiber density will be the number of fibers divided by the surface area made up of these fibers, or $n/2\pi r m_1 l$. The 2D pump is formed of rings of compound fibers such that $m_2 l = 2\pi r$. A stack of these rings makes up the cylindrical surface of the pump. The fiber density in this case is $n/2\pi r L = n/m_2 lL$.

The volume of the 1D pump is

$$V_1 = \pi r^2 L = \pi r^2 m_1 l \tag{6.4}$$

If the fibers contract at rate dl/dt the rate of volume change of the pump, i.e., flow rate out of the pump, is

$$dV_1/dt = \pi r^2 m_1 (dl/dt) \tag{6.5}$$

and, upon substituting Eq. (6.4) into Eq. (6.5), we obtain

$$dV_1/dt = (V_1/l)(dl/dt) \tag{6.6}$$

The volume of the 2D pump may be written as

$$V_2 = \pi r^2 L = m_2{}^2 l^2 L/4\pi \tag{6.7}$$

The rate of change of volume for a fiber contraction rate dl/dt is

$$dV_2/dt = m_2{}^2 L 2l(dl/dt)/4\pi \tag{6.8}$$

Upon substituting Eq. (6.7) into Eq. (6.8), we may write

$$dV_2/dt = 2(V_2/l)(dl/dt) \tag{6.9}$$

We have taken a constant fiber length l and constructed two different pumps of equal volume, $V_1 = V_2$, from which we see that one arrangement gives twice the flow rate of the other for the same rate of fiber shortening.

The 3D pump is not worth working out because it is unlikely that the cardiac fibers are so neatly orthogonal as the sketch indicates.

In the heart the fibers parallel to the long axis of the ventricle lie inside the cavities (*endocardium*) and circumferential fibers are outside. The longitudinal axis is activated first and the long axis shortens. This behavior is similar to that of the 1D pump. The internal pressure generated by this action expands the heart slightly, which will lengthen the circumferential fibers, possibly moving them to a higher length–tension position on the curve of Fig. 6.22b. Their subsequent contraction makes the heart behave as a 2D pump with twice the flow rate of the 1D pump. In this way the longitudinal fibers act to position the circumferential fibers.

Although this simple exercise in pump flow does not explain the physiological basis of Starling's (1918) law, it does indicate that measurements on single fibers combined with flow rate and pressure data are not necessarily sufficient information to lead to an explanation, and that the arrangement of the fibers in the cardiac muscle must be considered.

REFERENCES

Burger, H. C. (1968). "Heart and Vector." Gordon and Breach, New York.

Chou, T., and Helm, R. A. (1967). "Clinical Vectorcardiography." Grune and Stratton, New York.

Ciba Foundat. Symp., 24th (1974). "The Physiological Basis of Starling's Law of the Heart." Elsevier, Excerpta Medica, North-Holland Publ., Amsterdam.

Gabe, I. T. (1974). Starling's Law of the heart and the geometry of the ventricle, *in* "The Physiological Basis of Starling's Law of the Heart" (*Ciba Foundat. Symp., 24th*). Elsevier, Excerpta Medica, North-Holland Publ., Amsterdam.

Geselowitz, D. B., Langner, P. H., and Mansure, F. T. (1962). Further studies on the first derivative of the electrocardiogram including instruments available for clinical use, *Am. J. Heart* **64**, 805.

Gordon, A. M., Huxley, A. F., and Julian, F. J. (1966). The variation in isometric tension with sarcomere length in vertebrate muscle fibers, *J. Physiol.* **184**, 170.

Grishman, A., and Scherlis, L. (1952). "Spatial Vectorcardiography." Saunders, Philadelphia, Pennsylvania.

Guyton, A. C. (1955). Determination of cardiac output by equating venous return curves with cardiac response curves, *Physiol. Rev.* **35**, 123.

Guyton, A. (1971). "Textbook of Medical Physiology," 4th ed. Saunders, Philadelphia, Pennsylvania.

Katz, L. N. (1955). Analysis of the several factors regulating the performance of the heart, *Physiol. Rev.* **35**, 91.

Massie, E., and Walsh, T. J. (1960). "Clinical Vectorcardiography and Electrocardiography." Year Book Publ., Chicago, Illinois.

Mommaerts, W. F. H. M., Abbott, B. C., and Whalen, W. J. (1960). Selected topics on the physiology of the heart, *in* "The Structure and Function of Muscle" (G. H. Bourne, ed.), Vol. II. Academic Press, New York.

Pipberger, H. V., Dunn, R. A., and Berson, A. S. (1975). Computer methods in electrocardiography, *Ann. Rev. Biophys. Bioeng.* **4**, 15.

Sarnoff, S. J. (1955). Myocardial contractility as described by ventricular function curves; Observations on Starling's Law of the heart, *Physiol. Rev.* **35**, 107.

Schaefer, H., and Haas, H. G. (1962). Electrocardiography, *in* "Handbook of Physiology: Circulation, Section 2," Vol. 1. American Physiological Society, Washington, D.C.

Simonson, E., Tuna, N., Okamoto, N., and Toshima, H. (1966). Diagnostic accuracy of the vectorcardiograam and electrocardiogram, *Am. J. Cardiol.* **17**, 829.

Sjöstrand, F. S., and Andersson-Cedergren, E. (1960). Intercalcated discs of heart muscle, *in* "The Structure and Function of Muscle" (G. H. Bourne, ed.), Vol. I. Academic Press, New York.

Starling, E. H. (1918). "The Linacre Lecture on the Law of the Heart." Longmans, Green, London.

Starling, E. H., and Evans, L. (1962). "Principles of Human Physiology" (H. Davson and M. G. Eggleton, eds.), 13th ed. Lea and Febiger, Philadelphia, Pennsylvania.

CHAPTER 7

Circulation: Fluid Flow in Elastic Tubes

INTRODUCTION

The blood circulatory system is one of extreme complexity. In essence it is a high-pressure piping system (arteries) that distributes the output from the pumping action of the heart through the capillaries to a low-pressure return system (veins). As long as these generalities only are considered, it is a simple process to estimate the work of the heart and the blood flow rate. This will be done first. But as one approaches the details, the problem becomes more complex. For example, blood flowing in the capillaries behaves as a non-Newtonian fluid. This has been discussed in Chapter 2. In addition to the complexities of the fluid are the complications of the arteries. They are elastic but are composed of at least two kinds of fibers, each with its own elastic constant. The arteries have many branching points, which not only change the flow characteristics but can reflect a portion of the wave pulse.

In addition, the flow is not steady but pulsates (*pulsatile flow*), and a portion of the energy from the heart is stored elastically in the arterial system to maintain flow while the heart is being refilled. Mathematical models of this kind of system will be introduced briefly with suitable references for those wishing to pursue this approach.

CIRCULATORY SYSTEM

There are certain simple physical principles that can be applied to the circulatory system, because the heart can be substituted by a mechanical pump and all of the conduits could be assembled outside of the living system.

The heart is a physical pump but, because of the constant need for gas exchange in the lungs, nature has designed it as a four-chamber system, two to supply blood to the pulmonary system and two for the rest of the body. Each of the two systems can be considered to have two pumps in series, a storage and supply pump, called the *atrium*, and a main pump, called the *ventricle*. The pumping capacity of the ventricle is designed to pump all of the blood accumulated in the atrium. The duration or act of the ventricular pumping is called the *systole* (contraction) and the period between the pumping times when the ventricle is being filled from the atrium is called the *diastole* (dilation).

The pumping system on the right side of the heart—the right atrium and right ventricle—supplies the pulmonary arteries, and the left side of the

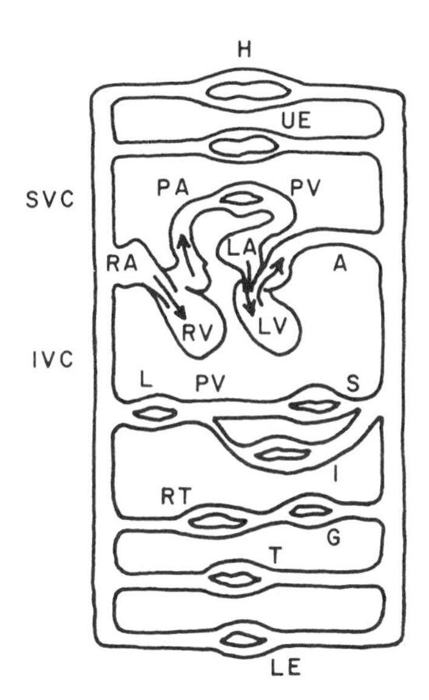

FIG. 7.1 Schematic diagram of circulation. H, capillaries in the head; UE, capillaries in the upper extremities; SVC, superior vena cava; IVC, inferior vena cava; RA, right atrium; RV, right ventricle; PA, pulmonary artery; PV, pulmonary vein; LA, left atrium; LV, left ventricle; A, aorta; L, capillaries in liver; PV, portal vein; S, capillaries in the spleen; I, capillaries in the intestines; RT, capillaries in the renal tubules; G, capillaries in the glomerulus of the kidney; T, capillaries in the trunk; LE, capillaries in the lower extremities. [Adapted from Green (1944) *in* Medical Physics (O. Glasser, ed.), Vol. I, Copyright © 1944 by Year Book Medical Publishers, Inc., Chicago. Used by permission.

heart supplies the remainder of the body. It has been ascertained that the left side of the heart does about six times the work of the right side.

The heart and the major parts of the circulatory system are schematically diagrammed in Fig. 7.1. It is seen that the circuit can be represented as series and parallel elements. The sum of the resistances of these elements is called the *total peripheral resistance* (TPR), and by the law of parallel resistances we may write

$$\frac{1}{\mathrm{TPR}} = \frac{1}{R_{\mathrm{H}}} + \frac{1}{R_{\mathrm{UE}}} + \frac{1}{R_{\mathrm{SIL}}} + \frac{1}{R_{\mathrm{G+RT}}} + \frac{1}{R_{\mathrm{T}}} + \frac{1}{R_{\mathrm{LE}}} \tag{7.1}$$

In principle, circuit theory can be applied to estimate the resistance of any section or organ by measuring the pressure drop between two points of an artery (input pressure minus output pressure) and the current of flow rate. Capacitance storage, which arises from the elasticity of the blood vessels, will be considered later.

WORK OF THE HEART

The heart does work to raise the pressure in the arteries as well as to impart kinetic energy to the blood. It also does work against friction within the blood, against the arterial walls, and against whatever may resist the expansion of the arteries. We may use Bernoulli's equation to estimate the work of the heart. Consider the diagram of Fig. 7.2 of the heart and aorta with pressure manometers.

Recall that the Bernoulli equation was derived on the basis of conservation of mechanical energy in a fluid system,

$$\begin{aligned} \Delta W &= \Delta \mathrm{KE} + \Delta \mathrm{PE} \\ &= \tfrac{1}{2}\rho({v_2}^2 - {v_1}^2) + \rho g(h_2 - h_1) + (P_2 - P_1) \end{aligned} \tag{1.6}$$

and thus the work done by the heart is that done to raise the height of the fluid, increase the velocity of the fluid, and increase the pressure of the fluid by forcing it into elastic arterial walls. These terms are velocity change, height change and pressure change.

Let us assume some average values, such as the difference in height from the ventricle to the aortic arch as 15 cm, the velocity of the blood at this point as 40 cm/sec, and a blood density of 1 g/cm^3. The initial velocity before systole is effectively zero. Note that arterial pressure at the aorta for a healthy man ranges from 80 mm-Hg during diastole to 120 mm-Hg during systole. However, the minimum pressure is maintained by the elastic stress of the arteries and the closing of the aortic valve; the venous drain pressure is effectively zero and the left atrial pressure prior to contraction can also

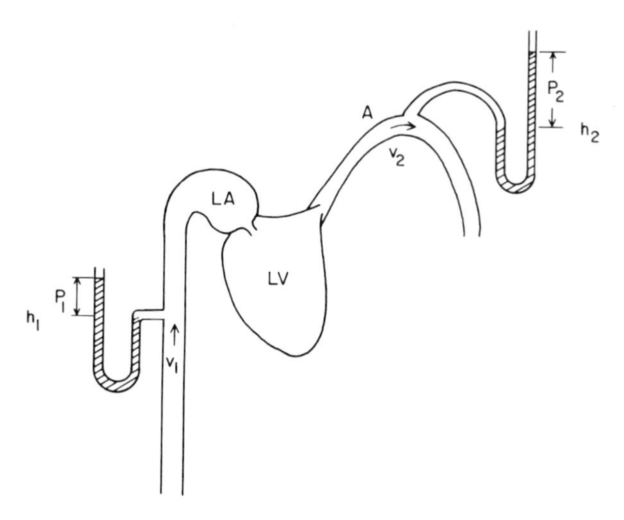

FIG. 7.2 Schematic of left side of heart. v_1, P_1, h_1, are the velocity, pressure and height of the blood entering the heart and v_2, P_2 and h_2 the same quantities for blood leaving the heart. A, aortic arch; LA and LV, left atrium and left ventricle (resp.). Adapted from Green (1944) *in* Medical Physics (O. Glaser, ed.), Vol. I. Copyright © 1944 by Year Book Medical Publishers, Inc., Chicago. Used by permission.

be taken as zero. Thus the pressure difference is 120 mm-Hg = 12 cm-Hg = 12/76 atm = 12/76 × 1 × 10^6 dyne/cm^2 = 0.16 × 10^6 dyne/cm^2. Therefore, for each cm^3 of blood expelled our terms have the magnitude of:

$$\begin{aligned}
\text{height change} &= mg(h_2 - h_1) = 1\text{ g} \times 980\text{ cm/sec}^2 \times 15\text{ cm} \\
&= 0.15 \times 10^5\text{ dyne cm}, \\
\text{pressure change} &= (m/\rho)(P_2 - P_1) = 1\text{ g}/(1\text{ g/cm}^3) \times 0.16 \times 10^6 \\
&\quad \times 10^6\text{ dyne/cm}^2 = 1.6 \times 10^5\text{ dyne cm}, \\
\text{velocity change} &= \tfrac{1}{2}m(v_2{}^2 - v_1{}^2) = \tfrac{1}{2} \times 1\text{ g }(40\text{ cm/sec})^2 \\
&= 8 \times 10^2\text{ dyne cm.}
\end{aligned}$$

We see that the work of velocity change is negligible and the work of height change is only 10% of that of pressure change. Thus the work per cm^3 of blood pumped is 1.7 × 10^5 dyne cm and, assuming an average quantity per stroke of 60 cm^3, the work of the left ventricle is about 10^7 dyne cm. To this must be added the work of the right ventricle, which pumps into the pulmonary artery, whose average pressure is about 20 mm-Hg or 1/6 the aorta's. The height change is about half that of the aortic arch so we may consider it negligible. The total heart work is then $(1 + \frac{1}{6})10^7$ dyne cm = 1.16 × 10^7 dyne cm. Note that since pumping against the aortic pressure is the main term in the work, the above calculations indicate that a body with hypertension (high blood pressure) requires an increased work load by the heart directly proportional to the ratio of the aortic pressure to a normal aortic pressure.

The average power of the heart can be obtained by dividing the average stroke work by the ejection time. This time is the interval during which the aortic valve is open, about 0.2 sec. The average power is therefore

$$\text{average horsepower} = \frac{(1.16 \times 10^{7}\ \text{dyne cm}) \times (1.34 \times 10^{-10})}{0.2\ \text{sec}} \simeq 0.008$$

where 1 erg/sec = 1.34×10^{-10} hp.

ZERO PRESSURE POINT

The return system through the veins is only slightly dependent on arterial pressure because the pressure through the capillaries drops essentially to zero. The veins have unidirectional passive valves much like flap valves in a water pump. Movements of the body muscles compress the veins and force the blood toward the larger veins with the passive valves preventing a return flow. Thus, someone standing motionless has an increased head pressure on the veins in the leg of more than 1 m of blood and the veins distend. Years of continuous standing sometimes leads to a permanent distention of these veins, resulting in a condition called varicose veins. With normal motion the venous blood returns to the right atrium of the heart, which then drains into the right ventricle during diastole. During systole the ventricle pumps the amount made available to it by the atrium through the pulmonary system into the left atrium. This in turn drains into the left ventricle to be pumped through the rest of the body's arteries. The size of the stroke volume of the heart adjusts itself to accommodate the volume of blood returned to it, as we have discussed in Chapter 6.

The pressures that we have discussed are relative or gauge pressures and we must specify an internal reference or zero point. There is one point in the body that is not affected by hydrostatic pressure of the blood, the tricuspid valve. The right atrium pumps blood from the return veins through the tricuspid valve into the right ventricle. (Recall from Chapter 6 that the left side has a similar valve called the mitral.) The reason for the lack of hydrostatic effects at the tricuspid is as follows. If the pressure at the trisucpid rises above normal the right ventricle fills more and the heart pumps either more rapibly or with greater volume than normal, and thereby reduces the pressure at the tricuspid. If the pressure falls below normal the right ventricle does not fill as much and the pumping decreases.

It is apparent from this that while the arterial pressure is everywhere positive, the venous pressure in a standing human must be negative above this zero point. Because of this, the veins in the neck collapse almost completely in a standing person. In the skull, however, the veins are in a noncollapsible chamber and are therefore at a negative pressure, differing at different heights in the skull. Thus in surgery if a vein in the skull is entered without its being at the same height as, or lower than, the tricuspid valve, air can be sucked into the circulatory system.

FLOW RATE

If we know the amount of substance that enters or leaves a stream and the concentration difference resulting from such an injection or removal of the substance, the flow rate of the stream can be calculated. Consider for example a long pipe with flowing water that has two taps far apart. If we inject a salt solution in one tap at a rate of I g/sec of salt and after a while take samples from the downstream tap, there will be a salt concentration of C g/l in the extracted sample. Under steady state conditions, if F in liters per second is the flow rate of the water, we may write

$$I = C \times F$$

and thus the flow rate is calculable. This is done in the bloodstream with indicator solutions, although sometimes an instantaneous injection of an indicator is used. In this latter case, because some paths in the body are shorter than others, the single pulse becomes a distribution function. This is because a second or third recirculation of parts of the indicator that has gone through the shorter paths may overlap with the first circulation of part of the indicator. Relatively simple analytic methods are available (see Zeirler, 1962) for this situation.

The steady state case is readily applied to gas transport by the blood. Suppose oxygen is consumed at the rate of 240 cm^3/min. If arterial blood contains 200 cm^3/liter and venous blood 160 cm^3/liter, each liter of blood would take up 40 cm^3 of oxygen in the lungs. It would therefore require a flow rate of 6 liter/min to transport 240 cm^3 of oxygen to the tissues in 1 min. We can generalize this expression to

$$F = O/(A - V) \tag{7.2}$$

where F is the flow rate, O the oxygen consumption rate, and A and V the oxygen content of the arterial and venous blood, respectively. This same principle applies to any foreign substance injected into the blood stream. Equation (7.2) is called Fick's law by physiologists. To avoid confusion, since physicists and chemists mean Eqs. (3.4) or (3.5) when speaking of Fick's laws, Eq. (7.2) should perhaps be called Fick's *zeroth* law.

An injection technique with substances opaque to x rays is in common practice among heart specialists because it reveals not only flow rate but also valve malfunction is indicated by regurgitation of blood back into the heart.

STROKE VOLUME FROM PULSE ANALYSIS

Following diastole, the left ventricle contracts and the pressure opens the aortic valve. This permits blood to flow into the aorta, which, being elastic, distends. Blood then flows into other major arteries, which also distend. As the pressure from the left ventricle begins to drop, because the ventricular contraction is nearly complete, the pressure in the aorta drops and the aortic valve, which acts as a check-valve, closes to prevent back flow. This point is noticeable in an accurate pulse curve and is known as the *incisura*. This is the onset of diastole but, since the arteries are elastically distended, they maintain a pressure that slowly decreases as the blood drains through the arterioles and capillaries into the venous system and back to the heart. Clearly, the greater the blood flow the greater the rate of pressure decrease. The stroke volume can be ascertained if the amount of blood stored in the arteries by their distension is determined. The pioneering experiments of this technique by Hamilton and Remington (1947) showed the information that is contained in the details of the pulse shape (Fig. 7.3).

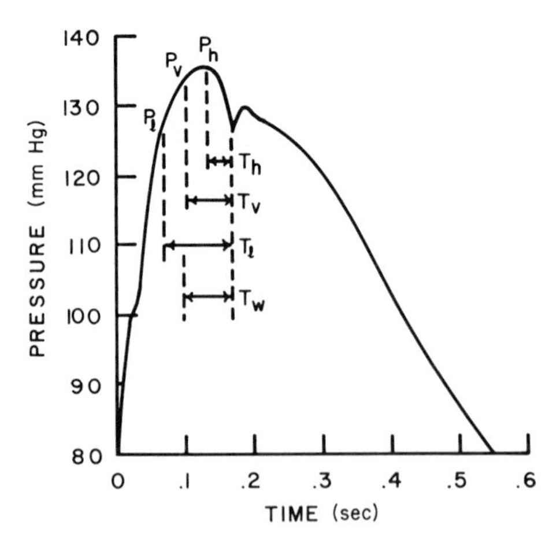

FIG. 7.3 Arterial pressure pulse. Note the incisura at 0.17 sec. [From Hamilton and Remington (1947).]

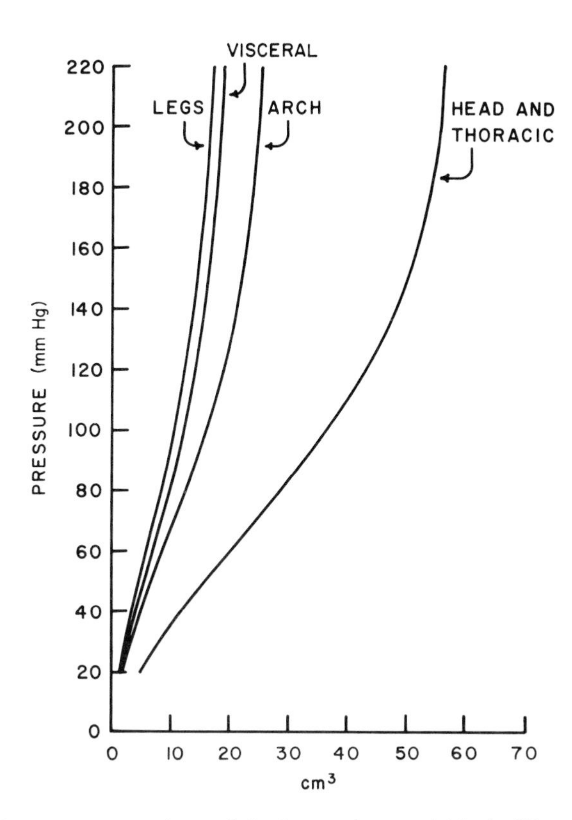

FIG. 7.4 Pressure versus volume of the four major arterial beds. [From Hamilton and Remington (1947).]

The essence of their approach is as follows. The arterial storage system is divided for simplification into four major groups; the head and thorax, the aortic arch, the viscera, and the legs. If their capacities are known as a function of pressure, as well as the transmission time of the pressure pulse from the heart, their mean pressures and hence mean volumes relative to the incisura time can be found. The sum of these volumes is called the *uptake*. If reasonable estimates can be made of the drainage or loss of blood during systole, the sum of this drainage plus the uptake is the stroke volume.

Various earlier data on sacrifice animals of the volume of the arterial system as a function of pressure are plotted in Fig. 7.4.

The pulse wave transmission times were measured and symbolized: T_h = time to reach head, T_v = time to reach viscera, T_ℓ = time to reach leg, and T_w = weighted transmission time for drainage. This latter time is required to estimate drainage but, since the uptake in each of the arterial systems is different in volume, an average drainage time weighted to the relative volumes is more accurate than a simple average time.

There is now sufficient information for the calculation of the uptake. The procedure was to construct transmission times from the incisura in the aortic pulse and to read the corresponding pressures in the four major storage groups. These times and the corresponding pressures are shown in Fig. 7.3, all being measured from the incisura. The corresponding volumes of the arterial beds at these pressures were obtained from Fig. 7.4. Thus the total arterial uptake or stroke volume is calculable when a reasonable value for drainage loss during systole is included. The stroke volumes on experimental animals of a wide range in size was also determined by dye injection and found to be in close agreement with this method.

Although many assumptions are implicit, it should be noted that consistently good results were obtained with relatively simple calculations. It seems that considerable information is available in the pulse shape and in the transmission velocity and, most importantly, the arterial system may be approximated adequately by four representative volumes. This degree of accuracy with such crude approximations confirms our introductory remark that if one does not seek out details then a relatively simplistic approach is adequate. However, following this pioneering work a much greater degree of sophistication has been achieved. An example of a modern technique using a radioactive isotope will be discussed in Chapter 9.

ELASTIC TUBES

The arteries are in reality quite elastic and, as we have seen, can elastically store energy of the blood. The kinetic energy of the blood distorts the elastic walls of the arteries, thereby converting some of the kinetic energy to potential energy, and we would therefore anticipate that the more elastic the blood vessel the lower will be the blood velocity. We shall first define our terms, noting that from this point on we are considering only thin-walled elastic tubes; the mathematics are a bit more complicated for thick-walled tubes.

The modulus of elasticity Y of a substance is defined as

$$Y = \frac{F/A}{\Delta L/L} \tag{7.3}$$

which expresses Hooke's law that the linear distortion $\Delta L/L$ is proportional to F/A, the force per unit cross-sectional area. We define the tensile stress F/A as σ_τ and therefore

$$Y = \sigma_\tau L/\Delta L. \tag{7.3'}$$

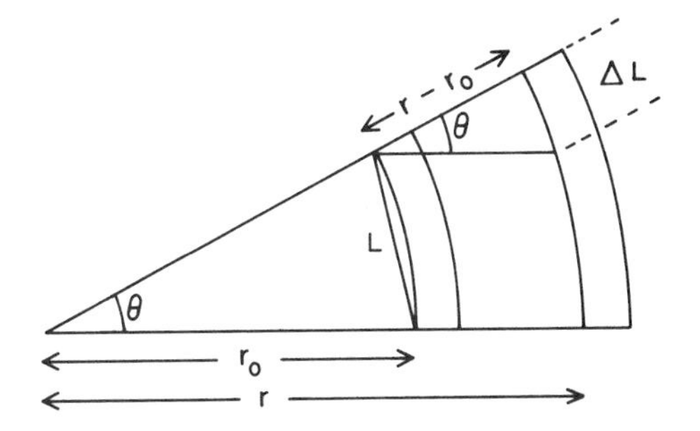

FIG. 7.5 Cross section of a wall element of a thin-walled elastic tube in relaxed and stretched condition.

A cross section of the wall in relaxed and stretched conditions of an elastic tube is shown in Fig. 7.5. It is seen that for small angles $\Delta L = (r - r_0)\theta$ and $L = r_0\theta$. Substitution of these relations into Eq. (7.3′) yields

$$\sigma_\tau = Y(r - r_0)/r_0 \tag{7.4}$$

We showed that the law of Laplace for a tube (Eq. (1.15)) can be written as

$$P_{TM} = \sigma_\tau w/r \tag{1.15}$$

where σ_τ is the tensile stress and w the wall thickness. P_{TM} is the transmural pressure, i.e., the inside pressure minus the outside pressure.

Upon substitution of Eq. (1.15) into Eq. (7.4) we obtain

$$Pr/w = Y(r - r_0)/r_0$$

in which we drop the subscript TM for convenience of writing. Since $r \simeq r_0$ we may write

$$P \simeq Yw(r - r_0)/r_0^2 \tag{7.5}$$

This equation is a simplified form of the actual relation of P and r. Texts on elasticity theory give

$$P = (2Yw/r_0^2)(r - r_0)/(2 - \psi) \tag{7.5′}$$

where ψ is Poisson's ratio (the amount of thickness decrease of the wall per unit strain), which is of the order of 0.25. Thus the pressure would be larger than that of Eq. (7.5) by about 10%. It is not necessary to achieve this accuracy in physiological systems because a tube such as an artery does not strictly obey Hooke's law.

WAVE PULSES IN ELASTIC TUBES

Consider a homogeneous nonviscous incompressible fluid in a thin-walled elastic tube, as in Fig. 7.6. Let a thin element $ABCD$ of length dx be displaced a distance ξ from the undisturbed position $A'B'C'D'$ when a wave passes through the tube. With no wave the radius of the element is r' and the fluid pressure is p' relative to the outside pressure p_0. The pressure on the fluid arises from the distension of the tube and is a function of the radius. Let the instantaneous pressure be $(p - p') = P$ on the cross-sectional face AB where p is the instantaneous absolute pressure at AB. The excess force on face AB is the excess pressure times the area $P\pi r^2$, where r is the radius of the AB surface. The excess force in the opposite direction on area CD is the sum of the excess pressure on AB plus the gradient of the area and excess pressure times the distance dx, or

$$\pi r^2 P + (\partial/\partial x)(\pi r^2 P)\, dx$$

Thus the accelerating force on element $ABCD$ is the vector sum of the forces on $ABCD$, or

$$-(\partial/\partial x)(\pi r^2 P)\, dx$$

The mass of the element is the density ρ times the volume, or

$$\pi r^2 \rho\, dx,$$

and the acceleration is $\partial \xi^2/\partial t^2$.

Upon substituting these terms into $F = ma$ we obtain the equation of motion

$$-(\partial/dx)(r^2 P) = \rho r^2(\partial^2 \xi/\partial t^2) \tag{7.6}$$

The original width of the element at $A'B'C'D'$ is

$$[1 - (\partial \xi/\partial x)]\, dx$$

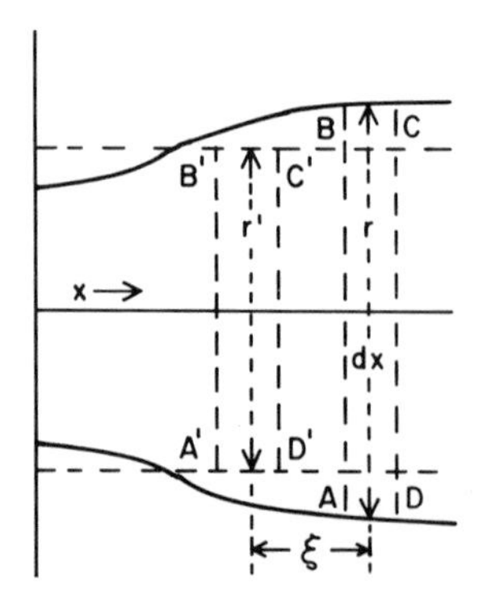

FIG. 7.6 Diagram of a portion of an elastic tube in which a wave is traveling from left to right.

and, since the fluid is incompressible, the original and final volumes must be equal:

$$\pi r^2 \, dx = \pi r'^2 [1 - (\partial\xi/\partial x)] \, dx$$

or

$$r^2 = r'^2 [1 - (\partial\xi/dx)] \tag{7.7}$$

Substitution of Eq. (7.7) into Eq. (7.6) gives

$$-\frac{\partial}{\partial x}\left[r'^2\left(1 - \frac{\partial\xi}{\partial x}\right)P\right] = \rho r'^2\left(1 - \frac{\partial\xi}{\partial x}\right)\frac{\partial^2\xi}{\partial t^2}$$

Since r' is not a function of x it cancels, and upon taking the derivative on the left-hand side and rearranging we obtain

$$\left(1 - \frac{\partial\xi}{\partial x}\right)^{-1}\left[-\frac{\partial P}{\partial x} + \frac{\partial P}{\partial x}\frac{\partial\xi}{\partial x} + P\frac{\partial^2\xi}{\partial x^2}\right]\frac{1}{\rho} = \frac{\partial^2\xi}{\partial t^2}$$

The first two terms of a binomial expansion of $[1 - (\partial\xi/\partial x)]^{-1}$ are $[1 + (\partial\xi/\partial x)]$. Multiplying through and dropping the terms that involve the square of $\partial\xi/dx$, Eq. (7.6) can now be written as

$$\frac{\partial^2\xi}{\partial t^2} = \frac{1}{\rho}\left[P\left(1 + \frac{\partial\xi}{\partial x}\right)\frac{\partial^2\xi}{\partial x^2} - \frac{\partial P}{\partial x}\right] \tag{7.8}$$

which is the equation for the pulse wave.

THE MOENS–KORTEWEG EQUATION

If we assume as before that the tube obeys Hooke's law and continue to neglect the Poisson ratio we may use Eq. (7.5) for the functional relation of P and r:

$$P \cong Yw(r - r')/r'^2 \tag{7.5}$$

where r_0 now is equal to r' because the tube is partially distended owing to the pressure P' of the fluid. Substituting from Eq. (7.7) for r gives

$$P \cong \frac{Yw[r'(1 - (\partial\xi/\partial x)^{1/2} - r']}{r'^2}$$

Binomial expansion of the square root gives

$$P \cong \frac{Yw[r'(1 - \frac{1}{2}(\partial\xi/\partial x)) - r']}{r'^2} \qquad \text{or} \qquad P \cong (Yw/2r')(\partial\xi/\partial x)$$

which we can substitute into Eq. (7.8) and, again dropping terms which involve the square of $\partial\xi/\partial x$ as well as the cross term, we obtain the wave equation

$$\partial^2\xi/\partial t^2 = (Yw/2\rho r')(\partial^2\xi/\partial x^2) \tag{7.9}$$

The dimensionality of the two sides requires the constant on the right-hand side to be v^2 and the wave velocity (Appendix B) is therefore

$$v = (Yw/2\rho r')^{1/2} \tag{7.10}$$

As before, if the Poisson ratio is to be included, Eq. (7.5′) would be used instead of Eq. (7.5) and, by inspection, it is seen that the velocity would be

$$v = [(Yw/2\rho r')2/(2 - \psi)]^{1/2} \tag{7.10a}$$

with a resulting 10% increase in the velocity.

The relation of Eq. (7.10) is known as the Moens–Korteweg equation although it was derived earlier by Thomas Young. In experiments with elastic tubes Moens found that the expression should be modified to

$$v = K(Yw/2\rho r')^{1/2} \tag{7.11}$$

where K is a dimensionless constant of about 0.9 for standing waves in elastic tubes. Others found that in dog aortas $K = 0.6$–0.7 and in humans about 0.8 for diastolic pressures above 70-mm Hg; at low pressure K is not a constant.

Pulse wave velocity data for the thoracic aorta in humans are shown in Table 7.1. It is seen in these data that loss of elasticity by the arteries with age increases the modulus Y and therefore increases the velocity, as expected. What Eq. (7.10) does not predict however is the nonlinear pressure dependence of the data. Since the density of blood does not vary sufficiently with pressure to account for this there is a clear indication of nonlinearity in the elastic modulus.

TABLE 7.1[a, b]

	Mean blood pressure (mm Hg)				
Age	7.5	50	100	150	200
20–24	3.47	3.29	3.57	4.50	5.74
36–42	3.28	3.49	4.91	6.70	8.94
71–78	3.02	4.65	7.57	11.12	14.73

[a] Data compiled by Hamilton (1944).
[b] Pulse wave velocity in m/sec.

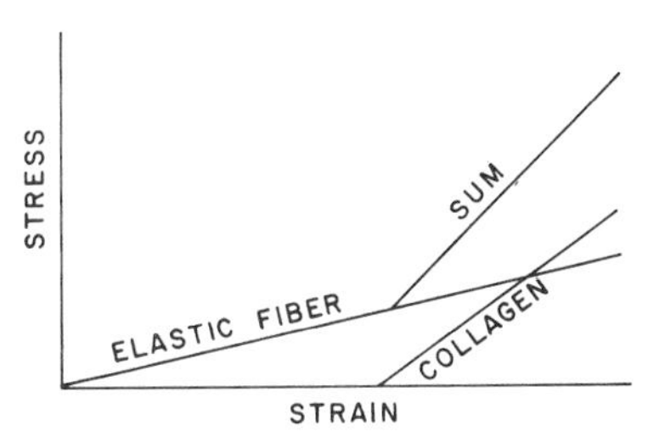

FIG. 7.7 Schematic stress–strain curves of the elastic fibers and collagen fibers of muscle.

It is now known that the blood vessel walls are made of a combination of elastic fibers and collagen fibers, each with its own modulus of elasticity. These are symbolically represented by stress–strain curves in Fig. 7.7. However, the actual situation is not so simple. The fibers do not obey Hooke's law but are viscoelastic. Thus, the rate of stretch is important and, since the molecules are not simple ones, there is a spectrum of relaxation times. In addition, arterial walls have a component of smooth muscle whose role must be considered. As an example, Fig. 7.8 shows the pressure–radius relationship of a dog's jugular vein. The closed circles are for a condition of smooth muscle tone. The second inflation was done after the vessel was irrigated with xylocaine to relax the muscle. Although both curves tend toward the same maximum radius, the slope, which is related to the elastic modulus, is steeper at low pressures for the unrelaxed smooth muscle.

The arrangement of the various fibers in a blood vessel wall is somewhat like an automobile tire of rubber and nylon cord. The rubber is quite elastic,

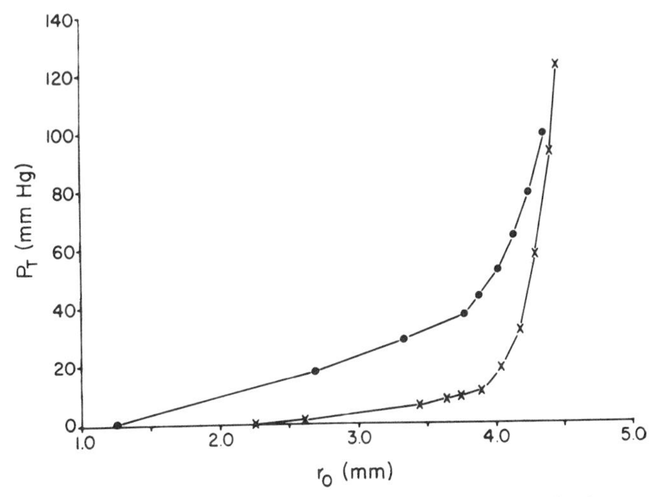

FIG. 7.8 Pressure–radius relationship of the external jugular vein of a dog *in vivo*. Closed circles are for the condition of smooth muscle tone, crosses for condition of smooth muscle relaxed with xylocaine. [From Bergel (1964).]

low modulus, but is kept from blowing out by the high modulus nylon fibers. The combination has the form of the sum curve of Fig. 7.7, which is similar to that of Fig. 7.8. Thus a proper model of two elastic constants of a blood vessel should, in principle, yield the proper blood velocity behavior. However, there are still further complicating features that have been considered by investigators, such as the damping of the wave due to fluid, elastic wall internal friction, and wave reflections from branching. Many of these complications are discussed in detail in the references.

MODELING OF PULSATILE ARTERIAL FLOW

In the eighteenth century Hales measured arterial blood pressure and attributed the changes to the expansion of the elastic walls by the blood volume injected by the heart. He thus completed the description of blood circulation begun by Harvey.

In its simplest form circulation consists of a pump, the heart, which forces blood into a system of branching tubes. The pump does not operate on a steady flow basis but in pulses. These periodic pulses send the blood into elastic tubes, which stretch in the nonlinear fashion described above. This *pulsatile* flow through the arteries is damped by the elasticity of the arteries and the media in which they lie, and the pulses die out by the time they reach the capillaries. The blood returns through the venous system by a more or less steady flow. The pulsatile flow is disturbed further by interthoracic changes in pressure due to respiration and by reflected waves from the branching points of the arterial system. As we have seen, the blood itself is not a simple fluid and has anomalous viscous properties because it is a colloid of suspended particles.

In order to investigate the nature of pulsatile fluid flow it is therefore necessary to construct a simplified model. The earliest simplification was the "windkessel," or air chamber, model. This was proposed in the nineteenth century by E. H. Weber, who used the statement, "What the air effects in the fire engine, the elastic walls effect in the human body." In those days fire engines had an air chamber in line with the pump. The air compressed elastically to store pressure excursions of the pump and thereby smoothed the flow of water. This model therefore introduces the capacitance of the air chamber, which has its equivalence in electrical capacitance.

The windkessel model in its simplest form consists of a pump, a resistance line, and a hollow container as a storage volume. In this model, the heart is the pump which exerts pressure P_{H}, the resistance is R, the arterial storage volume is V_{S}, and the flow volume is Q. This is analogous to an RC electrical circuit, shown in Fig. 7.9, in which $P_{\mathrm{H}} = \mathscr{E}$, the voltage, $V_{\mathrm{S}} = C$, the capaci-

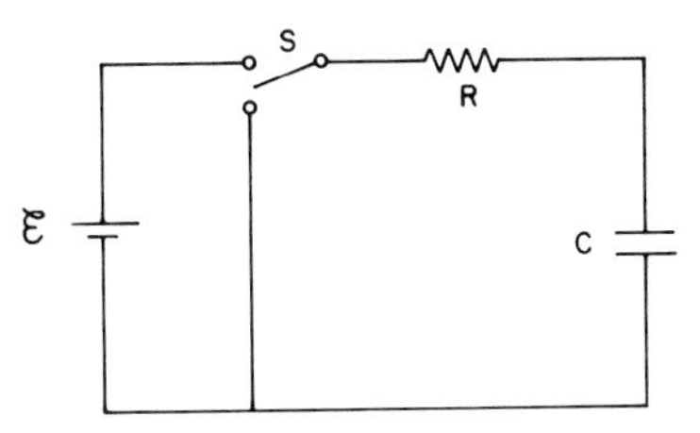

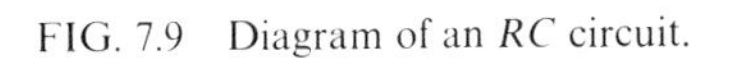
FIG. 7.9 Diagram of an RC circuit.

tance, and $Q = q$, the quantity of charge. The solution of this circuit is based on the conservation of energy in that the electrical work, $\mathscr{E}\,dq$, done by the source in an incremental time dt, where q is the electric charge, must equal the energy dissipated by the resistor plus the energy stored by the capacitor. When the switch is closed.

$$\mathscr{E}\,dq = i^2R\,dt + d(q^2/2C) = i^2R\,dt + (q/C)\,dq$$

but, since $i = dq/dt$

$$\mathscr{E} = R(dq/dt) + (q/C) \tag{7.12}$$

If $\mathscr{E}$ is considered to go instantaneously from 0 to its full value $\mathscr{E}$ it may be treated as a constant. The solutions of Eq. (7.12) (Section D.3 of Appendix D) is

$$q = C\mathscr{E}\,(1 - e^{-t/RC}) \tag{7.13}$$

If, when the capacitor is fully charged, the switch is moved to the lower terminal so as to disconnect the voltage source, Eq. (7.12) is written as

$$0 = R\,dq/dt + q/C$$

whose solution is (Section D.1)

$$q = q_0 e^{-t/RC} \tag{7.14}$$

Note that these differential equations and their solutions are the same functionally as the viscoelastic ones, Eqs. (2.6) and (2.8). From Eq. (7.13) at $t = \infty$ the fully charged capacitance will have charge $q_0 = C\mathscr{E}$ and then Eq. (7.14) may be written as

$$q = C\mathscr{E}e^{-t/RC} \tag{7.15}$$

We may now return to our physiological model and rewrite Eqs. (7.13) and (7.15) as

$$Q = V_S P_H(1 - e^{-t/RV_S}) \tag{7.16}$$

$$Q = V_S P_H e^{-t/RV_S} \tag{7.17}$$

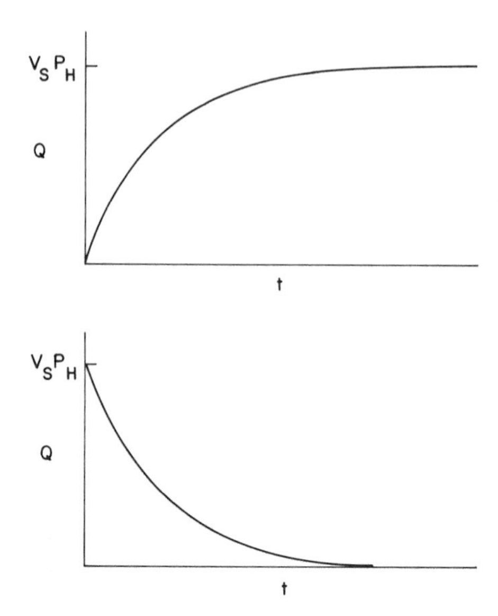

FIG. 7.10 Ascending and descending exponential behavior of arterial flow in the windkessel model. Q, flow volume; V_S, blood vessel storage capacity; P_H, heart pump pressure (assumed to be constant).

If we assume that the aorta is linearly elastic in the range of normal physiological pressures, we may write that its pressure is proportional to the volume of blood stored in it, or P_A is proportional to Q. Therefore Eqs. (7.16) and (7.17) may be written as aortic pressure variations with time while the heart is considered to be a constant pressure pump of magnitude P_H, which is either on or off. A plot of Eqs. (7.16) and (7.17) is shown in Fig. 7.10. In this figure both the increasing and decreasing exponential curves are plotted with the same time constant RV_S. This need not be the case since the contraction of the heart and the pulse wave in the arteries are two unrelated processes.

A measurement of the aortic pressure was shown in Fig. 6.3. This can be approximated by an ascending and a descending exponential with an ascending time constant 1/10 that of the descending one. This is shown in Fig. 7.11. It is seen that the windkessel model is a reasonable one for the aorta.

What is true for the aorta, however, does not apply downstream. If we measure the pressure gradient at two points along an artery, when the pressure wave hits the first point its pressure is higher than that at the second point and the pressure gradient slopes downstream. When the pressure wave hits the second point, its pressure is higher than that at the first point and the gradient slopes upstream. This oscillation of the pressure gradient

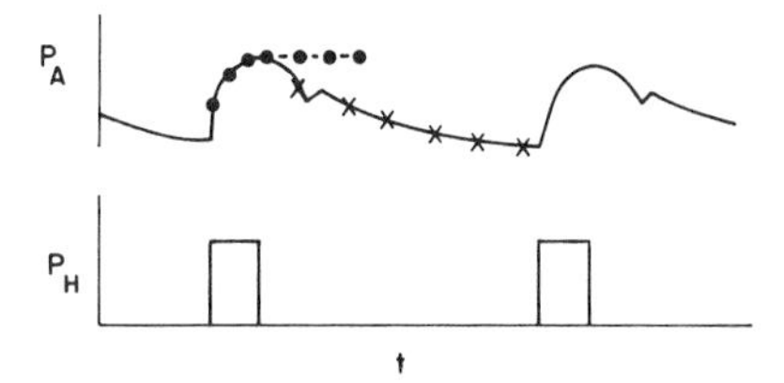

FIG. 7.11 P_A is the aortic pressure with time from Fig. 6.3. Circles and dashed curve represent a rising exponential with a time constant 1/10 the descending exponential, which is represented by crosses. Note that the incisura is for reference and is not a result of the equations. Lower curve is heart pump pressure, assumed to be constant and either on or off.

gives rise to a traveling wave, and the simple Poiseuille flow, Eq. (1.11), is not applicable. Furthermore, the mass of the blood will resist movement when pressure is first applied, because of its inertia. Once it starts moving its momentum will tend to keep it moving. In addition, the elasticity of the arteries plays a role in creating a phase difference between the pressure and the flow rate. This is seen in Fig. 7.12, in which is shown the flow velocity and the pressure measured simultaneously in the femoral artery.

A mathematical theory of such pulsatile flow was developed by Womersley (1955, 1957, 1958) and computer solutions were obtained. This theory, however, is still an oversimplification. The superposition of reflected waves and respiratory pressures complicates the situation still further. Figure 7.13 shows the pressure waveform and flow rate characteristics in different

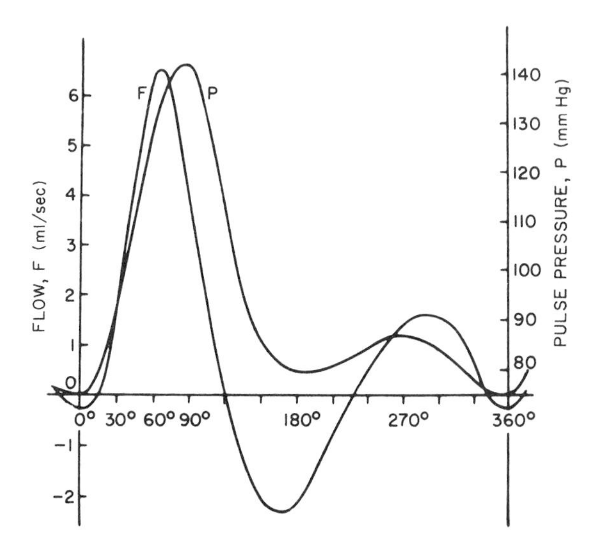

FIG. 7.12 Flow velocity pulse F and arterial pressure pulse P recorded simultaneously in the femoral artery of a dog plotted versus phase angle. [From McDonald (1955).]

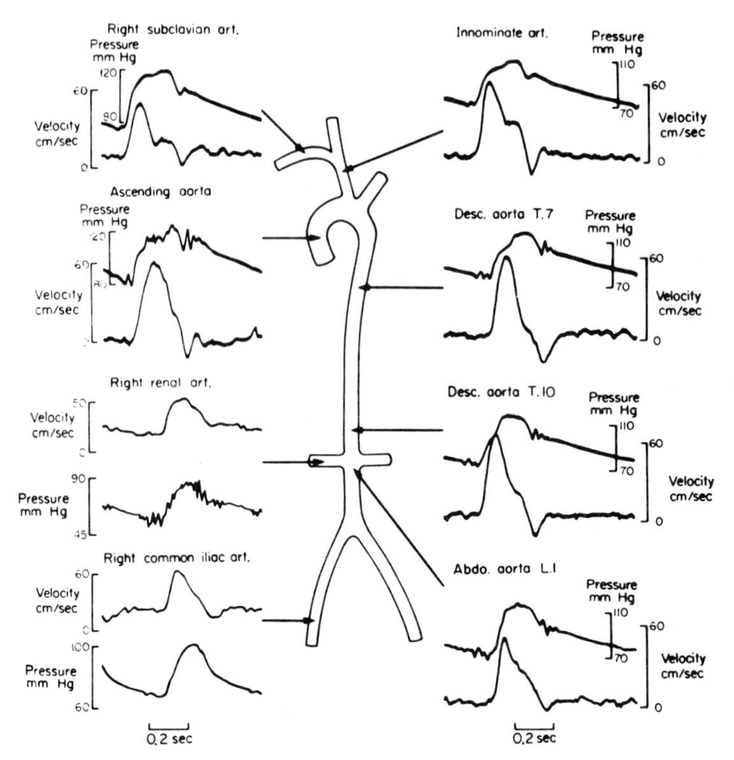

FIG. 7.13 Simultaneous pressure and blood velocity patterns at different points in the arterial system. [From Bergel and Schultz (1971).

arterial parts and the analysis of such waveforms will be very difficult. A number of analytic approaches have been tried and they are reviewed in the book by McDonald (1960) and those edited by Attinger (1964) and by Bergel (1972b). However, a complete analysis of such complex flow and pressure patterns remains to be made.

REFERENCES

Anliker, M. (1972). Toward a nontraumatic study of the circulatory system, *in* "Biomechanics" (Y. C. Fung, N. Perrone, and M. Anliker, eds.). Prentice-Hall, Englewood Cliffs, New Jersey.

Attinger, E. (ed.) (1964). "Pulsatile Blood Flow." McGraw-Hill, New York.

Bergel, D. H. (1964) *in* "Pulsatile Blood Flow" (E. O. Attinger, ed.). McGraw-Hill, New York.

Bergel, D. H. (1972a). The properties of blood vessels, *in* "Biomechanics" (Y. C. Fung, N. Perrone, and M. Anliker, eds.), Prentice-Hall, Englewood Cliffs, New Jersey.

Bergel, D. H. (ed.) (1972b). "Cardiovascular Fluid Dynamics," Vols. 1 and 2. Academic Press, New York.

Bergel, D. H., and Schultz, D. L. (1971). Arterial elasticity and fluid dynamics, *Progr. Biophys.* **22**, 1.

Blesser, W. B. (1969). "A Systems Approach to Biomedicine." McGraw-Hill, New York.

Burton, A. (1962). Physical principles of circulatory phenomena: The physical equilibria of the heart and blood vessels, *in* "Handbook of Physiology: Circulation, Section 2," Vol. 1. American Physiol. Association, Washington, D.C.

Frank, O. (1899). Die Grundform des arteriellen Pulses, erste Abhandlung, mathematische Analyze, *Z. Biol.* **37**, 483.

Frank, O. (1927). Die Theorie der Pulswellen, *Z.* Biol. **85**, 91.

Fronek, A. (1972). Analysis of recent developments in blood flow measurement, *in* "Biomechanics" (Y. C. Fung, N. Perrone, and M. Anliker, eds.). Prentice-Hall, Englewood Cliffs, New Jersey.

Green, H. D. (1944) *in* "Medical Physics" (O. Glasser, ed.), Vol. 1. Year Book Publ., Chicago, Illinois.

Guyton, A. (1971). "Textbook of Medical Physiology," 4th ed. Saunders, Philadelphia, Pennsylvania.

Hamilton, W. F. (1944), *in* "Medical Physics" (O. Glasser, ed.), Vol. I. Year Book Publ., Chicago, Illinois.

Hamilton, W. F., and Remington, J. W. (1947). The measurement of the stroke volume from the pressure pulse, *Am. J. Physiol.* **148**, 14,

Hardung, V. (1962). Propagation of pulse waves in visco-elastic tubing, *in* "Handbook of Physiology: Circulation, Section 2," Vol. 1. American Physiological Society, Washington D.C.

Hardung, V. (1964). Input impedance and reflection of pulse waves, *in* "Pulsatile Blood Flow" (E. O. Attinger, ed.). McGraw-Hill, New York.

Iberall, A., Cardon, S., and Young, E. (1970). "On Pulsatile and Steady Arterial Flow: The GTS Contribution." General Technical Services, Upper Darby, Pennsylvania.

Kenner, T. (1972). Flow and pressure in the arteries, *in* "Biomechanics" (Y. C. Fung, N. Perrone, and M. Anliker, eds.). Prentice-Hall, Englewood Cliffs, New Jersey.

King, A. L. (1947). Waves in elastic tubes: Velocity of the pulse wave in large arteries, *J. Appl. Phys.* **18**, 595.

McDonald, D. A. (1960). "Blood Flow in Arteries." Williams and Wilkins, Baltimore, Maryland.

McDonald, D. A. (1955). The relation of pulsatile pressure to flow in arteries, *J. Physiol.* **127**, 533.

Rushmer, R. (1961). "Cardiovascular Dynamics." Saunders, Philadelphia, Pennsylvania.

Skalak, R. (1972). Mechanics of the microcirculation, *in* "Biomechanics" (Y. C. Fung, N. Perrone, and M. Anliker, eds.). Prentice Hall, Englewood Cliffs, New Jersey.

Warner, H. R., Swan, H. J. C., Connolly, D. C., Tompkins, R. C., and Wood, E. H. (1953). Quantitation of beat-to-beat changes in stroke volume from the aortic pulse contour in man, *J. Appl. Physiol.* **5**, 495.

Womersley, J. R. (1955). Oscillatory motion of a viscous liquid in a thin-walled elastic tube. I: The linear approximation for long waves, *Phil. Mag.* **46**, 199.

Womersley, J. R. (1957). "An Elastic Tube Theory of Pulse Transmission and Oscillatory Flow in Mammalian Arteries." Wright Air Development Center Tech. Rep. WADC-TR 56-614.

Womersley, J. R. (1958). Oscillatory flow in arteries: The reflections of the pulse wave at junctions and rigid inserts in the arterial system, *Phys. Med. Biol.* **2**, 313.

Zierler, K. L. (1962). Circulation time and the theory of indicator dilution methods for determining blood flow and volume, *in* "Handbook of Physiology: Circulation, Section 2," Vol. 1. American Physiological Society, Washington, D.C.

CHAPTER 8

Ultrasonic Probes: Scanning and Echocardiography

INTRODUCTION

Ultrasound procedures now in use are essentially pulse–echo sonar. The ultrasound is generated by a piezoelectric transducer and directed into the body by contact with the skin. As ultrasound travels through the body it is reflected by interfaces between tissues of varying densities, the larger the density change the greater the reflection. The reflected signal is received either on a separate transducer or on the same one if the signal is pulsed. Because of the high reflectivity of bone, when scanning soft tissue the probe must be placed at a point in the body from which the beam can be directed to strike the region under investigation without striking a bone.

If a moving part is to be studied, such as heart action, properly timed reflected pulses, displayed on an oscilloscope and photographed, can yield a time sequence. This scope figure can be compared with the electrocardiograph and become an auxiliary diagnostic tool. This technique is called *echocardiography*.

Some of the ultrasonic energy is absorbed by the body tissues and is converted into heat. If the heat is not dissipated fast enough tissue damage can result.

Current predictions of some investigators NSF (1973) are that the medical applications of ultrasound will equal or surpass those of x rays within the next decade. For this reason the present chapter introduces topics of beam propagation, focusing, reflection, and damage. Because of the complexity of the basic mechanisms we will first discuss some of the applications. With this order the reasons for the analysis and the options available will be apparent.

SCANNING AND ECHOCARDIOGRAPHY

The principle of piezoelectricity was discussed in Chapter 5; pressure exerted on opposite faces of a piezoelectric crystal causes a voltage difference between the faces. The converse occurs if a voltage difference is created between opposite faces; the crystal changes dimensions. Crystals are highly responsive in that the changes occur almost as fast as the voltage can be made to change.

Usually a transducer is made of a composite of barium titanate, which has a large piezoelectric coefficient, formed into a disk about 1 to 1.5 cm diameter and 0.1 to 0.2 cm thick. The faces are silver-plated to provide the necessary contacts for the voltage signal. Barium titanate has a low mechanical damping; in other words, it rings like a bell when set into motion. Such ringing would prevent it from responding to the echo signals. Therefore, its motion must be damped. This is usually accomplished by placing the piezoelectric transducer at one end of a cylinder filled with fine tungsten powder imbedded in resin, which scatters and absorbs the sound transmitted into it and effectively damps any ringing.

In operation the transducer is placed against the skin with a gel as an interfacing compound and electrical pulses are fed from a high-frequency generator to the transducer electrodes. The pulse frequency is in the megacycle range and the duration of a pulse is a few microseconds; thus each pulse contains only a few oscillations. The repetition rate of these pulse bursts is a few hundred per second. The ultrasonic waves penetrate the body tissues and are reflected from tissue boundaries. In the time between the pulses the transducer becomes a listening device, in that a reflected pulse causes a mechanical change in the dimensions of the transducer, which in turn causes a voltage signal. This voltage signal is put into the y axis of an oscilloscope. When the generator pulses the transducer it also is arranged to make a y-axis signal on the oscilloscope. The x axis is time and any reflected pulses will appear in the y direction at different times in the x direction. The times are readily converted to distances because the velocity is known. An example of this mode of operation is shown in Fig. 8.1. This is a

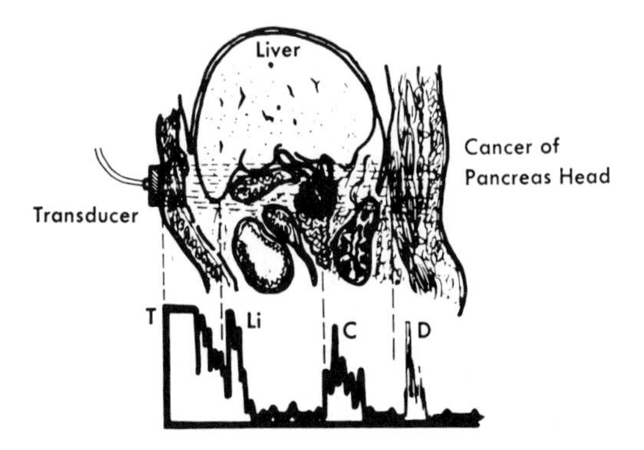

FIG. 8.1 Ultrasonic diagnosis of cancer in the region of the bile duct. Upper sketch is schematic of the area investigated by a horizontal beam. Lower part is oscilloscope trace of reflection intensities versus time. T, transmitted pulse; Li, liver margin echo; C, cancerous echo; D, diaphragm echo. [From Wagai *et al* (1965), in "Ultrasonic Energy" (E. Kelley, ed.), The University of Illinois Press, Urbana, Illinois. Copyright 1965 by the University of Illinois.]

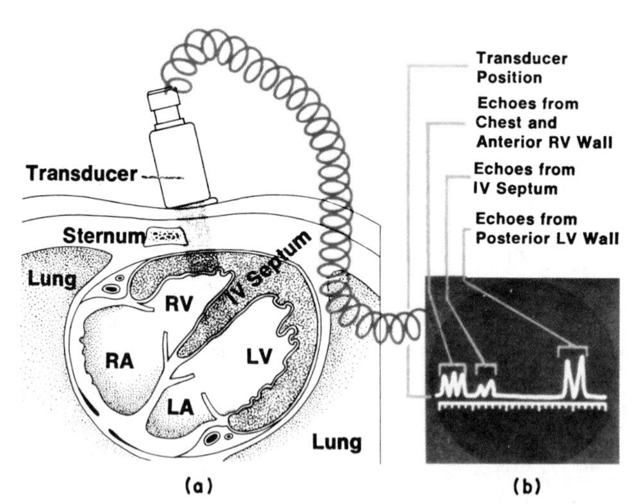

FIG. 8.2 (a) Production of an A mode by an ultrasonic transducer. Several echoes are obtained as the beam is partially reflected from different interfaces. (b) The echoes are displayed by height on an oscilloscope trace proportional to echo intensity and distance is measured along the x axis, which corresponds to time of the echo. (c) An echocardiograph displayed in A, B, and M modes. The B mode converts the peaks to intensity. The M mode, or time–motion mode, is created by sweeping the B mode up the oscilloscope screen. The motion of the echo-producing interfaces to and from the transducer produces waveforms. [From "Directions in Cardiovascular Medicine," Vol. 1 Hoechst Pharmaceuticals, Somerville, New Jersey, 1970. Based on data of Gramiak and Shah (1971).]

schematic of an ultrasonic diagnosis of cancer in the region of the bile duct. The oscilloscope trace is shown with the relation of echo times on the x axis to depth of the reflected signal.

Ultrasonics has become extraordinarily useful in the examination of the heart. The heart is constantly in motion but, with respect to the speed of the pulses and the response of the electrical system, this motion is very slow.

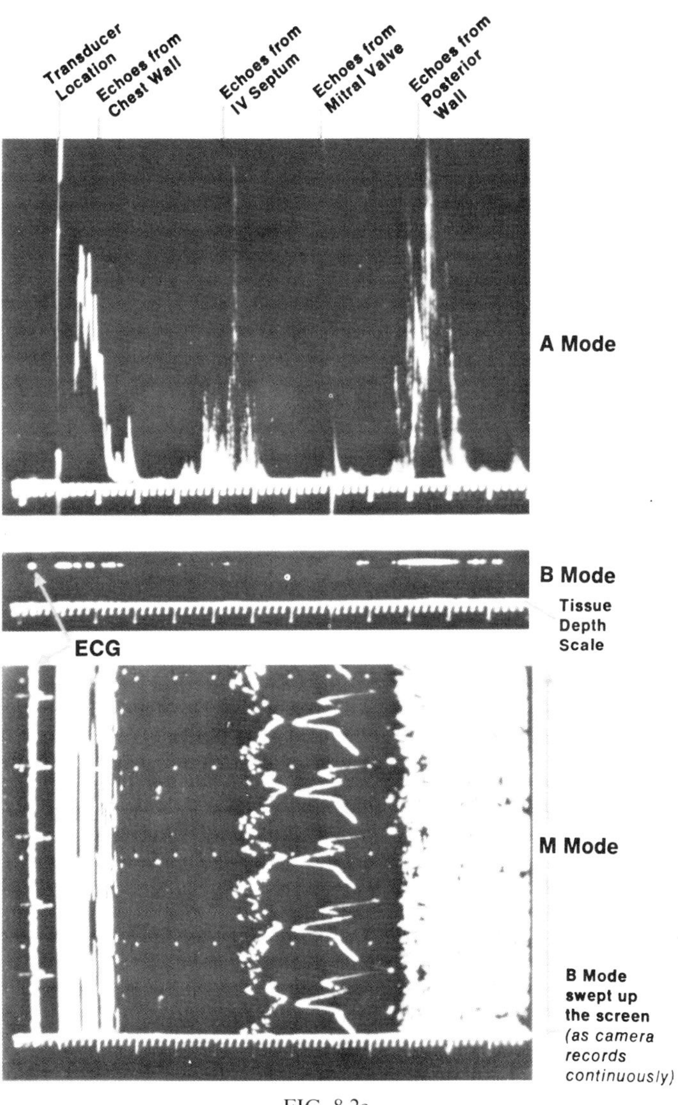

FIG. 8.2c

The reflected pulses from the heart are displayed on the oscilloscope. Such a display of reflected pulse height on a time axis is called the A mode of operation. Instead of having the reflected pulse magnitude made proportional to the magnitude of the y deflection, however, it can as readily be made proportional to the intensity or brightness of a spot on the oscilloscope. This is called the B mode. Note that while the A mode is an amplitude modulated display, the B mode is an intensity modulated display. The A mode as used in Fig. 8.1 is useful for static structures. The B mode is used for ultrasonic tomography or the motion pattern of reflecting structures. When the B mode pattern is swept across the oscilloscope screen it is known as the M mode. If the examination is of the heart this technique is called *echocardiography*. Note that the changes are too fast for the eye to follow on the screen and a moving picture of the screen is usually taken for later slow-motion examination. Such data are also generally taken simultaneously with EKG (sometimes called ECG) and phonograms. Figure 8.2 shows a typical arrangement for obtaining an echocardiograph of the heart in one direction. The photos in Fig. 8.2c right show the A, B, and M mode data.

The M mode is very useful in heart valve examination. Mitral valve stenosis, or constriction, can arise from poor mobility of the mitral leaflets or calcification in the vicinity. Figure 8.3 shows an echocardiographic analysis of the mitral valve. Although other structures are seen, the motion of the mitral valve is quite distinctive. The degree of stenosis can be determined from the rate of diastolic descent of the leaflet. The rate of descent is

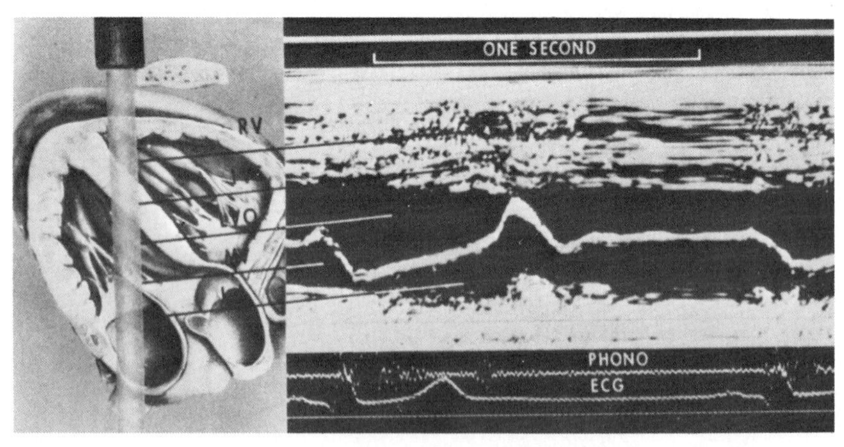

FIG. 8.3 Echocardiograph of a normal heart with the beam directed at the mitral valve. The reflection symbols are: RV, right ventricular cavity; IVS, interventricular septum; LVO, left ventricular outflow tract; MV, mitral valve; LA, left atrium. [From Gramiak and Shah (1971).]

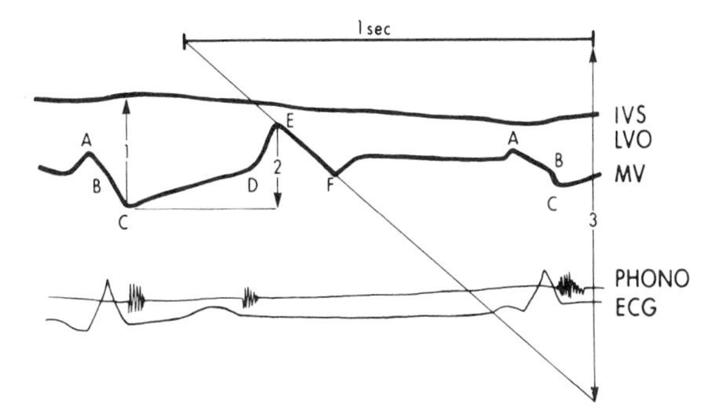

FIG. 8.4 A measurement of the rate of descent of the mitral leaflet from an echocardiograph. E–F is the diastolic descent and its slope is measured graphically in millimeters per second. [From Gramiak and Shah (1971).]

obtained from the E–F slope as shown in Fig. 8.4 and is measured in millimeters per second. Schematic examples of various conditions are shown in Fig. 8.5, where the upper curve is a simultaneous EKG.

It should be noted, however, that the application of such a diagnostic tool is not immediate upon its development. Not only must its use be learned but the signals must be interpreted. This can only be done by comparing it to more familiar signals or by surgical and autopsy examinations. A mass

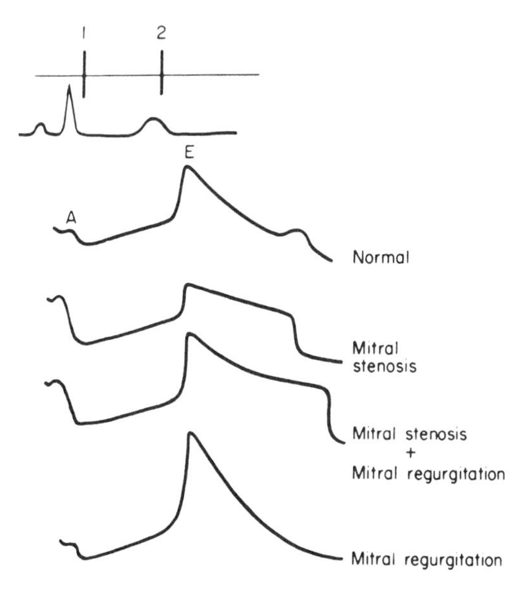

FIG. 8.5 Schematics of mitral valve echocardiography. The upper curve is a simultaneous EKG. [From Segal *et al.* (1967).]

of data must be accumulated over long years by dedicated investigators before the application of a new technique can be relied upon.

It is seen that an ultrasonic probe is an extremely useful diagnostic tool. It has limitations, however, and these must be considered in planning its utilization.

The design of a probe for clinical usage has three important variables; power, frequency, and transducer diameter. In the next sections we shall derive the fundamental equations on which designs are based. We shall first show that beam intensity is given by the relation

$$I = \overline{p^2}/\rho c$$

where $\overline{p^2}$ is the average of the square of the pressure of the wave, ρ is the density of the medium and c the velocity of sound in the medium. Using this relation we will then show that the useful depth of an ultrasonic beam x_{max} is given by

$$x_{max} \cong a^2/\lambda$$

where a is the transducer radius and λ the wavelength. Finally, we shall derive the relation that the beam diverges by an angle β given by the relation

$$\sin \beta = 0.61\lambda/a$$

THE WAVE EQUATION

In order to develop the necessary concepts we must first review some properties of simple harmonic motion and the pulses of waves generated in condensed media.†

Suppose we have a tube of a fluid in the x direction and have some type of colored cross-sectional markers in the fluid at positions x_1, x_2, x_3, etc. When a sound wave passes through the tube these markers will be temporarily displaced a distance ξ, as in Fig. 8.6, so that at a given instant the plane at x_1 will be at position $x_1 + \xi(x_1)$, the plane originally at x_2 will be at position $x_2 + \xi(x_2)$, etc. This, of course, assumes no turbulence. The velocity of any plane is given as $d\xi/dt$, which is also the average plane velocity. Thus ξ depends on both x and t, that is, the quantity $\xi(x, t)$ measures the average displacement due to a sound wave of those molecules whose average position was originally x. We must find the relation between ξ, x, and t. To do this we use Newton's second law, the conservation of matter, and the assumption that the compression is so rapid that no heat escapes from the compressed region, i.e., the compression is adiabatic.

† This is similar to the derivation of the wave equation in Appendix B but is redone here to make this chapter independently understandable.

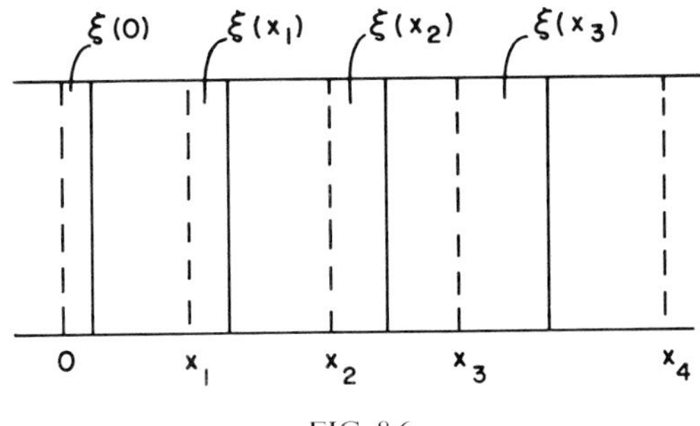

FIG. 8.6

Call the equilibrium pressure P_0 and the excess pressure due to compression P. If the equilibrium density is ρ_0 the actual density is a function of x and t, or $\rho(x, t)$. If we call δ the relative change in density then

$$\delta = (\rho - \rho_0)/\rho_0$$

or

$$\rho(x, t) = \rho_0(1 + \delta) \tag{8.1}$$

We can find a relation for δ by recognizing that, since there is no turbulence, the fluid between any two planes remains between those planes. Consider a single region of Fig. 8.6 as shown in Fig. 8.7. At equilibrium the mass of the fluid between the planes x and $x + dx$, if the cross-sectional area is A, is the product of the density and the volume:

$$m = \rho A\, dx \tag{8.2}$$

When the planes are displaced, however, the volume is changed and, since the number of molecules is unchanged, the density must change. Figure 8.7 indicates that the displacement of the first plane is $\xi(x)$ and that of the second

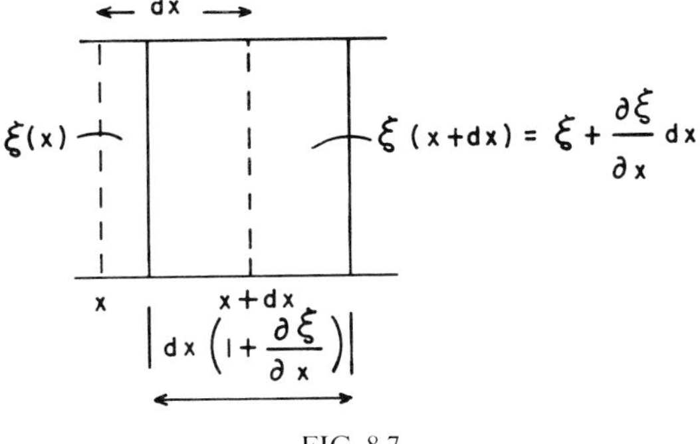

FIG. 8.7

plane is $\xi(x + dx) = \xi(x) + (\partial\xi/\partial x)\,dx$ because $\partial\xi/\partial x$ is the gradient of ξ with x. The volume enclosed by these planes is seen in Fig. 8.7 to be

$$A\{dx + [\xi(x + dx) - \xi(x)]\}, \qquad \text{or} \qquad A\,dx + A(\partial\xi/\partial x)\,dx$$

The density of the fluid between the planes must be changed so that the original mass $\rho A\,dx$ is unchanged, that is,

$$\rho(x, t)[A\,dx + A(\partial\xi/\partial x)\,dx] = \rho A\,dx \tag{8.3}$$

If we substitute Eq. (8.1) we have

$$\rho A\,dx = \rho_0(1 + \delta)A\,dx\,[1 + (\partial\xi/\partial x)] \tag{8.4}$$

For small changes in density and displacement we can neglect the product $\delta(\partial\xi/\partial x)$ and write from Eq. (8.4)

$$\delta = -\partial\xi/\partial x \tag{8.5}$$

which is a form of the equation of continuity.

We may now use Newton's second law, $F = ma$, to develop the wave equation. The force on the fluid in the volume between the planes of Fig. 8.7 is the vector sum of the unbalanced forces on the left plane and right plane. From the left this force is

$$[P_0 + P(x)]A$$

from the right it is

$$[P_0 + P(x + dx)]A = [P_0 + P(x) + (\partial P/\partial x)\,dx]A$$

and their difference is equal to the mass of the fluid $\rho A\,dx$ times the acceleration $\partial^2\xi/\partial t^2$:

$$-A(\partial P/\partial x)\,dx = \rho A\,dx\,(\partial^2\xi/\partial t^2)$$

or

$$\partial P/\partial x = -\rho(\partial^2\xi/\partial t^2) \tag{8.6}$$

This is the wave equation, where ρ is the equilibrium density. We wish now to have a different form of the wave equation without P and ρ.

VELOCITY OF THE LONGITUDINAL WAVE

We will now consider the relationship of the velocity of sound in a fluid, its density, and its bulk modulus B.

Figure 8.8 represents a tube of fluid with a piston at one end. At $t = 0$ the piston is set in motion toward the right with a speed v. The lower part of the

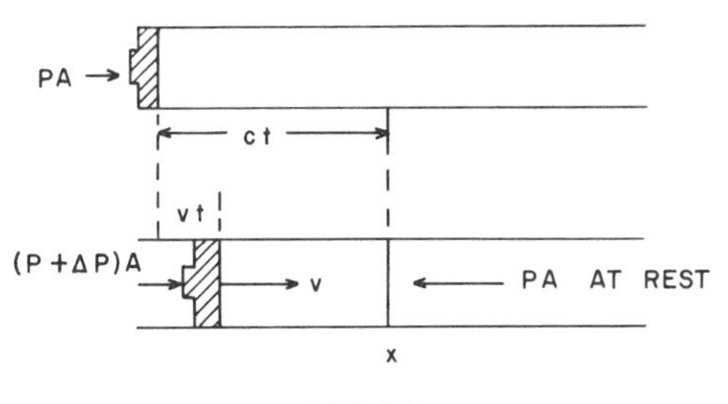

FIG. 8.8

figure shows the situation at time t where all parts of the fluid to the left of point x are in motion with velocity v while to the right of point x the fluid is still at rest. The boundary between the moving and stationary fluid travels to the right with velocity c, the velocity of sound in the fluid. In time t the piston has moved a distance vt and the boundary has moved a distance ct. The quantity of fluid set in motion in time t is that originally occupying the volume Act, its mass is ρAct and it has acquired a momentum of

$$\text{longitudinal momentum} = \rho Actv \tag{8.7}$$

We may use the definition of bulk modulus and the fact that the original volume of the moving fluid Act has been decreased by an amount Avt to write

$$B = \frac{\text{change in pressure}}{\text{fractional change in volume}} = \frac{\Delta P}{Avt/Act} \tag{8.8}$$

from which

$$\Delta P = B(v/c)$$

The force from the left on the moving fluid is $(P + \Delta P)A$ and from the right is PA. Thus the net force is ΔPA and the impulse is ΔPAt:

$$\text{longitudinal impulse} = \Delta PAt = B(v/c)At$$

Since impulse is equal to momentum change we equate this with Eq. (8.7),

$$B(v/c)At = \rho Actv$$

from which

$$c = (B/\rho)^{1/2} \qquad \text{or} \qquad B = c^2\rho \tag{8.9}$$

where c is the velocity of the pressure wave.

If we now define the ΔP as pressure due to the wave we may write

$$p \equiv \Delta P = -B(\Delta V/V) \tag{8.8}$$

in which a minus sign is included because as the pressure increases the volume decreases. Upon substituting Eq. (8.9) into Eq. (8.8) we obtain

$$p = -\rho c^2(\Delta V/V)$$

and, since for the volume Adx of Fig. 8.8 there has been a change in volume $Ad\xi$, the change in volume per unit volume is $\partial\xi/\partial x$ and we write

$$p = -\rho c^2(\partial\xi/\partial x) \tag{8.10}$$

Upon differentiating we obtain

$$\partial p/\partial x = -\rho c^2(\partial^2\xi/\partial x^2) \tag{8.11}$$

and substituting Eq. (8.11) into Eq. (8.6) gives

$$\partial^2\xi/\partial x^2 = 1/c^2(\partial^2\xi/\xi t^2) \tag{8.12}$$

This is another form of the wave equation, in which c is the velocity of sound in the fluid. It is the same as that derived in Appendix B.

INTENSITY

The intensity is defined as the power transmitted in the direction of the wave. Instantaneous power is the excess pressure p against a surface multiplied by the velocity of the fluid wave, i.e., multiply Eq. (8.10) by $\partial\xi/\partial t$:

$$p(\partial\xi/\partial t) = -\rho c^2(\partial\xi/\partial x)(\partial\xi/\partial t)$$

The left-hand side of this equation is the instantaneous power transmitted. The average value of the power is the intensity I of a wave.

If we have a travelling wave, we may write it as

$$\xi = \xi_0 \cos(kx - \omega t)$$

then

$$\partial\xi/\partial x = -\xi_0 k \sin(kx - \omega t)$$

and, since $p = -\rho c^2(\partial\xi/\partial x)$ (Eq. (8.10)), we may write

$$p = [\rho c^2 k\xi_0] \sin(kx - \omega t) \tag{8.13}$$

where the term in brackets is the pressure amplitude p_{max}. Substitution of $k = 2\pi f/c = \omega/c$, where f is frequency and $\omega = 2\pi f$, gives the useful relation

$$p_{max} = \omega\rho c\xi_0 \tag{8.14}$$

The time rate of change of ξ is

$$d\xi/dt = \omega\xi_0 \sin(kx - \omega t)$$

The intensity I defined above is the average value of the product

$$I = \overline{p(\partial\xi/\partial t)} = \overline{\rho c^2 k \xi_0{}^2 \omega \sin^2(kx - \omega t)}$$

and we may substitute Eq. (8.13) to obtain

$$I = \overline{p^2/\rho c} \tag{8.15}$$

the intensity in terms of pressure, density, and velocity of sound. We note here that sometimes intensity is expressed in terms of $p_{\max}$. Since the average value over a cycle of $\sin^2$ is $\frac{1}{2}$, Eq. (8.15) becomes

$$I = p^2{}_{\max}/2\rho c \tag{8.15$'$}$$

THE DESIGN OF SOURCES OF ULTRASONIC WAVES

In order to use reflections of sound waves in the body as a detection device of its structure and of how this structure varies with time, it is desirable to generate a parallel beam of sound waves to act as an ultrasonic "searchlight." A vibrating disk at the surface of the body is a suitable source of such a beam of sound waves. The variables in the design and operation of this disk-shaped source are its radius, the frequency of vibration, and the amplitude of vibration or power level at which the source is to be operated. We wish to choose these variables so that the beam generated yields the maximum resolution with the minimum damage to the tissues.

Consider first the question of resolution. Clearly, a parallel beam of sound waves will have a finer resolution the smaller its diameter, thus suggesting the use of a small diameter source. However, it is not so simple as this since a beam of sound waves emanating from a disk-shaped source does not remain parallel but diverges, and this divergence depends upon the diameter of the disk and its frequency of vibration. We shall explore the question of divergence in some detail below, and it will be seen that the higher the frequency the smaller the divergence. However, as will be discussed, increasing the frequency increases the absorption of energy by the tissues and hence the destructiveness of the beam. Thus a balance must be struck between competing factors in the choice of disk diameter and frequency of operation. To illustrate the considerations to be made in making this choice let us consider first beam divergence, then the absorption of energy from the beam by tissue.

THE PROPAGATION OF SPHERICAL ACOUSTIC WAVES

In the previous sections, the wave equation (Eq. (8.6) or Eq. (8.12)) was derived for propagation in one dimension (plane waves). Although simple in form, one-dimensional propagation occurs in only a limited number of situations. For the purposes of discussion here we must consider the propagation of diverging spherical waves in which the acoustic energy spreads over an increasingly larger area as the wave propagates away from its source.

The properties of such waves are deduced by a three-dimensional version of the considerations in the early part of the chapter. [The details of the derivations are rather complex and the interested reader may consult Kinsler and Frey (1962, Chapter 7).] It is found by solving the three-dimensional wave equation that spherical pressure waves of the form

$$p = (A/r)e^{i(\omega t - kr)} \tag{8.16}$$

may propagate through the material, where A is a constant. Note that this solution is similar to Eq. (8.13) but that the amplitude decays as $1/r$. It is also found that particle velocity accompanying such a pressure wave is proportional to the pressure:

$$d\xi/dt = p(r)/Z(r) \tag{8.17}$$

where the proportionality function Z, the acoustic impedance, is given by

$$Z(r) = [(\rho ck)r/(1 + k^2r^2)^{1/2}]e^{i\theta} \tag{8.18}$$

and

$$\theta = \arctan(1/kr)$$

If we have a vibrating source in the material, the waves emanating from the source may be found by properly choosing the constant A in Eq. (8.16) and summing up the waves originating from various parts of the source. This is simply illustrated by the waves emanating from a pulsating sphere of radius a, vibrating with a frequency ω and an amplitude ξ_0. The particle velocity at the surface of the sphere is

$$(d\xi/dt)(\text{sphere}) = \xi_0 e^{i\omega t} \tag{8.19}$$

(see for example the discussion just before Eq. (8.15)). If we assume that the particles of surrounding medium follow the motion of the sphere perfectly, the motion in the medium is described by Eqs. (8.16)–(8.18), choosing A so that

$$(d\xi/dt)(\text{sphere}) = (d\xi/dt)(\text{medium}) \qquad (r = a)$$

and upon substituting Eqs. (8.16)–(8.18) into Eq. (8.19) we write

$$\xi_0 e^{i\omega t} = (A/a)e^{i(\omega t - ka)}e^{-i\theta}[(1 + k^2a^2)^{1/2}/(\rho ck)a],$$

where θ is to be evaluated at $r = a$. Hence

$$A = [e^{ika}(\rho ck)/(1 + k^2a^2)^{1/2}](S/4\pi)e^{i\theta}\xi_0$$

where $S =$ the surface of the sphere, $4\pi a^2$. When the diameter of the sphere is so small that $ka \ll 1$ (i.e., $\omega a/c \ll 1$), $e^{ika} \to 1$. Further, since $\theta = \arctan(1/kr) = \arctan\infty = 90°$, $e^{i\theta} = \cos\theta + i\sin\theta = i$, we may write

$$A \cong i\rho ck\xi_0(S/4\pi)$$

Thus, the acoustic pressure wave generated by the small vibrating sphere is, from Eq. (8.16),

$$p = (i\rho ck/4\pi r)(S\xi_0)e^{i(\omega t - kr)} \tag{8.20}$$

The intensity of the radiation from the pulsating sphere is obtained by inserting the pressure (8.20) into Eq. (8.15):

$$I = (1/32\pi^2)(\rho ck^2/r^2)(S\xi_0)^2 \tag{8.21}$$

The intensity decays as $1/r^2$ as one goes away from the source.

RADIATION OF ACOUSTIC WAVES FROM A DISK-SHAPED SOURCE

The form of the pressure wave generated by a small vibrating sphere of area S (Eq. (8.20)) can be used to obtain the pressure wave radiated by a disk-shaped piston. It may be shown that an increment of area dS (of an extended surface S) moving at a velocity ξ perpendicular to the area contributes an increment of pressure dp,

$$dp = i(\rho ck/2\pi r)(\xi_0\, dS)\exp[i(\omega t - kr)] \tag{8.22}$$

at distance r. This is just Eq. (8.20) with S replaced by dS and multiplied by a factor of 2. The total pressure at any point in the medium is simply the sum of the incremental pressures produced at all the incremental areas of the surface.

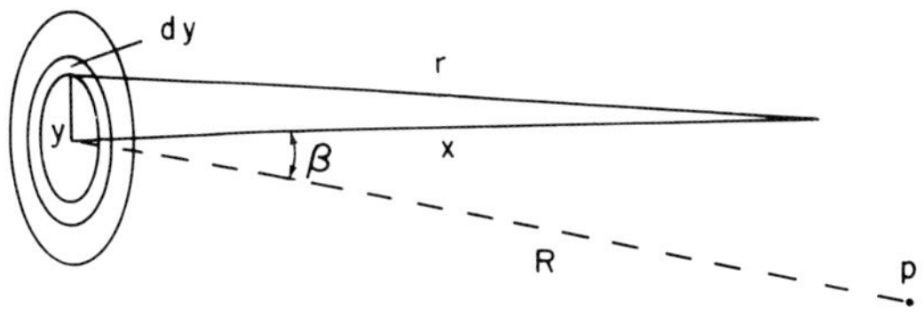

FIG. 8.9

The pressure at a point x on the axis of the disk shown in Fig. 8.9, replacing r in Eq. (8.22) by $x^2 + y^2$, is

$$p(x) = \frac{i\rho ck\xi_0 \exp(i\omega t)}{2\pi} \int_0^a \frac{2\pi y \exp[-ik(x^2 + y^2)^{1/2}]}{(x^2 + y^2)^{1/2}} dy \tag{8.23}$$

Integrating, one obtains

$$p(x) = -\rho c\xi_0 \exp(i\omega t)[\exp[-ik(x^2 + a^2)^{1/2}] - \exp(-ikx)] \tag{8.24}$$

Taking the real part of this equation and inserting it in Eq. (8.15) results in an intensity of†

$$I(x) = 2\rho c(\xi_0)^2 \sin^2(k/2)[(x^2 + a^2)^{1/2} - x)] \tag{8.25}$$

The variation of intensity with distance from the disk is shown in Fig. 8.10. Note that the intensity oscillates as the distance from the piston increases up to a certain critical distance, beyond which the intensity decreases monotonically to zero. The distance of the last minimum in Eq. (8.25) can be obtained as follows. For a large x approximation the $(x^2 + a^2)^{1/2}$ term can be expanded binomially keeping the first two terms. The argument of Eq. (8.25) becomes

$$(k/2)\{x[1 + \tfrac{1}{2}(a^2/x^2)] - x\} = (ka^2/4x)$$

† Equation (8.15) requires the average of the real part. Define

$$\theta_1 = k(x^2 + a^2)^{1/2} - \omega t \qquad \text{and} \qquad \theta_2 = kx - \omega t$$

Then

$$(e^{-i\theta_1} - e^{-i\theta_2}) \rightarrow (\cos\theta_1 - \cos\theta_2)^2 = \cos^2\theta_1 - \cos\theta_1 \cos\theta_2 + \cos^2\theta_2$$

For simplicity, define $\theta_1{}^0 = k(x^2 + a^2)^{1/2}$ and $\theta_2{}^0 = kx$. Trigonometric identities yield

$$\cos\theta_1 = \cos(\theta_1{}^0 - \omega t) = \cos\theta_1{}^0 \cos\omega t - \sin\theta_1{}^0 \sin\omega t$$

$$\cos\theta_2 = \cos(\theta_2{}^0 - \omega t) = \cos\theta_2{}^0 \cos\omega t - \sin\theta_2{}^0 \sin\omega t$$

Adding and factoring these yields

$$\cos\omega t(\cos\theta_1{}^0 - \cos\theta_2{}^0) - \sin\omega t(\sin\theta_1{}^0 - \sin\theta_2{}^0)$$

and upon squaring and averaging over one cycle we obtain

$$\overline{\cos^2\omega t}\{\quad\}^2 + \overline{\sin^2\omega t}\{\quad\}^2$$

The average over a cycle of $\cos^2$ and $\sin^2$ is $\frac{1}{2}$. Again, using a trigonometric identity one obtains

$$\tfrac{1}{2}[-2\sin\tfrac{1}{2}(\theta_1{}^0 + \theta_2{}^0)\sin\tfrac{1}{2}(\theta_1{}^0 - \theta_2{}^0)]^2 + \tfrac{1}{2}[2\cos\tfrac{1}{2}(\theta_1{}^0 + \theta_2{}^0)\sin\tfrac{1}{2}(\theta_1{}^0 - \theta_2{}^0)]^2$$

$$= 2\sin^2\tfrac{1}{2}(\theta_1{}^0 - \theta_2{}^0)[\sin^2(\theta_1{}^0 + \theta_2{}^0) + \cos^2(\theta_1{}^0 + \theta_2{}^0)]$$

$$= 2\sin^2\tfrac{1}{2}[k(x^2 + a^2)^{1/2} - kx]$$

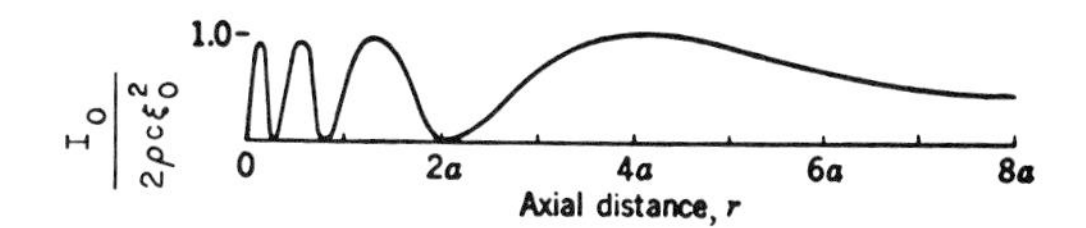

FIG. 8.10 Axial intensity as a function of radial distance r in the vicinity of a vibrating piston for $a = 8\pi/k$. [From Kinsler and Frey (1962).]

The last maximum occurs when

$$ka^2/4x_{max} \cong \pi/2$$

or

$$x_{max} \cong (ka^2/2\pi) = a^2/\lambda \tag{8.26}$$

When $x \gg x_{max}$, it is readily shown by Taylor expansion of Eq. (8.25) that

$$I(x) \cong (1/8\pi^2)(\rho ck^2/x^2)(S\xi_0)^2 \quad \text{for } x \gg x_{max} \tag{8.27}$$

where S is the area of the disk. Note the similarity with Eq. (8.21). At large distances the intensity begins to decay with distance much like the divergent spherical wave from the spherical source. In fact a detailed study of the radiation from the disk-shaped source shows that for $x \lesssim 2x_{max}$, the radiation is "beamlike" and beyond this the beam begins to diverge into a cone of radiation.

As an example of the depth of "beamlike" behavior, the velocity of sound in water or blood is 1.5×10^3 m/sec or 1.5×10^5 cm/sec. If we have a 2-cm diameter transducer, $a = 1$ cm, and for a 10^6-Hz frequency, $\lambda = 0.15$ cm; then by Eq. (8.26) the length of the beam before it diverges is 6.7 cm. We can extend this depth by increasing the diameter of the transducer, but then the probe does not have sufficient resolution. We could also extend it by reducing the wavelength, i.e., increasing the frequency f, but this has another drawback, which we will now consider.

BEAM DIVERGENCE

Following Pohlman (1949, 1951), the region of beamlike behavior up to $x = 2x_{max}$ is called the "near zone." The length of this near zone is proportional to the frequency. Beyond it the beam widens into a cone, which causes a reduction in intensity as the distance from the transducer increases.

The intensity at a point P (Fig. 8.9) a distance R away from the transducer at an angle β to the axis of the transducer is given by Eq. (8.27) but modulated by a directivity function $D(\beta)$ (see Kinsler and Frey, 1962):

$$I(R, \beta) \cong (1/8\pi^2)(\rho ck^2/R^2)(S\xi_0)^2 D^2(ka \sin \beta) \tag{8.28}$$

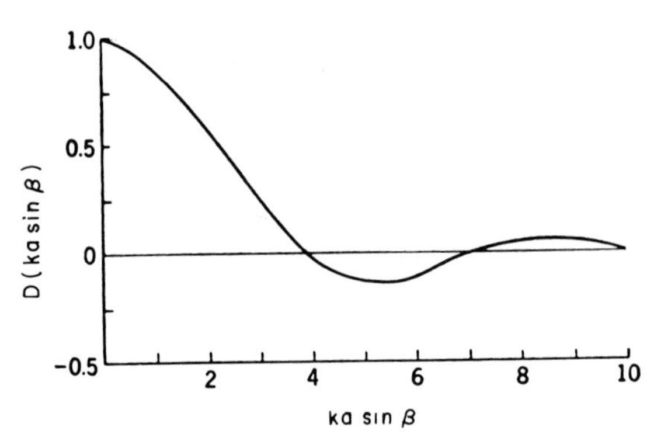

FIG. 8.11 Directivity function for a circular piston. [From Kinsler and Frey (1962).]

The directivity function D as a function of its argument $ka \sin \beta$ is shown in Fig. 8.11. This angular variation results in a lobelike distribution of acoustic intensity, illustrated in Fig. 8.12. A photograph of an actual beam is shown in Fig. 8.13 in which the characteristics of the early maxima and final maximum of Fig. 8.10 are seen in the central beam as well as the lobes.

It is seen in Fig. 8.11 that the directivity function becomes zero at a value of

$$ka \sin \beta = 3.83$$

This determines the angular divergence of the major lobe as

$$\sin \beta = 3.83/ka = (3.83/2\pi)(\lambda/a) = 0.61(\lambda/a) \qquad (8.29)$$

This divergence gives some flexibility in beam depth because by a proper choice of wavelength and disk size the angle β of divergence will be small. Equations (8.26) and (8.29) indicate that a well-directed beam can be obtained only if the wavelength is considerably less than the diameter of the crystal.

Suppose, for example, the ratio of crystal radius to wavelength is 10 to 1. Then from Eq. (8.26)

$$x_{\max} = a^2/\lambda = a^2/(a/10) = 10a$$

and from Eq. (8.29)

$$\sin \beta = 0.61(\lambda/a) = 0.61(a/10)/a = 0.061 \qquad \text{or} \qquad \beta = 3.5°$$

which is a well-focused beam even somewhat beyond $x_{\max}$. Note that β is a polar angle so that the beam width is actually 2β.

If we take an average sound velocity in blood and muscle as 1500 m/sec we can calculate for different frequencies the length of the cylindrical part

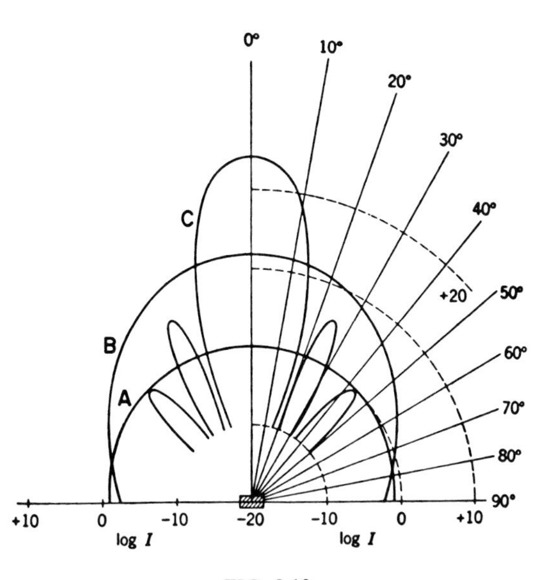

FIG. 8.12

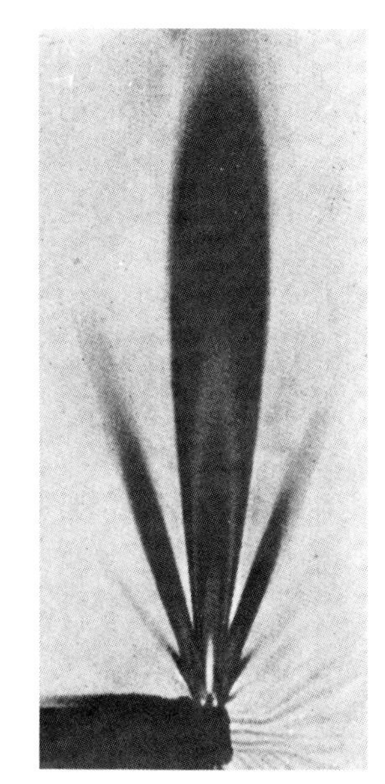

FIG. 8.13

FIG. 8.12 Polar radiation patterns of a flat piston as plotted for three different frequencies. log I is plotted against θ for (A) $\lambda = 8a$, (B) $\lambda = 2a$, and (C) $\lambda = a/2$. [From Kinsler and Frey (1962).]

FIG. 8.13 Photograph of ultrasonic beam from a disk of radius 2λ. [From Heidemann and Osterhammel (1937).]

of the beam, x_{max}, and the angle of divergence. Three examples are shown in Table 8.1 for a transducer of 0.6-cm radius. This table shows that, although the angle of divergence is small, it is not zero. Consider, for example, the 2.5×10^6-Hz frequency. The beam is cylindrical for 6 cm, after which it begins to diverge at an angle of 3.5°. Thus the beam will be doubled in diameter at a distance $x_{max} + 5$ cm or about 11 cm. This doubling in beam diameter has two deleterious effects; it reduces the linear resolution by a factor of 2, and the power by a factor of 4. Equations (8.18)–(8.20) indicate

TABLE 8.1

Frequency (f)	Wavelength (λ)	x_{max}	β
1×10^6 Hz	0.15 cm	2.4 cm	8.8°
2.5×10^6 Hz	0.06 cm	6 cm	3.5°
5×10^6 Hz	0.03 cm	12 cm	1.75°

that the situation theoretically improves as one goes to higher frequency, as shown by the β and x_{max} for 5×10^6 Hz in Table 8.1. However, in practice it does not because of absorption, as will be shown in the next section.

ABSORPTION OF ENERGY

The complicating feature that limits the range of frequency selection is absorption. Suppose we have the energy being absorbed by friction losses between molecules in the medium.

The change in particle amplitude ξ is caused by an absorption coefficient α and, if ξ_x is the remaining amplitude of the pressure wave at a point x, then the incremental amplitude change will obey the relation

$$-d\xi = \alpha\xi_x \, dx$$

which, by integration, gives the amplitude at any point x as

$$\xi_x = \xi_0 e^{-\alpha x}$$

where ξ_0 is the original amplitude. The intensity I at point x is, from Eq. (8.15), proportional to the square of the pressure amplitude, and therefore

$$I_x = I_0 e^{-2\alpha x} \tag{8.30}$$

Classically, if one considers only a kinetic energy term for energy loss by motion of the molecules, the absorption coefficient should be proportional to the square of the frequency f since $v_{max} = 2\pi f\xi_0$ and the kinetic energy varies with v^2. However, there are other loss terms, such as heat loss, and the adiabatic approximation is not completely valid. Further, in viscous materials there are large shearing forces so that density changes are not in phase with pressure changes. Curiously, experiments on biological tissues show that the ratio α/f is constant with frequency. This is seen in Fig. 8.14. The spread of the curves arises from the fact that the data collected did not cite temperature, which is an important variable.

The data of Fig. 8.14 suggest that we can represent the experimental results in biological tissues as $\alpha = bf$ where b is a constant. The higher the frequency the greater the absorption and thus Eq. (8.30) may be written as

$$I_x = I_0 e^{-2bfx} \tag{8.30'}$$

It is seen that since f is in the exponent there is not a very broad range of frequencies available for useful penetration within reasonable power limits.

Note that $1/\alpha$ is the depth of penetration of the beam. Pohlman (1949, 1951) gives example values of α and $1/\alpha$ for some materials at a frequency of 10^6 Hz shown in Table 8.2.

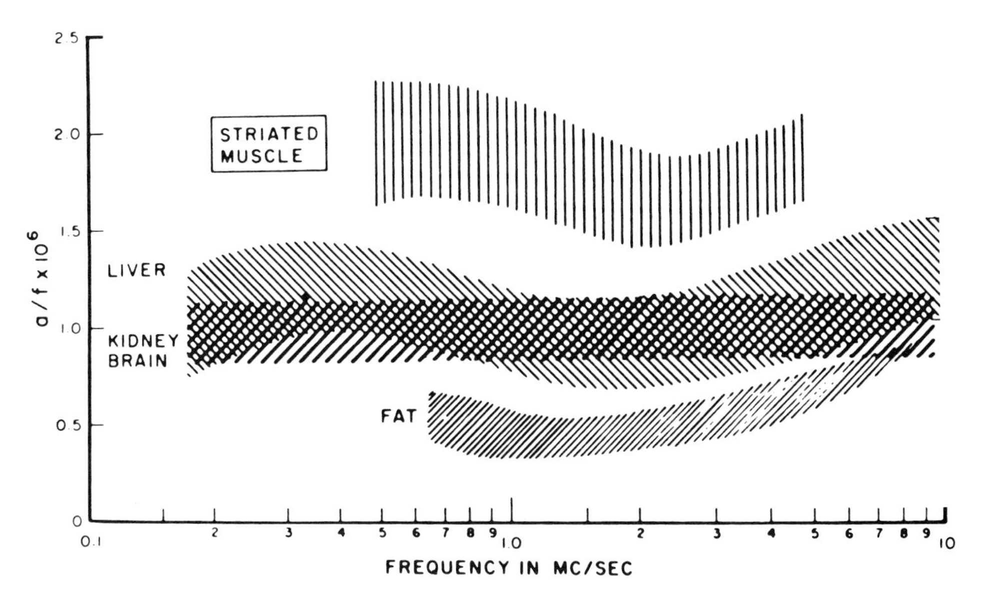

FIG. 8.14 Acoustic amplitude absorption coefficient α/f in dB/cm versus frequency for several mammalian tissues. [From Goldman and Hueter (1956).]

Pohlman has also introduced a useful term for tabulation purposes called the *half value* layer, that is, the thickness of body components in which the intensity falls to one half of its original value for a given frequency. For example, Table 8.3 shows *in vitro* measurements for the half value layer of heart muscle as a function of frequency. Thus, the higher the frequency the less the penetration of the beam.

The taking of meaningful scanning data requires a compromise of resolving power and penetration depth. These are determined empirically for a given patient and the organ to be examined. The transducer is usually about 1-cm diameter and the practical frequency range is from 0.5×10^6 to 5×10^6 Hz.

TABLE 8.2

Material	α (cm^{-1})	$1/\alpha$ (cm)
Water	0.0003	3000
Blood plasma	0.007	130
Whole blood	0.02	50
Skeletal muscle	0.2–0.25	4–5
Liver	0.17	6
Kidney	0.22	5
Fatty tissue	0.13	8

TABLE 8.3

f (Hz)	Half value layer (cm)
0.8×10^6	2.4
1.5×10^6	1.1
2.4×10^6	0.8
4.5×10^6	0.4

Before discussing power levels it is useful to understand that it is the reflected wave that is picked up and displayed on an oscilloscope. Without going through the somewhat lengthy derivation of impedance matching we will quote the result (see Kinsler and Frey, 1962) that the reflected intensity I from a beam of initial intensity I_0, when it is reflected in perpendicular incidence, is

$$I = I_0\left(\frac{\rho_1 c_1 - \rho_2 c_2}{\rho_1 c_1 + \rho_2 c_2}\right)^2 \tag{8.31}$$

where the ρ's and c's are the respective densities and velocities of sound of the two media. Typical values for humans are given in Table 8.4. It is seen from Table 8.4 that bone gives such a high reflectivity when next to any other tissue that one cannot "see" beyond it. Thus in tissue scanning or echocardiography the transducer can only be placed between ribs, on the neck aiming into the thorax, or on the abdominal region.

TABLE 8.4

Tissue	ρ (gm/cm^3)	Sound velocity (m/sec)
Blood	1	1.56×10^3
Fat	0.928	1.47×10^3
Muscle	1.058	1.568×10^3
Bone (skull)	1.85	3.36×10^3

DAMAGE TO TISSUES

We have seen that the sound pressure above atmospheric pressure is given by

$$p_{\max} = \omega \rho c \xi_0 \tag{8.14}$$

and, by Eq. (8.15′), the intensity is

$$I = p^2_{\max}/2\rho c \tag{8.15′}$$

The literature on biological investigations of the effects of ultrasound cites results in cgs units, so for convenience we will use that system. Therefore, the pressure is in units of dyn/cm^2 and the intensity in W/cm^2 (1 W/cm^2 = 10^7 erg/sec cm^2). Consider the following example for order of magnitude estimates. In normal therapeutic use at $f = 1 \times 10^6$ Hz the rule of thumb is an intensity maximum of 4 W/cm^2. From Table 8.4 for blood or muscle, $\rho c = 1.5 \times 10^5$ gm/sec cm^2 and the sound pressure is 3.5×10^6 dyn/cm^2 = 3.5 atm. These values substituted into Eq. (8.14) yield $\xi_0 = 3.7 \times 10^{-6}$ cm. The acceleration of a harmonically vibrating system is by definition

$$a = d^2x/dt^2$$

and, if $x = \xi_0 \sin(kx - \omega t)$,

$$a = -\omega^2 \xi_0 \sin(kx - \omega t)$$

The maximum acceleration occurs at the largest possible value of the sin, i.e., unity:

$$a_{\max} = -\omega^2 \xi_0$$

Since $\omega = 2\pi f$, for $f = 1 \times 10^6$ Hz and $\xi_0 = 3.7 \times 10^{-6}$ cm

$$a = 1.5 \times 10^8 \text{ cm/sec}^2$$

Note that the acceleration of gravity is about 1×10^3 cm/sec^2 and thus the acceleration of individual particles is greater than 100,000 times the acceleration of gravity.

These pressure differences occur every half wavelength and in this example, 10^6 Hz in muscle or blood, one half wavelength is about 0.075 cm. Thus there are strong forces trying to separate the cohesiveness of the medium. In a fluid in which the forces of attraction are not very strong these pressure differences cause a phenomenon called *cavitation*, i.e., rarefaction voids. If these voids contain gas the phenomenon is known as *pseudocavitation*. Experiments on frogs and mice have indicated that cavitation and resulting tissue damage can occur at 10^6 Hz for intensities of 20 W/cm^2.

An even more serious source of tissue damage is heating. Clearly, energy dissipation of a beam operating at the level of watts results in considerable heat, which must be conducted away thermally from the tissue. A temperature rise will occur until a steady state is reached in which the heating rate is equal to the cooling rate (at reasonable levels of intensity). Such temperature rises have been investigated with imbedded thermocouples and found to be of the order of several degrees. In experiments on humans the patient can report discomfort. Thus low-level power is used for the therapeutic effects of mild internal heating of tissue and the level of discomfort can be controlled.

TABLE 8.5

f (Hz)	I (W/cm^2)
1×10^6	80
2.5×10^6	40
5×10^6	20

During examination of internal organs by methods such as echocardiography sufficient power is required to obtain a reflection signal well above the noise level. The requirements, determined experimentally, are shown in Table 8.5. These values are well into the discomfort and damage levels. However, there is no reason for continuous operation at such high frequencies to observe a slowly moving mass such as the myocardium. Therefore the operation is one of pulsing. Typical values are about 200 pulses per second with a pulse duration of 5×10^{-6} sec. Thus the total transmission time is $200 \times 5 \times 10^{-6}$ sec $= 10^{-3}$ sec and the power dissipation in the body is therefore 0.08 W/cm^2 at 1×10^6 Hz, etc., clearly within the safety region.

RESOLUTION IMPROVEMENTS

As we have discussed above, the constraint on resolution, i.e., fineness of detail, results from the beam divergence. This was given by Eq. (8.29),

$$\sin \beta = 0.61(\lambda/a) \tag{8.29}$$

where β is the divergence angle and a the radius of the beam. Thus, while the wavelength λ is fixed by tissue absorption, the smaller a is made to increase the resolution the greater the beam spread at distances of interest.

An improvement in this situation has been recently developed by Whittingham (1976). He has constructed an ultrasonic probe in which the transducer is composed of 100 very small elements that are independently controlled. Each transducer sends and receives in 0.0005 sec and a scan by all 100 is accomplished in 0.05 sec, or 20 scans per second. Within this time of 0.05 sec the separate transducer elements are controlled to pulse in groups of 20. When the picture begins, transducers numbered 1 through 20 pulse together. Half a millisecond later transducers 2 through 21 pulse together, followed in another half millisecond by group 3 through 22, etc. In this way there is always a broad radiating surface that moves along by the width of one of the small transducers. It is reported that this system produces oscilloscope tracings of considerably greater detail than the conventional single transducer.

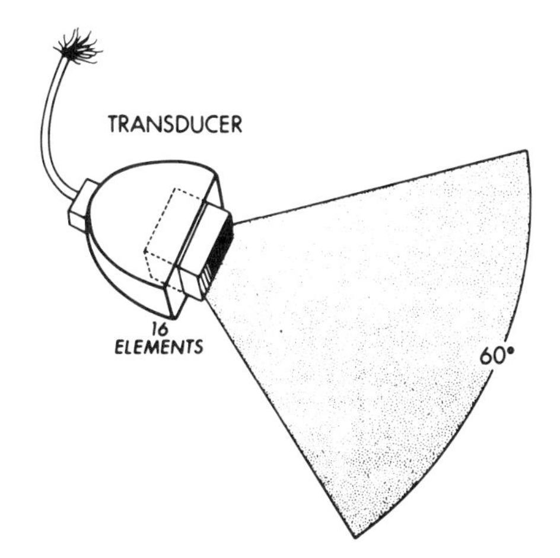

FIG. 8.15 Diagram illustrating THAUMASCAN transducer and scan format. [From von Ramm and Thurstone (1975).]

Another technique for achieving improved resolution in a planar scan is called THAUMASCAN, an acronym for *two-dimensional, high-resolution, actual time, ultrasound, multielement, array scanner*. It is designed to operate in "real time", that is, the organ is viewed while scanning rather than on a film at a later time.

The arrangement is 16 transducers in a linear array that view an angle of 60°, as illustrated in Fig. 8.15. The operation is based on Huygen's Principle of addition of wavelets. Figure 8.16 is a diagram of 5 transducers, for simplicity in a linear array. A small computer, PDP-11, is programmed to send the voltage pulses to the transducers with a delay between each. As seen in Fig. 8.16, the transmitted wavefront will be at an angle whose magnitude depends on the relative delay between the signals. Consider now the more complicated situation of Fig. 8.17. This method of transmitting the pulses causes the wavefront to focus, the focal length depending upon the initial

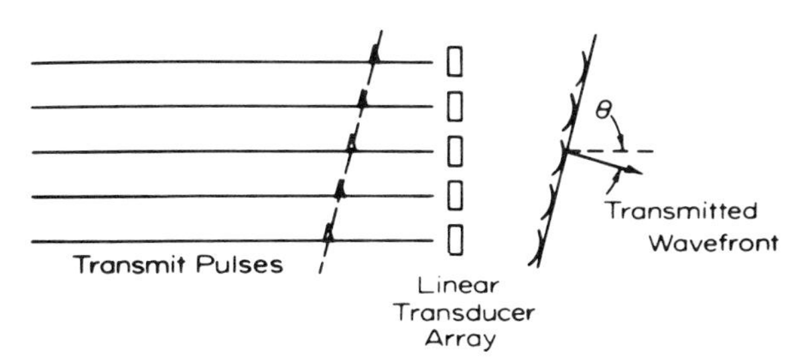

FIG. 8.16 Schematic representation of transmitted beam steering. [From von Ramm and Thurstone (1975).]

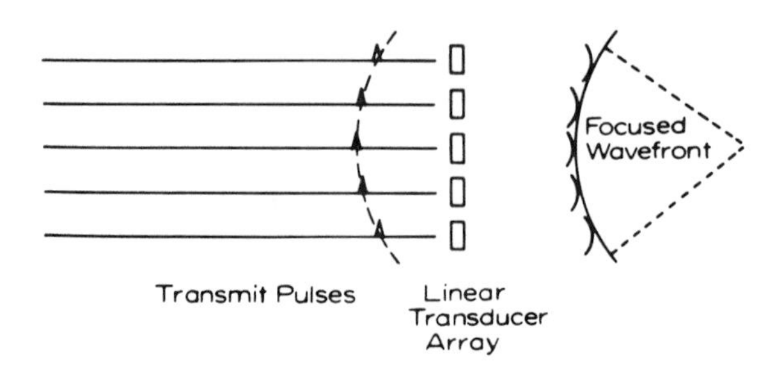

FIG. 8.17 Schematic representation of transmitted beam focusing. [From von Ramm and Thurstone (1975).]

curvature of the pulses. Upon reflection from the target the echoes arrive at the transducer at different times, as shown in Fig. 8.18, which requires a phasing of the received signal so that the display corresponds to the orientation of the transmitted signal. This is accomplished by switchable delay lines in the receiver circuit. With such delay lines the receiver can be focused at a given range and then this focus can be changed, or "tracked", in synchrony with the range of returning echoes. This is done by having a delay controller in conjunction with the delay lines. This controller determines the total delay time associated with each transducer channel. The time of the returning echo is proportional to the distance of the reflecting tissue. Immediately after the transducers have transmitted the signal the receive electronics are focused to receive at short range and then the focal distance is increased in synchrony with the increasing range of target echoes. This moving focus reception can be superimposed on a moving angle transmission and reception, as in Fig. 8.16, to cover a flat plane (tomograph) in the body. The display on an oscilloscope screen is wedge-shaped with a 60° angle. The investigators report a range resolution of 1.5 mm and an azimuthal resolution from 2 to 4 mm throughout the field of view.

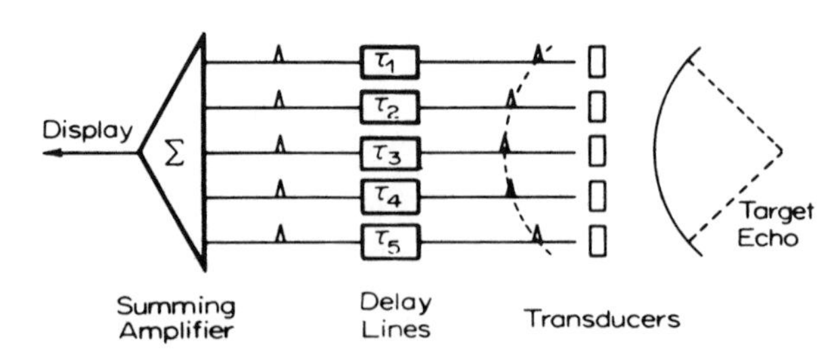

FIG. 8.18 Schematic representation of reflected beam focus mode. [From von Ramm and Thurstone (1975).]

GRAY SCALE AND IMAGE PROCESSING

In the ultrasonic probe described in the preceding sections, the transducer is activated by an echo and the resulting signal is displayed on an oscilloscope screen. A type of information processing, called "leading edge," was used in the early devices. In this, an amplitude threshold could be set and any signal above this amplitude would be recorded. Leading edge processing readily gives the outline of an organ where the acoustic impedance is large at the interface. If additional information on the internal structure of the organ was desired, repreated scans had to be made, with the threshold being lowered for each scan. The physician had to accumulate the data, either by photographing the oscilloscope image or by remembering it, and then form a composite picture in his own mind of the internal structure of the organs. In addition, to study an organ properly the beam had to be swept across the organ in both x and y directions, with consideration of the fact that a portion of an organ curving away will not reflect with the same intensity as a portion whose plane is normal to the incident beam.

The development of the gray scale for image processing eliminated the need for repeated scans at different amplitude settings. Figure 8.19 illustrates a hypothetical A mode (see Fig. 8.2) operation with four regions of gray for simplicity of explanation. The lowest, black, is the noise and artifact level and is eliminated. As the amplitude of the echo increases, that falling within the next darkest range appears on the oscilloscope as a very dark gray (assume the scope with no signal is black) and as the amplitude increases the signal goes through lighter shades up to white. The "first generation" of gray scale display required that a trained operator make numerous sectored sweeps across the outer layer of the body. The system took a time exposure photograph of a nonstorage oscilloscope display in which the camera shutter was open for the duration of the scan. Since each transmitted

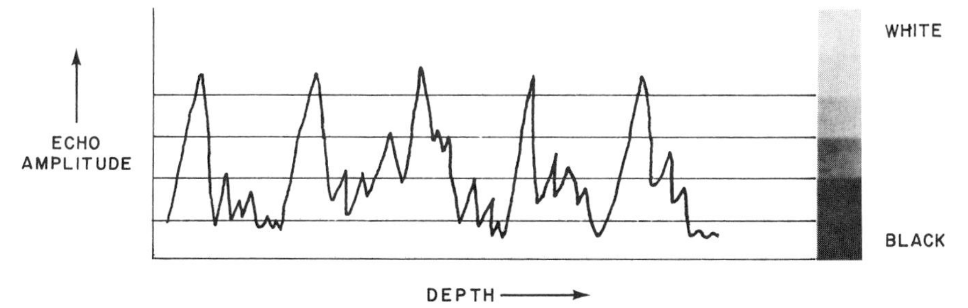

FIG. 8.19 Hypothetical A mode display with resulting gray scale. Four shades of gray are shown for ease of explanation; in reality, the transition from black to white is continuous. [From Hileman and McLain (1975).]

pulse is represented by a spot of light on the oscilloscope and then on the film, stopping the scan momentarily, lingering too long at one spot, or repeating an area would result in excessive brighness in some areas.

A "second generation" of devices has been developed in which the image of the scan is recorded in a storage device. The storage device is an amplitude detector, so the storage of an echo is a function of the amplitude of the echo alone and not how many times the same echo is received. It is, of course, desirable to see the resulting picture as it is being made, without waiting for a film to be processed. To accomplish this there have recently been developed scan converter display systems. These can gather and store information in one mode, an x–y scan, and display in another mode, such as on a television screen. The scan converters contain an electron tube at one end and an electron gun at the other, similar to a cathode ray tube, although the target is not a phosphor screen. The electron beam is modulated by the signal, as in a cathode ray tube, and the image is stored on the target in the form of points of electric charge. In the scan converter mode the charge-stored image can be read nondestructively and displayed on a video screen. This is not a new innovation but has been used for years in aerospace and other commercial applications. However, the combination of this with the amplitude-

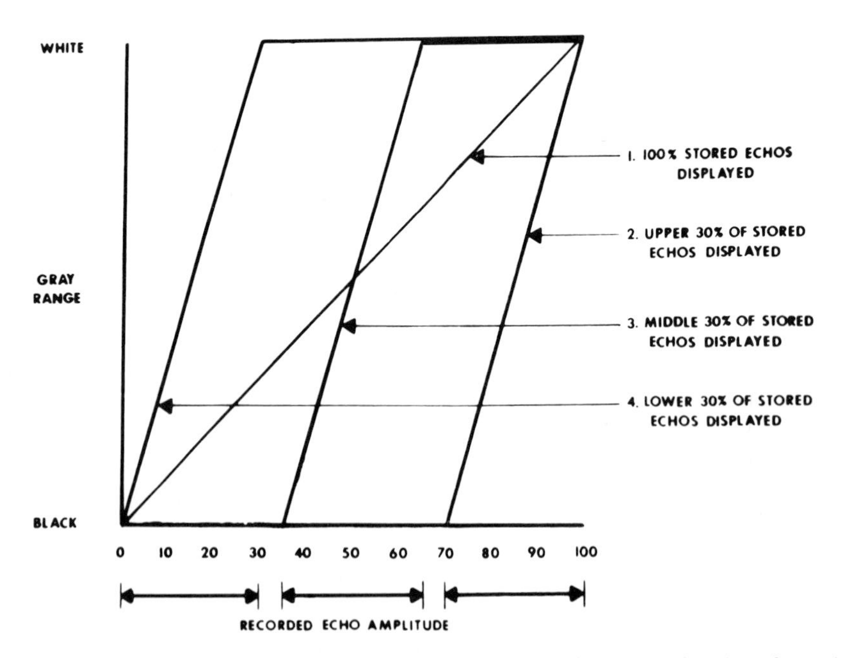

FIG. 8.20 Graph relating output of video intensity (shade of gray) as a function of stored echo amplitude. [From Hileman and McLain (1975).]

dependent signal mentioned above renders possible the storage of a true gray scale image.

There are two other advantages to modern image storing and converting techniques. One is that any portion of the stored image can be made to fill the video screen, thus allowing a magnification. This, of course, has obvious limitations because of the resolution of the original signal. The other is gray scale expansion. The amplitude range of the ultrasonic signals is about 50 dB (dB = decibel = 10 × log ratio of two intensities) while the brightness range of cathode ray tube phosphors is about 20 dB. Furthermore, human vision can differentiate only about 10–15 different levels of gray displayed on a cathode ray tube and those only when arranged in successive steps. Although a logarithmic amplifier can compress 60 dB of amplitude information into 20 dB of display, clearly information is lost. A quantified display of a region of amplitude can be expanded into the full gray scale. Figure 8.20 illustrates this capability. In line 1 all amplitude information is displayed on a gray scale which ranges from black to white. Line 2 represents the expansion of the upper 30% of the information into the full black to white range. Lines 3 and 4 are for the middle and lowest 30%, respectively, expanded into the full range.

With these developments ultrasonography has left the technique-dependent era and entered into an era equivalent to radiography.

REFERENCES

Baum, G., and Greenwood, I. (1960). Ultrasonography—an aid in orbital tumor diagnosis, *Am. Med. Assoc. Arch. Ophthal.* **64**, 180.

Bergmann, L. (1954). "Der Ultraschall, 6. Auflage." Hirzel Verlag, Stuttgart.

Busey, H. W., and Rosenblum, L. H. (1975). Physical aspects of gray scale ultrasound, *in* "Ultrasound in Medicine" (D. White, ed.), Vol. 1. Plenum Press, New York.

Duchak, J. M., Chang, S., and Feigenbaum, H. (1972). The posterior mitral valve echo and the echocardiographic diagnosis of mitral stenosis, *Am. J. Cardiol.* **29**, 628.

Dunn, F. (1965), Ultrasonic absorption by biological materials, *in* "Ultrasonic Energy" (E. Kelley, ed.). Univ. of Illinois Press, Urbana, Illinois.

Edler, I. (1961). Ultrasound cardiography, *Acta Med. Scand. Suppl.* 370.

El'piner, I. E. (1964). "Ultrasound: Physical, Chemical and Biological Effects." Consultants Bureau, New York.

Feigenbaum, H. (1972). Clinical applications of echocardiography, *Progr. Cardiovasc. Disease* **14**, 531.

Fry, W. J., Wulff, V. J., Tucker, D., and Fry, F. J. (1950). Physical factors involved in ultrasonically induced changes in living systems: I, Identification of non-temperature effects, *J. Acoust. Soc. Am.* **22**, 867.

Fry, W. J., Tucker, D., Fry, F. J., and Wulff, V. J. (1951). Physical factors involved in ultrasonically induced changes in living systems: II, Amplitude duration relations and the effect of hydrostatic pressure for nerve tissue, *J. Acoust. Soc. Am.* **23**, 364.

Goldman, D. E., and Hueter, T. F. (1956). Tabular data of the velocity of sound in mammalian tissues, *J. Acoust. Soc. Am.* **28**, 35.

Gramiak, R., Shah, P. M., and Kramer, P. H. (1969). Ultrasound cardiography: Contrast studies in anatomy and function, *Radiology* **92**, 939.

Gramiak, R., and Shah, P. M. (1971). Cardiac ultrasonography, *Radiol. Clin. North Am.* **9**, 469.

Gustafson, A. (1967). Correlation between ultrasound cardiography hemodynamics and surgical finding in mitral stenosis, *Am. J. Cardiol.* **19**, 32.

Halliday, D., and Resnick, R. (1962). "Physics for Students of Science and Engineering." Wiley, New York.

Heidemann, E., and Osterhammel, K. (1973). Optische Untersuchung der Richtcharakteristik von Ultraschallquellen, *Z. Phys.* **107**, 273.

Hertz, C. H. (1967). Ultrasonic engineering in heart diagnosis, *Am. J. Cardiol.* **19**, 6.

Hileman, R. E., and McLain, J. A. (1975). Image improvement with second generation gray scale, *in* "Ultrasound in Medicine" (D. White, ed.), Vol. 1. Plenum Press, New York.

Kelley, E. (ed.) (1965). "Ultrasonic Energy." Univ. of Illinois Press, Urbana, Illinois.

Kikuchi, Y. (1956). Early cancer diagnosis through ultrasonics, Lecture, *2nd ICA Congr. HD-9, Cambridge, Massachusetts.*

Kinsler, L. E., and Frey, A. R. (1962). "Fundamentals of Acoustics." Wiley, New York.

National Science Foundation (1973). "Prospectives of Ultrasonic Imaging in Medical Diagnosis: Report of the NSF Survey Team on Ultrasonic Imaging." National Science Foundation, Washington, D.C.

Pohlman, R. (1939). Über die Absorption des Ultraschalls im menschlichen Gewebe und ihre Abhängigkeit von der Frequenz, *Physik. Z.* **40**, 159.

Pohlman, R. (1949). Die Ultraschalltherapie in ihrer heutigen Entwicklung, *Schwerz. Med. Wschr.* **79**, 754.

Pohlman, R. (1951). "Die Ultraschalltherapie." Verlag Hans Huber, Bern.

Richardson, E. G. (1952). "Ultrasonic Physics," Elsevier, Amsterdam.

Segal, B. L., Likoff, W., and Kingsley, B. (1967). Echocardiography: Clinical application in combined mitral stenosis and mitral regurgitation, *Am. J. Cardiol.* **19**, 42.

Toshio, W., Yazawa, R. M., Ito, K., and Kikuchi, Y. (1965). Ultrasonic diagnosis of intracranial disease, breast tumors, and abdominal diseases, *in* "Ultrasonic Energy" (E. Kelley, ed.). Univ. of Illinois Press, Urbana, Illinois.

Von Ramm, O. T., and Thurstone, F. L. (1975). Thaumascan: Design considerations and performance characteristics, *in* "Ultrasound in Medicine" (D. White, ed.), Vol. 1. Plenum Press, New York.

Wagai, T., Miyazawa, R., Ito, K., and Kikuchi, Y. (1965). Ultrasonic diagnosis of intracranial disease, breast tumors, and abdominal diseases, *in* "Ultrasonic Energy" (E. Kelley, ed.). Univ. of Illinois Press, Urbana, Illinois.

Whittingham, T. (1976). A multiple transducer system for heart, abdominal and obstetric scanning, *in* "Ultrasonics in Medicine" (E. Kazner, M. DeVlieger, H. R. Müller, and V. R. McCready, eds.), Int. Cong. Ser. 363, Excerpta Medica, Amsterdam.

CHAPTER 9

Nuclear Medicine: Tracers and Radiotherapy

INTRODUCTION

The atomic nucleus is composed of *protons* and *neutrons*. The protons have a single positive charge equal and opposite to the charge of an electron; the neutron has no charge and has essentially the same mass as the proton. The number of protons in a nucleus is represented by the symbol Z and is called the atomic number. This characterizes the name and chemical properties of the atom. Similarly, the number of neutrons is represented by the symbol N, and the mass number A is the sum $Z + N$. The mass of a neutron is 1.6749×10^{-27} kg and that of a proton is 1.6726×10^{-27} kg. For the present purposes we can take the mass of a nucleus as $(Z + N) \times 1.67 \times 10^{-27}$ kg. A neutral atom has its nucleus surrounded by electrons equal in number to the protons in its nucleus. The mass of an electron is only 9.1×10^{-31} kg and this mass need not be considered in comparison with that of a nucleus for our present purposes. The diameter of a neutral atom is the order of 10^{-10} m while that of the nucleus is 10^{-14} m.

ISOTOPES

Each element is characterized by the number of protons in the nucleus; if that number is changed the nucleus is that of a different element. However, nuclei of a given element can have various numbers of neutrons and these varieties of a given element are called *isotopes* of that element. Each isotope of an element is also called a *nuclide*. The conventional symbolic form is as follows. If X is the chemical symbol of the element then nuclides are identified as

$$_{Z}X^{A},$$

although sometimes A is written as a superscript to the left of X. Thus the three isotopes of hydrogen are

$$_{1}H^{1},\quad _{1}H^{2},\quad _{1}H^{3}$$

where the latter two are called *deuterium* and *tritium*, respectively. Note that the Z is not necessary because the chemical symbol specifies Z. It is usually omitted because the chemical characteristics of an element are not altered by changes in the numbers of neutrons. That is, all isotopes of an element have the same chemical behavior.

Not all isotopes are stable. The repulsive force between protons in nuclei is compensated for by the presence of neutrons. No nucleus with Z greater than that of hydrogen has fewer neutons than protons and as Z becomes greater than 20 the ratio of neutrons to protons progressively increases. When an isotope is unstable it may emit one or more particles. This is called *radioactive decay*. Sometimes the resulting nucleus may decay further. An unstable, or *radioactive*, nucleus can emit neutrons, electrons, positrons, alpha particles, or gamma rays. *Positrons* are positively charged electrons, *alpha particles* are helium nuclei, $_{2}He^{4++}$ (the superscript $++$ sign means that it is an ion with twice the charge of a proton, i.e., two electrons are missing from the helium atom), and *gamma rays* are quanta of high-energy radiation. Although it may seem strange, the electons emitted, called *beta rays*, actually come from inside the nucleus and not from the orbital electrons.

Some particles from nuclear decay cause ionization of other atoms that they strike or flashes of light when they strike phosphors. Such ionizations and light flashes can be detected by instruments called counters. Other particles will cause a nuclear reaction with other nuclei and their subsequent decay can be counted. The particles emitted and the energies of the emitted particles are characteristic of the individual type of nuclide. These two are measureable quantities and have been tabulated (Lederer *et al.*, 1967).

RADIOACTIVE DECAY

Many useful artificial isotopes are made by bombarding a nucleus with high-energy particles to make an unstable isotope. Reactors supply high-energy neutrons, and cyclotrons can be used to accelerate charged particles such as protons, deuterons, and alpha particles. A typical example would be the following. If naturally occurring stable carbon $_6C^{12}$ is bombarded by high-energy deuterons, $_1H^2$, it produces nitrogen $_7N^{14}$. This reaction is written in the form of a chemical equation

$$_6C^{12} + {}_1H^2 \longrightarrow {}_7N^{14}$$

Although $_7N^{14}$ is usually stable, when produced in this way it possesses sufficient internal excitation so that it is unstable. When this excited state decays it emits a neutron, thereby becoming nitrogen $_7N^{13}$. This is written as

$$_7N^{14} \longrightarrow {}_7N^{13} + n$$

A large class of reactions occur in this manner and are written as

$$a + X \longrightarrow Y + b$$

which means that particle a interacts with nucleus X to yield nucleus Y plus particle b. The shorthand notation for this is

$$X\,(a, b)\,Y$$

and for the above case the reaction would be written

$$_6C^{12}\,(d, n)\,{}_7N^{13}$$

(where d = deuteron, the ionic form of deuterium). However, $_7N^{13}$ is unstable and has a half-life of 10 min,† emitting a positron and thereby reducing its positive charge from 7 units to 6 while maintaining a constant nuclear mass, A. This reaction is written

$$_7N^{13} \longrightarrow {}_6C^{13} + {}_{+1}e^0$$

in which $_{+1}e^0$ is the symbol for a positron.

Just as orbital electrons in atoms have excited states from which they decay to the ground state with the emission of a photon, so do nuclei. In the case of nuclei the photons have large energies and are called *gamma rays* (γ rays). Often in decay of a radioactive nucleus the emitted particle does not carry away the total available energy, and the final nucleus is not then

† A discussion of decay rates, including a precise definition of half-life, is given in the section on radioactive decay rate and Eqs. (9.1)–(9.3).

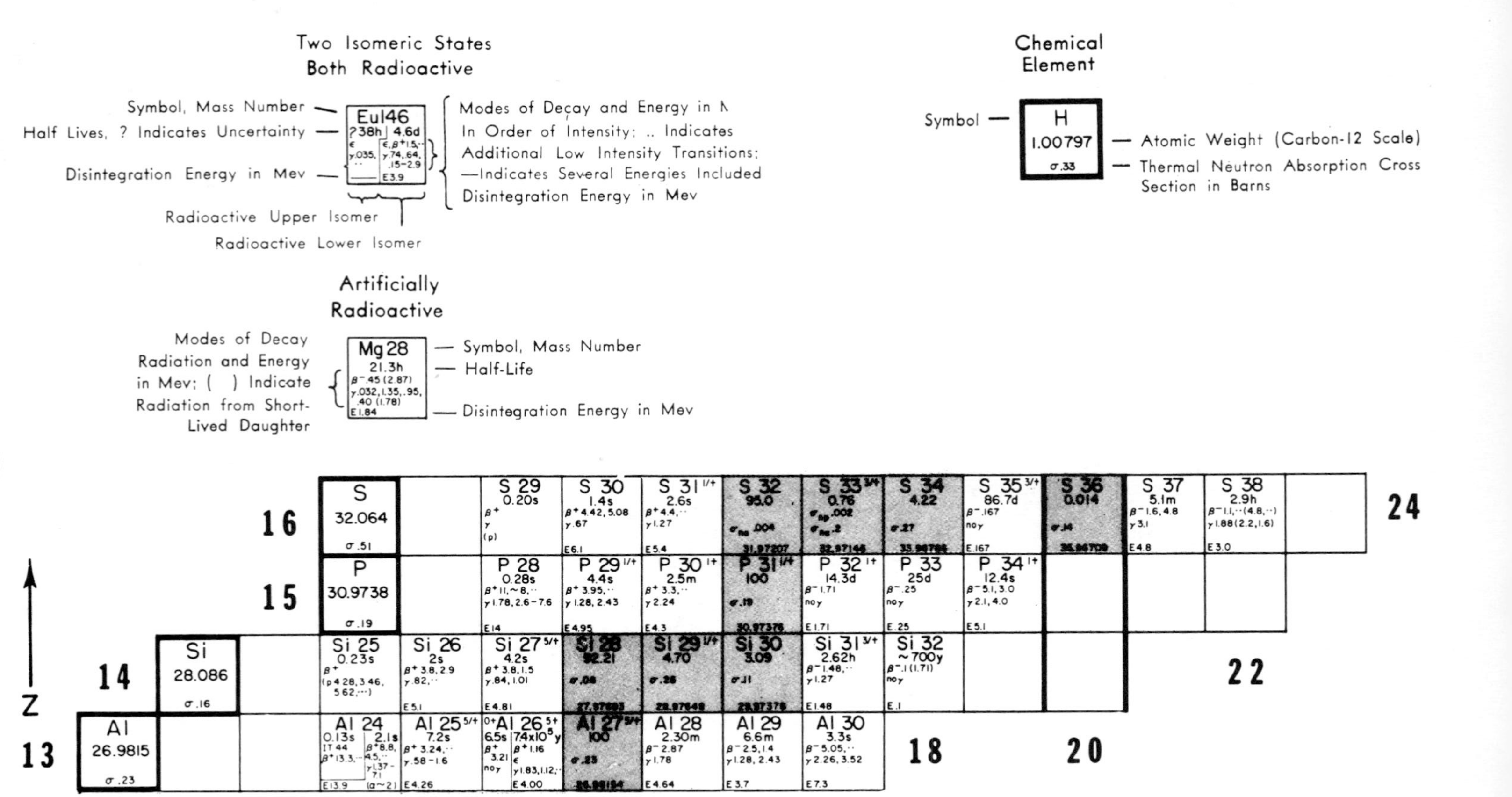

FIG. 9.1 Section of a chart of the nuclides. [From "Chart of the Nuclides," 1977 Revision General Electric Co., Knolls Atomic Power Laboratories, Schenectady, New York.]

in its lowest energy, or ground state. A γ ray or an x ray cascade may then be emitted to attain the ground state.

An excited nucleus can be created either as one stage of a multistage decay or by excitation of a different nucleus with an accelerated particle. Often the decay of an excited nucleus to a lower energy state occurs promptly, but in some situations, in which the decay is forbidden by considerations beyond our scope (nuclear spin), the time to decay is comparatively long. Nuclei that have measurable decay times are called *isomers* and the excited states that possess those lifetimes are called *metastable states*. Such nuclei are usually written with an m to indicate metastable. For example, Tc^{99m} (6 hr) means that a metastable state of technicium with an atomic mass of 99 has a 6-hr half-life.

A great deal of information can be given in a chart of the nuclides. Such charts are structured so that N increases from left to right and Z increases from bottom to top. A typical segment of this type of chart is shown in Fig. 9.1. The interpretation of the symbols and numbers is shown in the adjacent boxes. The shaded isotopes are the stable ones that are found in nature, and their relative abundances are indicated. The unshaded isotopes are the unstable ones and their decay schemes are indicated.† Except for the thermal neutron capture cross sections σ, the method of creating these unstable isotopes is not indicated.‡

RADIOACTIVE DECAY METHODS

A nucleus in an excited metastable state, instead of decaying with the emission of γ rays, may give up its excited state energy in other ways. For example, by transferring it to its surrounding electrons. The orbital electrons that spend part of their time in or near the nucleus effectively bombard it. Under certain conditions the excited nucleus may transfer its energy to one of these electrons and thereby decay to a state of lower energy. The electron now has this energy and will leave the atom with a kinetic energy equal to the energy of the excited nucleus minus the electron's binding energy to the atom. This electron is called a *conversion electron* and the process is known as *internal conversion*. The energy level formerly occupied by the conversion electron is subsequently filled by an atomic electron from a higher energy level by the emission of an atomic photon. Since internal conversion usually occurs from an inner atomic shell, this photon is in the energy range of

† Decay energies are given in MeV (= million electron volts = 10^6 eV) or keV (= kilo electron volts = 10^3 eV). 1 eV = 1.6021 × 10^{-19} J (= joule).

‡ The unit of neutron cross section is b (= barn). 1 b = 10^{-28} m^2.

x rays. The atom is still shy one electron and it captures some free electron in the surrounding medium with the emission of a low-energy photon. The x ray released by the exchange of orbital electrons need not escape the atom as x rays, but may be gained by yet another atomic electron, which then leaves the atom. This is known as the *Auger process*. Subsequent electron rearrangement of the orbital electrons will then take place. Note that this Auger process is the orbital equivalent of the internal conversion process that takes place in the nucleus.

Unstable nuclei may be created by bombardment of stable ones with external particles or by chains of radioactive decay, during which some unstable isotopes may be formed. The nucleus can change its neutron-to-proton ratio by transforming one of its protons into a neutron or vice versa. Such a change is called an *isobaric process* since it does not change the total number of neutrons *and* protons, and therefore leaves the atomic number $A = N + Z$ unchanged. Isobaric processes fall into three categories (1) beta (β^-) decay, (2) positron (β^+) decay, and (3) electron capture.

Beta decay is the emission of an electron and a *neutrino* from the nucleus. The neutrino is a subatomic particle that has either an extremely small or zero mass, zero charge, and which interacts only very slightly with matter. We need not consider it further. As an example, β^- emission occurs when stable phosphorus ${}_{15}P^{31}$ is bombarded with neutrons. The nucleus captures a neutron and becomes ${}_{15}P^{32}$, which is unstable. This unstable nucleus can convert one of the neutrons into a proton by emitting an electron to become a stable isotope, sulfur, ${}_{16}S^{32}$:

$$ {}_{15}P^{31} + n \longrightarrow {}_{15}P^{32} \longrightarrow {}_{16}S^{32} + \beta^- + \text{neutrino} $$

The total energy carried off from the nucleus by this β^- decay process is 1.71 MeV. This energy is shared by the electron and the neutrino in a continuous way, with no fixed ratio. If one measures the number of electrons emitted and their energies, a distribution curve is obtained as shown in Fig. 9.2. Thus, there is no sharp β^- energy, but rather a distribution of energies extending from a minimum of zero energy up to the maximum available energy of 1.7 MeV, with an average of 0.7 MeV. This is an important

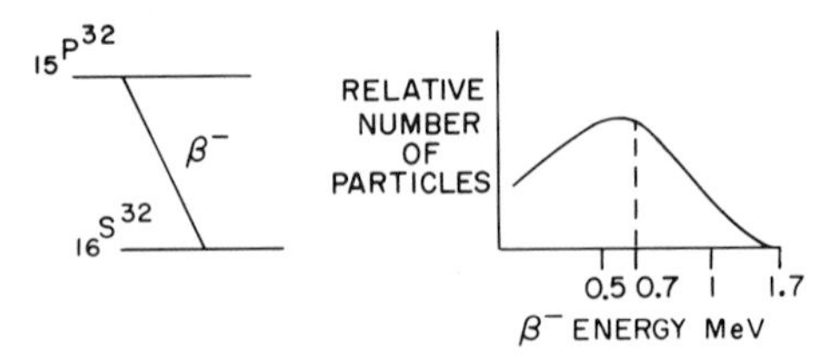

FIG. 9.2

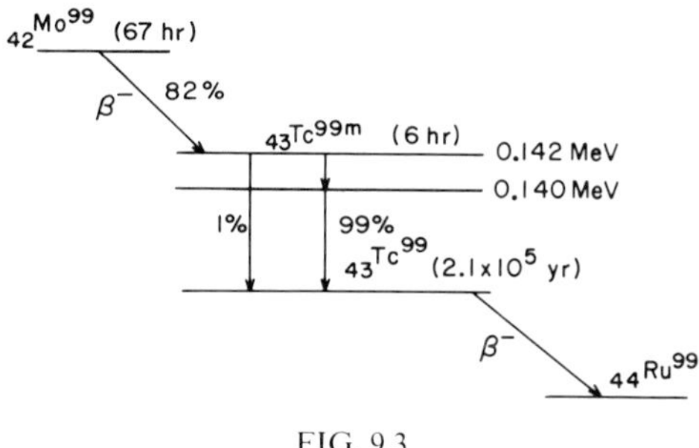

FIG. 9.3

point in designing a tracer detection system and in computing the radiation dosage to body tissue. Usually the average energy is used but one should be aware of the fact that other energies are involved.

Sometimes the decay schemes are more complex. Consider the decay of Mo^{99} to Tc^{99} as shown in Fig. 9.3. As indicated, 82% of the decays of Mo^{99} lead to the excited state of Tc^{99}. The remaining 18% go to excited states that are not indicated in the figure. The times in parenthesis refer to the half-lives of the indicated levels. To create this family of radioactive isotopes stable Mo^{98} is first bombarded with neutrons in a nuclear reactor to create the *parent* activity Mo^{99}. This isotope decays primarily (82%) to a *daughter* product Tc^{99m}. This is an excited nucleus, which emits a 0.142-MeV γ ray and decays to Tc^{99}. Tc^{99} then decays to Ru^{99} by β^- emission with a 200,000-year half-life, but for tracer purposes this slow decay rate has little significance. The Mo^{99} isotope is supplied commercially with an ion exchange resin column from which the Tc^{99m} can be "milked" daily since it has a different atomic number, and hence chemical characteristics, than molybdenum. Technicium is similar chemically to manganese and rhenium, and its pertechnetate ion ($Tc0_4{}^-$) resembles iodine in its biological distribution. Thus, by such a generator, the short-lived isotope Tc can be made available to a hospital that is not particularly close to an accelerator or nuclear reactor. Note that, as shown in Fig. 9.3, there are two possible γ-ray decay methods.

Positron decay, β^+, is the second way in which a nucleus with an unstable ratio of protons to neutrons can spontaneously convert into a stable form. It is similar to β^- decay, in that a neutrino is emitted simultaneously. An example is $_{11}Na^{22}$, shown in Fig. 9.4. In this case 90% of the decay of Na^{22} is by β^+ emission to an excited state of Ne^{22}. The energy lost by the nucleus goes into kinetic energy of the β^+ and the neutrino and, as with β^-, there is a continuous distribution of β^+ energies. Note that the arrow of the decay process slants to the left when Z decreases and to the right when Z increases, Figs. 9.2 and 9.3.

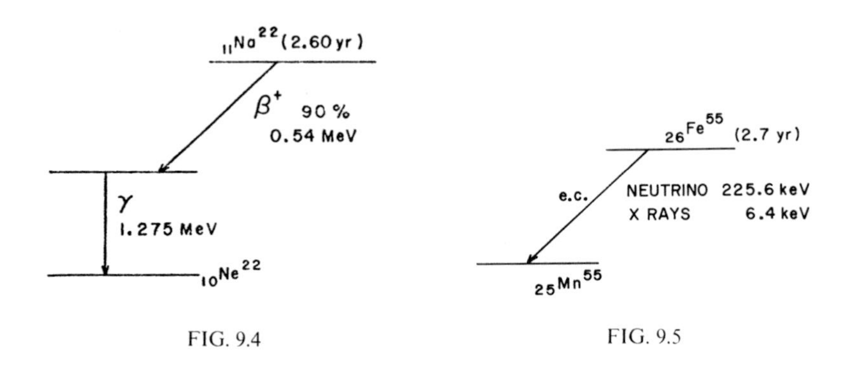

FIG. 9.4

FIG. 9.5

As indicated in Fig. 9.4, 90% of the decay of Na^{22} to Ne^{22} occurs by the simultaneous emission of a positron and neutrino. The remaining 10% occurs by the process of electron capture (not indicated in the figure).

The third way in which a proton in the nucleus can be converted into a neutron is by electron capture (e.c.). For reasons explained below this process is also referred to as K capture. It is electrically equivalent to the emission of a β^+, although energetically they are different and, therefore, their relative importance will vary depending on the energy available and the atomic number of the parent nucleus. The electron capture process occurs because the inner orbital electrons spend part of their time within the nucleus. If one is captured by the unstable nucleus a proton will be converted to a neutron and a neutrino will be ejected. This neutrino carries away the energy difference between the unstable nucleus and its new, stable form. An example is Fe^{55}, shown in Fig. 9.5. Since in this case the neutrino carries away only 225.6 keV of the 232-keV energy difference, the remaining 6.4 keV is released in the following manner. The innermost atomic electron that was captured (called the K electron in x-ray terminology) was bound in its orbit of the atom by an energy of about 22 keV. If a single free electron from outside the atom replaced this K electron directly an x ray of 22 keV would be emitted. However, as described before, a series of electron Auger effects usually takes place. The x ray with largest energy occurs for the transition of L electron into the K shell. Lower energy x rays are emitted for the transition of an M electron into the L shell and so on until the sum of the x ray energies equals the energy difference between the transition energy 232 keV and the neutrino energy.

An alpha particle is a helium nucleus. It has two neutrons and two protons, and is doubly ionized. It can be written as $_2H^{4++}$ but is usually represented by the symbol α. Some of the heavier unstable nuclei spontaneously emit α particles but, as will be seen in a later section, their range is so short that

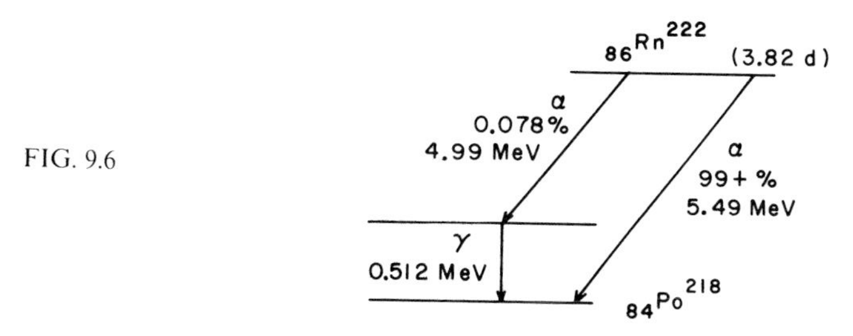

FIG. 9.6

they are absorbed by the body tissue and cannot be detected by an external counting device. Thus, they are not useful as tracers but, because they lose their energy in such a short range, they are useful for implantation in tumors. An example is radon-222, which is a product of radium decay, and is encapsulated in glass needles for surgical implantation. Its decay scheme is shown in Fig. 9.6. Most of the decay is by the emission of an α particle with 5.49 MeV of kinetic energy. In α decay the momentum is shared by only two bodies, the α particle and the recoiling nucleus. Therefore, all α particles have the same energy from a given nucleus. This is in contrast to the electron and positron emissions, in which the momentum is shared by three bodies—the charged particle, the neutrino, and the recoiling nucleus. Such a three-body decay results in a spectrum of energies of the charged particle, as illustrated in Fig. 9.2.

Some heavy isotopes eject a neutron during decay, the most prominent example being uranium. However, although neutrons are not clinically useful as tracers, controlled sources, such as accelerators or reactors, as well as neutron-emitting nuclides, such as californium-252, are being used for neutron radiation therapy.

RADIOACTIVE DECAY RATE

The number of decays or disintegrations per second of a quantity of radioactive material is given in units of the *curie*:

$$1 \text{ curie} = 3.7 \times 10^{10} \text{ disintegrations/sec}$$

The curie alone does not characterize the potential hazard of a source; this also depends on the nature and energies of the emitted particles, as discussed later.

It has been found experimentally that the intensity of the radioactive decay from a given sample decreases exponentially with time, $\exp(-\lambda t)$, where λ is called the decay constant. We can easily construct an equation that reflects this behavior. If we assume that a given sample has n radioactive nuclei of the same kind and that each has a constant probability λ for decay in a unit time, then $\lambda\,dt$ is the probability that any given nucleus will decay in time dt. Thus dn, the change in the number of available nuclei in time dt, will be the product of the number present at that time times the probability $\lambda\,dt$ that each will decay in dt, or

$$dn = -n\lambda\,dt, \tag{9.1}$$

where the minus sign means that n decreases with time. If we have n_0 undecayed nuclei at $t = 0$, integration of Eq. (9.1) will give the number n of undecayed nuclei at time t:

$$\int_{n_0}^{n} (dn/n) = -\lambda \int_0^t dt$$

$$n = n_0 e^{-\lambda t} \tag{9.2}$$

If we now differentiate this equation with respect to time we obtain the expression for the number of radioactive decays per second:

$$-dn/dt = \lambda n_0 e^{-\lambda t} \tag{9.2'}$$

This equation exhibits the desired exponential dependence on time.

The characteristic time of decay of material undergoing exponential decay is the *half-life*. This is the time in which half of the nuclei of a given group will have decayed. If we call this time $T_{1/2}$ then we may substitute into Eq. (9.2) and write

$$n_0/2 = n_0 e^{-\lambda T_{1/2}}, \qquad e^{\lambda T_{1/2}} = 2$$

Taking natural logarithms of both sides we obtain

$$\lambda T_{1/2} = \ln 2$$

$$T_{1/2} = (\ln 2)/\lambda = 0.693/\lambda \tag{9.3}$$

which is the connection between the half-life and the exponential decay constant. Another measure of the lifetime of a radioactive nucleus that is occasionally used is the *mean life* τ. This is the time required for the activity to decay to $1/e$ of its original value. $\tau = 1/\lambda = 1.443T_{1/2}$.

TRACER STATICS

Charged particles such as electrons, positrons, and α particles have a short range in body tissue. We shall see later in this chapter that they can damage and destroy cells. Such emitters are used for implantation in a tumor where they can affect the tumor cells but their range is too short to damage surrounding tissue. Because of this short distance they are not suitable for *in vivo* tracer studies. However, if tissue or fluid is extracted from the body then β emission can be counted. In general, for *in vivo* studies γ emitters are used. In this way individual counters can be placed over different parts of the body or a large number of counters covering the body can be used—this latter arrangement is one of several ways of *whole-body* counting. Whole-body counting can also be done by moving a single counter over the body, stopping at intervals to record the counts.

Let us consider an example of each type, following the treatment by Sheppard (1962). Suppose we wish to estimate the number of fish in a lake. We could catch some, tag them, and release them. Later we could catch a group of fish, count the fraction of tagged ones and thereby estimate the total number of fish in the lake. There are some obvious statistical questions, for example, how long do we wait to ensure a completely random distribution of the tagged fish with the total population? Clearly, the time must be long compared with the widest dimension of the lake divided by the mean swimming speed, but short compared with the average fish lifetime and reproduction time.

Similar considerations exist in an experiment to determine the total body content of potassium. We add K^{42}, a γ-ray emitter, take a blood sample, and measure both the total potassium content and the radioactive content. Before one can be certain of a correct answer a variety of auxiliary experiments must be performed to assure that sufficient time has elapsed for random dilution, that the dilution rate in blood is equal to the dilution rate in other tissues of the body, that potassium is not stored selectively in any body tissues, that any passed from the body during the waiting time is accounted for, etc. This technique is called *isotope dilution* and involves the above considerations related to *tracer statics*. The isotope dilution technique can also be used for time studies. For example, if we wish to learn the exchange rate for water in body tissues a person can ingest some water consisting in part of tritium oxide. If urine samples are checked over a period of days (and corrected for the decay that has taken place during the elapsed time) a plot of excreted water versus time would give an exchange rate in the body tissues.

As an example of the application of simple isotopic decay rate we consider the problem of human red blood cell survival. The nuclide Cr^{51} decays by electron capture (e.c.), with a half-life of 27.8 days, and 9 % of the decays are accompanied by 0.32-MeV γ rays, which are easily counted. Red blood cells tagged *in vitro* with chromium-51 are injected into a patient. Blood samples are taken daily, counted and plotted; a linear and a semilogarithmic plot are shown in Fig. 9.7. It is seen that in a normal individual the half-life of the cell is about 37 days. Note that in order to make such a plot the remaining activity must be determined by correcting for the half-life of the

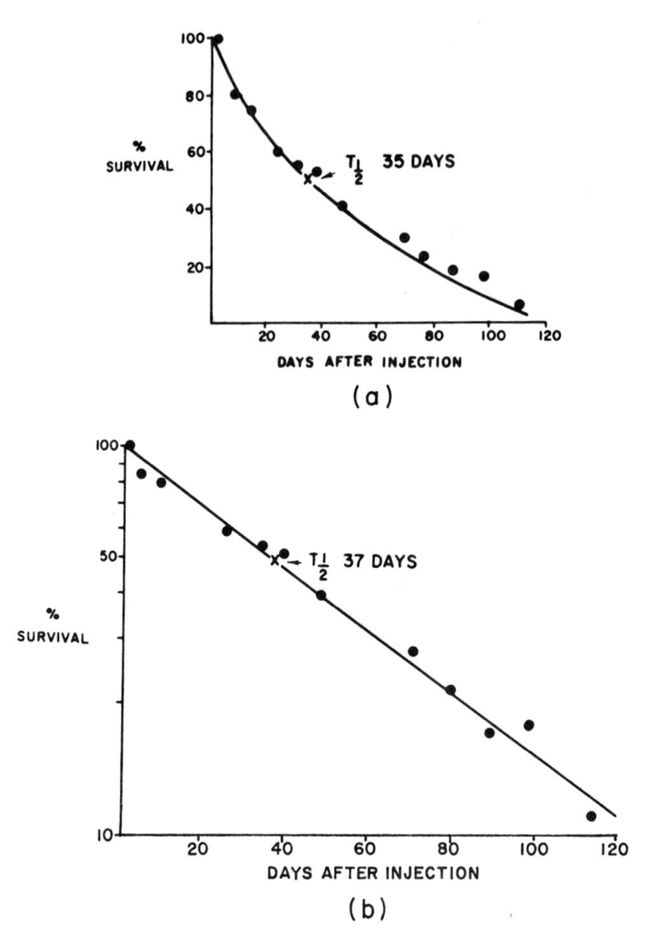

FIG. 9.7 Plots of percent of red blood cell survival versus time as measured by Cr^{51} tagging. (a) Data plotted on linear coordinates. (b) Percent survival plotted on semilog coordinate. [From Korst (1968).]

isotope. There is no experimental evidence of feedback of chromium from the spleen, where red blood cells are destroyed and dissolved. However, when the spleen malfunctions and destroys the red cells at a greater than normal rate, hemolytic anemia results. This condition is detectable by scanning the area of the spleen and comparing the counting rate with that of normal individuals. Figure 9.8 shows the percentage of red cell survival versus time for patients with hemolytic anemia, the shaded area being the range for normal patients. Steroid therapy or removal of the spleen can usually correct this condition.

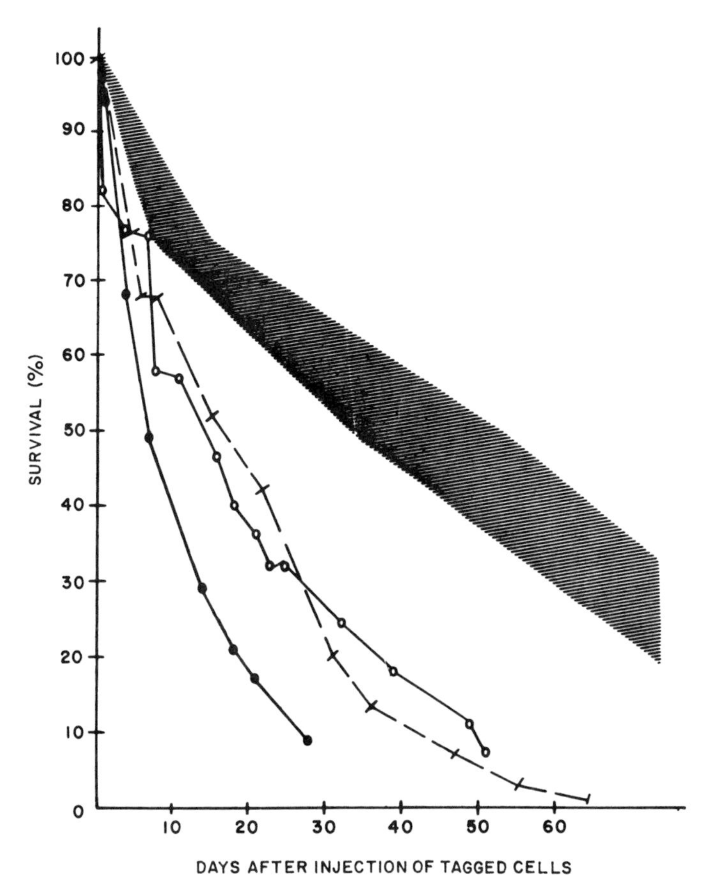

FIG. 9.8 Radiochromate-tagged red cell survival curves before splenectomy in three patients with acquired hemolytic anemia compared to the normal range (black area). [From Korst (1968).]

A combination of isotope dilution measurements and the following of an injected radioactive bolus can be used as a noninvasive technique for determining cardiac performance. This is described in the next section.

RADIOGRAPHIC MEASUREMENT OF CARDIAC PERFORMANCE

It is of considerable interest to determine cardiac function quantitatively in a rapid, noninvasive manner. We have seen in Chapter 7 the early attempts of Hamilton and his co-workers to measure stroke volume from pulse analysis. With the advent of radionuclide technology such measurements can now be done with a great deal of sophistication in a short time. The technique to be described was developed by Groch *et al.* (1976) and the resulting instrument, called the Gamma-cor RCG cardiac probe, is being marketed by Searle Radiographics.

A bolus (a mass) of 0.5-cm^3 volume of 1–1.5 mCi (= millicuries) Tc^{99m} human serum albumin is injected into an external vein such as the jugular. This bolus travels to the right side of the heart, is pumped through the pulmonary system, drains into the left ventricle, from which it is pumped into the aorta and onward into the body's arterial system and venous return.

The section of the heart of most interest is the left ventricle, and two important parameters are the *left ventricular ejection fraction* (LVEF), the fraction of blood in the ventricle ejected during systole, and the *cardiac output*, the quantity of blood pumped by the left ventricle per minute.

A crystal detector of γ rays, which is described in the next chapter, is placed behind a collimator, a hollow cylinder of lead, and this collimator–detector, called a probe, is placed directly over the center of the left ventricle. This center is located with an echocardiograph probe, described in Chapter 8. The major axis of the left ventricle is taken from the aortic root to the cardiac apex, and the midpoint of this axis, and therefore of the left ventricle, is taken to be beneath the mitral valve cusps. When this is located a mark is placed on the chest for placement of the probe. Concentric about the probe is another collimator–detector which is focused on the area immediately surrounding the left ventricle. A cross-sectional diagram through these concentric probes with the respective regions from which γ-ray emissions will be detected is shown in Fig. 9.9. Note that the left and right collimators are part of one collimator concentric to the central collimator, and signals from any part of the area with diagonal crosshatching go to the same recorder.

As the radioactive bolus enters the right heart the outer detector records a high level of activity (Fig. 9.9, heavy line). The bolus is then pumped into the pulmonary circulatory system and very little activity is seen in either detector. When the bolus returns from pulmonary circulation and enters the left ventricle, the beat-to-beat variation seen in the count rate (Fig. 9.9, thin line), which corresponds to left ventricular filling and ejection, is recorded by the central probe.

Typical recordings of the two detectors in Fig. 9.9 are shown graphically in Fig. 9.10 for analysis. An immediately evident parameter obtainable from this graph is the time for a labeled quantity of blood to leave the right side of the heart and arrive at the left side. During this time it has been passing through the pulmonary circulatory system. This pulmonary transit time (PTT) is approximately determined from the time at which the activity of

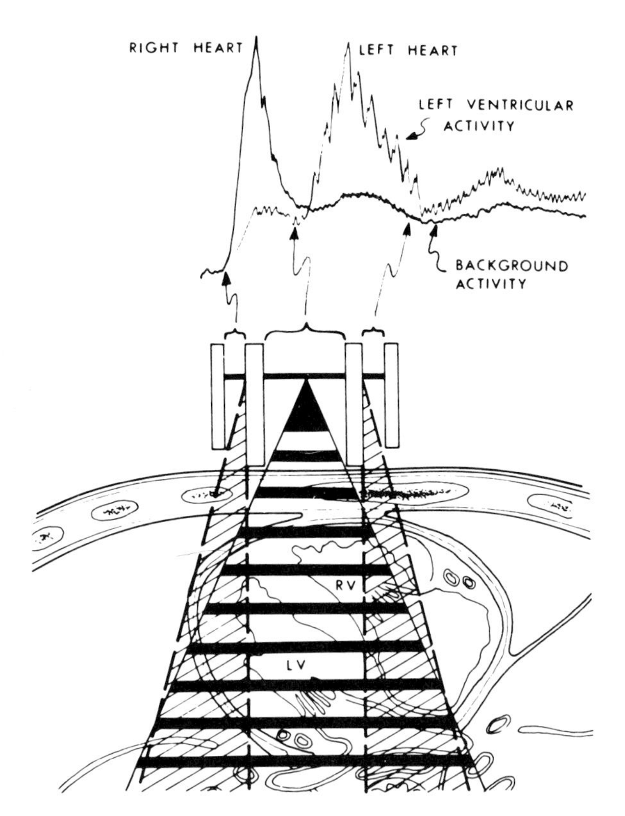

FIG. 9.9 Cross section of collimator–detector on chest over the center of the left ventricle. The detectors are two concentric cylinders; the diagonal lines show the field of view of the outer detector and the horizontal lines that of the inner detector. [From Groch *et al.* (1976).]

the right heart is 75% of its maximum to the time of the peak of activity in the left ventricle.

The heavy curve represents activity other than in the left ventricle, and therefore the activity in the left ventricle at any time is the difference between that recorded by the central probe (Fig. 9.10, thin line) and the outer probe Fig. 9.10, (heavy line), and this is represented as y. Since y is proportional to the quantity of blood in the left ventricle at any time, in a given beat the distance x represents the change in this quantity of blood between the beginning and end of systole. Groch *et al.* (1976) define the peak as end-diastole and the nadir as end-systole and define the left ventricle ejection fraction as

$$\text{LVEF} = \frac{(\text{counts in peak}) - (\text{counts in nadir})}{(\text{counts in peak}) - (\text{background})} = \frac{x}{y} \qquad (9.4)$$

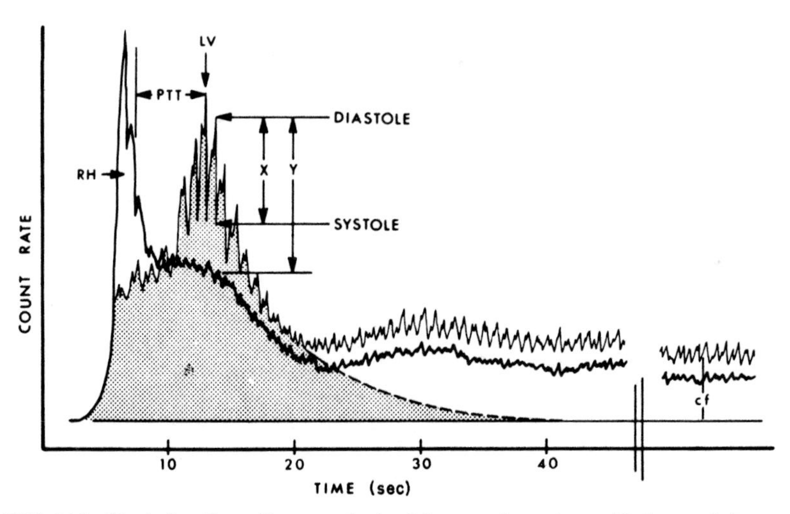

FIG. 9.10 Typical radiocardiogram obtained from probe system with time activity curves for left ventricle (thin line) and right heart and background (thick line). RH, right heart; LV, left ventricle; PTT, pulmonary transit time. [From Groch *et al.* (1976).]

Thus the left ventricle ejection fraction is determinable and the result should be the same for each stroke, although highest accuracy will be obtained from the beat with the highest count rate.

Cardiac output, blood volume per minute, can be obtained in the following way. If a known quantity I of an indicator is injected into the venous side of the circulatory system near the heart, it will soon begin to appear on the arterial side mixed with cardiac output Q. If $C_a(t)$ is the concentration of the indicator in arterial blood then $QC_a(t)$ is the amount of indicator entering the arterial side at any time t. If all of the quantities $QC_a(t)$ were collected on the arterial side all of the indicator I would be recovered. We may therefore write

$$Q\int_0^\infty C_a(t)\,dt = I \tag{9.5}$$

and the cardiac output Q is

$$Q = I\bigg/\int_0^\infty C_a(t)\,dt \tag{9.6}$$

If a radioactive tracer is used for the indicator and it is assumed that there is none in the lungs, then the radioactivity $R(t)$ at any time t measured on the chest over the heart may be written as the sum of the contributions of the four chambers:

$$R(t) = W_{RA}C_{RA}(t) + W_{RV}C_{RV}(t) + W_{LA}C_{LA}(t) + W_{LV}C_{LV}(t) \tag{9.7}$$

where W is the effective volume of each chamber seen by the counter, RA, RV, LA, and LV stand respectively for right atrium, right ventricle, left atrium, and left ventricle, and the C's are the respective concentrations. From the indicator–dilution relation, Eq. (9.5), we can write

$$\int_0^\infty C_{RA}(t)dt = \int_0^\infty C_{RV}(t)dt = \int_0^\infty C_{LA}(t)dt = \int_0^\infty C_{LV}(t)dt = I/Q \quad (9.8)$$

since no blood escapes from the path through the heart chambers. If Eq. (9.7) is multiplied by dt and integrated from 0 to ∞ then Eq. (9.8) may be substituted for the integrals with the results

$$\int_0^\infty R(t)dt = (W_{RA} + W_{RV} + W_{LA} + W_{LV})(I/Q) \quad (9.9)$$

Now let W_H be the effective counting volume for all four heart chambers,

$$W_H = W_{RA} + W_{RV} + W_{LA} + W_{LV} \quad (9.10)$$

and define area A of counts from 0 to ∞ of the radioactive bolus, indicated by the shaded area of Fig. 9.10. Since A is the total number of detected counts that have passed through the heart,

$$A = \int_0^\infty R(t)dt = W_H I/Q \quad (9.11)$$

or cardiac output Q is given by

$$Q = W_H I/A \quad (9.12)$$

and is calculable when W_H is evaluated.

If the tracer used remains in the bloodstream then in a few minutes an equilibrium (final) concentration C_f of activity will be present in the blood. A measure of activity when this equilibrium is achieved, R_e, made in the same position will be

$$R_e = W_T C_f \quad (9.13)$$

where W_T is the effective volume seen by the counter at equilibrium time. This will be larger than W_H because all of the blood now contains activity and small blood vessels in the chest wall, etc., will contribute. Independent measurements of cardiac output have shown that a reasonable correction factor is

$$W_H = 0.8 W_T$$

Therefore, $R_e = W_H C_f/0.8$ and cardiac output is

$$Q = 0.8 R_e I/C_f A \quad (9.14)$$

But, by definition, $I = C_f \times$ (total blood volume) at equilibrium and

$$Q = (0.8R_e/A) \times \text{(total blood volume)} \qquad (9.15)$$

If a blood sample is taken after equilibrium is achieved and counted, C_f is the result. Knowing I, the total blood volume can be determined. By measuring the area A of Fig. 9.10 the cardiac output Q can be calculated from either Eq. (9.14) or (9.15).

COMPARTMENT MODELING

The constituents or components of a living system can be divided into mathematical *compartments* that constitute distinguishable phases or pools. The boundaries of these compartments may be physical, such as the chambers of the heart, or a phase, such as the bicarbonate pool. In some cases the location may be changing, as in cell mitosis. Substances transfer in and out of compartments with differing reaction rates. If the compartments are sequential, i.e., in series or chainlike, the system is called *catenary*. If a system is in parallel, or if there is a system with a single central compartment that exchanges with multiple peripheral compartments that do not exchange with one another, the system is called *mammillary*. The mathematical analyses for multicompartment systems have been worked out and are given in the book by Sheppard (1962) and reviews such as those by Robertson (1962) and by Jacquez (1968).

The fundamental measurement in a radioactive tracer experiment is counts per unit time, usually per minute, in a counting device. The number of counts per minute depends on the amount of material being counted, and thus a mass must be specified. The counts per minute of a given mass, usually per milligram, is called the specific activity a. Thus, the measurement is effectively normalized. In general, the initial quantity of radioactive tracer is not a meaningful parameter per se, but the initial specific activity is.

Let us consider the simplest case of a two-compartment system, with compartments labeled 1 and 2. Let S be the total amount of traced substance in the system, i.e., moles, grams, etc., and let S_1 and S_2 be the amounts of S in compartments 1 and 2, respectively. S_1 and S_2 are in equilibrium but the relative concentrations of the tracer in S_1 and S_2 change. The number of counts due to tracer in compartments 1 and 2 are R_1 and R_2 and the specific activities are a_1 and a_2, respectively. For simplicity let the rate of exchange k be the same in both directions. From these definitions we may write that the counts of tracer R equals a, counts/min/g, times the number of grams S, or

$$R_1 = a_1 S_1 \qquad \text{and} \qquad R_2 = a_2 S_2$$

Since S_1 and S_2 are constant, the change of R with time is

$$dR_1/dt = S_1(da_1/dt) \qquad \text{and} \qquad dR_2/dt = S_2(da_2/dt)$$

Since the rate constant k is the same in both directions the rate of change of radioactivity in compartment 1, dR_1/dt, is the difference between the specific activity leaving and entering, or

$$dR_1/dt = -ka_1 + ka_2 \tag{9.16}$$

similarly, for compartment 2

$$dR_2/dt = -ka_2 + ka_1 \tag{9.17}$$

Upon substituting the previous relations we obtain

$$S_1(da_1/dt) = -ka_1 + ka_2 \tag{9.16'}$$

$$S_2(da_2/dt) = ka_1 - ka_2 \tag{9.17'}$$

These two equations may be added and the sum written as

$$\frac{d(a_1 - a_2)}{dt} = -k(a_1 - a_2)\left(\frac{1}{S_1} + \frac{1}{S_2}\right)$$

which integrates simply to

$$(a_1 - a_2) = (a_1 - a_2)_0 \exp\left[-k\left(\frac{1}{S_1} + \frac{1}{S_2}\right)t\right] \tag{9.18}$$

in which $(a_1 - a_2)_0$ is the initial specific activity difference. If all of the initial tracer is in compartment 1 at time = 0 call this $a(0)$.
The total activity is†

$$S_1a_1 + S_2a_2 = S_1a(0)$$

and, by definition, $S_1 + S_2 = S$. If this equation is solved simultaneously with Eq. (9.18) we obtain

$$a_1 = [a(0)/S][S_1 + S_2 \exp(-kSt/S_1S_2)] \tag{9.19}$$

and

$$a_2 = [a(0)/S]S_1[1 - \exp(-kSt/S_1S_2)] \tag{9.20}$$

† This analysis assumes that the lifetime of the tracer is much longer than the time of the experiment. If the tracer decays appreciably during the experiment similar, but more complex, expressions apply.

Equations (9.19) and (9.20) indicate that there is an exponential type of decrease in the specific activity in compartment 1, and a corresponding increase in compartment 2, due to the transport of tracer between the compartments.

The transport rate is usually the desired quantity. This may be determined from a plot of a_1 versus t, and from Eq. (9.16′) we may write

$$k = S_1 \frac{da_1/dt}{a_2 - a_1} \tag{9.21}$$

By plotting experimental data of a_1 versus t one can determine k from this analysis if S_1 is known. The rate constant k may vary with time but Eq. (9.21) gives the value at any time.

If there is a one-compartment system that is open, i.e., S_2 is an infinite reservoir, the above two-compartment equations can be used. In this case $a_2 \to 0$ (infinite dilution), and $1/S_2 \to 0$. Thus from Eq. (9.18) the expression for a_1 is

$$a_1 = a(0)\exp(-kt/S_1) \tag{9.22}$$

which is simple exponential behavior from which k can be determined from the slope of a semilogarithmic plot.

Compartment modeling can be extended to any number of any combinations of catenary or mammillary compartments and the methods of solution are given in the references cited above.

PATH OF NUCLEAR PARTICLES

There are three basic types of radiation: (1) photons, (2) charged particles, and (3) neutral particles. In the first category are light, x rays, and γ rays. Although light has little penetrating power, ultraviolet light does have sufficient energy to excite electrons near the surface of an exposed cell. x rays and γ rays not only excite electrons but they can give so much energy to an electron that that electron itself acts as a charged particle that can excite additional electrons within the material. Heavy charged particles, such as protons and α particles, also lose their energy by collision with electrons, exciting or ionizing atoms of the target material. The most common neutral nuclear particle is the neutron. It does not interact directly with the atomic electrons, but instead loses its energy through collision with or capture by atomic nuclei. In general, high-energy neutrons (say, above 1000 eV) are not captured, since capture cross section decreases with increasing energy; instead, they collide with a host nucleus, which is displaced from its original position. This nucleus, called a *primary knock-on*, leaves behind some of its electrons and behaves as a new massive charged particle within the material. The original neutron loses energy by successive collisions until it reaches an energy of less than an electron volt, at which time it has a high probability of capture by an atomic nucleus. The result of this neutron capture process depends on the stability of the isotope produced. If it is unstable, it will emit particles and these, having momentum, may cause damage by knocking other atoms from their position, or by breaking chemical bonds. Although the neutron has a very low kinetic energy before capture, its capture by a nucleus releases an excitation energy of about 7–8 MeV in the form of γ rays.

The loss of energy of a charged particle is not uniform but increases with decreasing velocity. Nor can the damage in the path be characterized as uniform. There are local "bursts" of energy that occur when a relatively large amount of energy is given to electrons in the path of the primary moving particle. The energetic electrons create their own path of energy dissipation. Such electrons are called *delta* (δ) *rays.*

ENERGY LOSS OF CHARGED PARTICLES

The passage of energetic particles through matter is frequently characterized by their total range, which may be determined experimentally in a straightforward way. However, the damage to cells along the particles' path is a function of the rate of energy loss per unit distance. Therefore, we first concentrate on understanding the details of energy loss and return later to the question of total range. Since atomic nuclei are very small, most of the travel of a charged particle is through a cloud of electrons, and it is through interactions with these electrons that a charged particle loses its energy. These electrons are either raised to excited states or completely removed from their nucleus, i.e., the atom is ionized. We will consider these relative numbers in the next section. We will now derive $-dE/dx$, the energy loss as a function of distance along the path of the energetic particle. The calculation of $-dE/dx$ may be done using classical physics, and was first carried through by Niels Bohr in 1915.

Consider a particle of mass m, velocity V, and charge Ze moving in the x direction past an electron at point A as in Fig. 9.11. The perpendicular

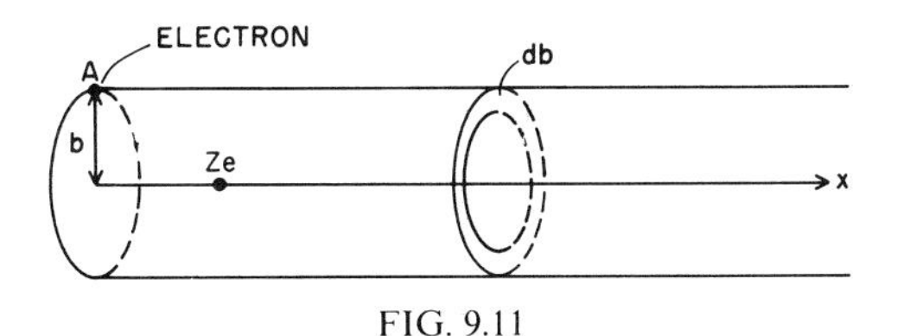

FIG. 9.11

distance b from the path to the electron at A is called the *impact parameter*. It is assumed that this electron is free and initially at rest and moves so slightly during the passage of the charged particle that its interaction with the electric field can be calculated at the unperturbed location of the electron.

With this assumption the net impulse parallel to the path of the charged particle is zero because, by symmetry, for each position of the charged particle to the left of A before passage there is a position to the right of A after the passage and so there are points with equal and opposite forces along the x direction.

An estimate of the magnitude of the perpendicular impulse I given to the electron may be obtained from the definition

$$\begin{aligned} I &= \int F dt \\ &= \text{electrostatic force} \times \text{time of collision} \\ &\cong (Ze^2/b^2) \times (b/V) \end{aligned}$$

For a more exact calculation consider a circular cylinder centered on the path of the charged particle with its surface passing through the electron in question as in Fig. 9.11. Let $\mathscr{E}$ be the electrostatic field intensity due to the charged particle. Then by Gauss's theorem

$$\int_{\text{surface}} \mathscr{E} \cdot d\boldsymbol{\sigma} = 4\pi Ze$$

where σ is the surface. If $\mathscr{E}_\perp$ is the component of the electric field perpendicular to the path then the flux is

$$\int_{-\infty}^{\infty} \mathscr{E}_\perp 2\pi b \, dx = 4\pi Ze$$

and therefore

$$\int_{-\infty}^{\infty} \mathscr{E}_\perp \, dx = 2Ze/b \tag{9.23}$$

The change of $\mathscr{E}_\perp$ with time at the electron is the same as holding the charged particle fixed and moving the electron along the cylindrical surface with velocity V. Therefore, using Eq. (9.23) we may write

$$\int_{-\infty}^{\infty} \mathscr{E}_\perp(t) \, dt = \int_{-\infty}^{\infty} (\mathscr{E}_\perp(x)/V) \, dx = (1/V) \int_{-\infty}^{\infty} \mathscr{E}_\perp(x) \, dx = 2Ze/Vb \tag{9.24}$$

The impulse to the electron is equal to its momentum change p. This impulse in the perpendicular direction is, as before,

$$I = \int F_{\perp}\, dt$$

$$= \int_{-\infty}^{\infty} \mathscr{E}_{\perp} e\, dt = 2Ze^2/Vb = p$$

where p is the momentum change of the electron, which we assume was at rest before the passage of the charged particle. Note that the earlier estimate for I was within a factor of 2 of this value. Since kinetic energy (KE) of a particle with momentum p is $p^2/2m$, the KE acquired by the electron is

$$\mathrm{KE} = p^2/2m = 2Z^2e^4/mV^2b^2 \tag{9.25}$$

The number of collisions dn/dx per unit path length with electrons in the range of b to $b + db$ is equal to the number of electrons per centimeter of length in the cylindrical shell bounded by an inside radius b and an outer radius $b + db$. If N is the number of electrons per cm^3 in the medium, the number of collisions with electrons per unit path length is

$$dn/dx = N2\pi b\, db \tag{9.26}$$

Thus, the energy lost per cm to electrons in this shell, $-dE(b)/dx$, is the product of the energy lost to one electron times the number of electron collisions in the cylindrical shell per cm of path length, or

$$-dE(b)/dx = (4\pi Z^2e^4N/mV^2)(db/b) \tag{9.27}$$

The total energy delivered from the charged particle to all shells within the cylinder with radii in the range b_{min} to b_{max} is the integral of Eq. (9.27), or

$$-dE/dx = (4\pi Z^2e^4N/mV^2)\ln(b_{max}/b_{min}) \tag{9.28}$$

We cannot evaluate this expression between the limits of 0 and ∞ because the energy lost per unit path length would be infinite. We must, therefore, choose physically meaningful values of b_{min} and b_{max}.

Let us now examine some of the physical parameters involved in the determination of b_{min} and b_{max}. As the charged particle moves past the electron the "time of collision" τ becomes an important parameter. This time of collision is the same order of magnitude as the time the particle will take in traveling a distance of length b, the impact parameter distance: $\tau = b/V$. If this collision time is very short compared with the time of vibration of an electron, the interatomic forces will not have time to act on the electron before the charged particle has passed out of range, and the energy transferred to the electron will be essentially the same as if the electron were free. On the other hand, if the time of collision is long compared with the time of vibration the electron will behave almost as if it were rigidly bound and the energy transferred will be negligible.

Let us first consider b_{min}. Classically, the maximum velocity that can be given to the electron in a head-on collision with a heavy particle is less than $2V$. This is readily seen if we apply the laws of momentum and energy conservation to the collision between two particles of masses M and m and initial velocities V_1 and v_1, respectively, in a head-on collision, with V_2 and v_2 their respective velocities after collision.

By conservation of momentum

$$MV_1 + mv_1 = MV_2 + mv_2$$

or

$$M(V_1 - V_2) = -m(v_1 - v_2) \tag{9.29}$$

By conservation of energy

$$\tfrac{1}{2}MV_1^2 + \tfrac{1}{2}mv_1^2 = \tfrac{1}{2}MV_2^2 + \tfrac{1}{2}mv_2^2$$

or

$$\tfrac{1}{2}M(V_1^2 - V_2^2) = -\tfrac{1}{2}m(v_1^2 - v_2^2) \tag{9.30}$$

Dividing Eq. (9.29) by Eq. (9.30) we obtain

$$\frac{V_1 - V_2}{V_1^2 - V_2^2} = \frac{v_1 - v_2}{v_1^2 - v_2^2}$$

and, upon factoring the denominators, we obtain

$$V_1 + V_2 = v_1 + v_2$$

In our present case v_1 is zero and v_2, the velocity of the electron after collision, is $V_1 + V_2$ or almost $2V_1$, where V_1 is the initial velocity of the ion. Therefore the kinetic energy given to the electron cannot exceed (dropping the subscript 1)

$$\text{KE}_{\text{max}} = \tfrac{1}{2}\text{m}(2V)^2 \tag{9.31}$$

and b cannot have values that imply greater energy transfers than $2mV^2$.

We have expressed in Eq. (9.25) the energy transferred per collision as a function of b. Values of b smaller than the simultaneous solution of Eq. (9.25) and (9.31),

$$\text{KE}_{\text{max}} = 2Z^2e^4/mV^2b^2 = 2mV^2 \tag{9.32}$$

must therefore be excluded in the integration. This determines b_{min} as

$$b_{min} = Ze^2/mV^2 \tag{9.32'}$$

where m is the mass of an electron and V is the velocity of the particle.

We now consider the determination of b_{max}. If b_{max} were allowed to become infinite, the energy of the charged particle would be instantaneously lost because of the contribution of unlimited small energies given to distant electrons. However, since these electrons are bound in orbits within material, infinitesimal energies cannot be given to them, only energies that correspond to excitation into the excited states. If the average excitation energy is J, the minimum kinetic energy from Eq. (9.25) is

$$\text{KE}_{min} = J = 2Z^2e^4/mV^2b^2$$

(Note that an empirical value for J is approximately 12.5Z eV, where Z is the atomic number of the atoms of the medium.) Solving for b_{max} gives

$$b_{max} = 2Ze^2/(2mV^2J)^{1/2} \tag{9.33}$$

We may therefore write the rate of energy loss of a charged particle moving at nonrelativistic velocities through matter by substitution of b_{max} and b_{min} into Eq. (9.28) to obtain

$$-dE/dx = (4\pi Z^2e^4N/mV^2) \ln 2mV^2/(2mV^2J)^{1/2} \tag{9.34}$$

Noting that $\ln x^{1/2} = \frac{1}{2} \ln x$ we may write Eqn. (9.34) in the more commonly used form as

$$-dE/dx = (2\pi Z^2e^4N/mV^2) \ln(2mV^2/J) \tag{9.35}$$

Note that the primary dependence of $-dE/dx$ on the velocity of the ion is through the factor $1/V^2$. Thus the energy loss rate increases rapidly as V decreases.

The quantity $-dE/dx$ is called the *linear stopping coefficient* or *stopping power* of the material through which the heavy charged particle passes. The assumptions in the derivation render it inapplicable to electrons, since it was assumed that the particle is much more massive than an electron and is undeflected by its collision with an electron. For charged particles having a velocity of less than about 10% of the velocity of light, i.e., where relativistic corrections are unimportant, Eq. (9.35) indicates that their rate of loss of energy is proportional to the density of electrons in the medium, the square of the charge number Z of the particle, and inversely proportional to the square of the velocity of the particle. The energy loss rate is independent of the mass of the incident particle and hence a measurement of the stopping power of a material for one nonrelativistic particle can be scaled to the stopping power for other charged particles through the factor $(Z/V)^2$ in Eq. (9.35).

RANGE OF CHARGED PARTICLES

The *range* of a particle is the distance it moves through a stopping substance before coming effectively to rest. If a particle has initial energy E_i its range R is

$$R = \int_{E_m}^{E_i} [dE/(-dE/dx)] \tag{9.36}$$

where E_m is the minimum detectable energy. The scaling mentioned above for the rate of energy loss, or stopping coefficient, is indicated clearly in Fig. 9.12, in which the ranges of protons, and α particles are shown as a function of their initial energies. The curves essentially superimpose with only a shift in the ordinate.

A more useful example of energy loss rate, which can be related to cell death, is that in protein, given in Table 9.1 from data calculated by Setlow

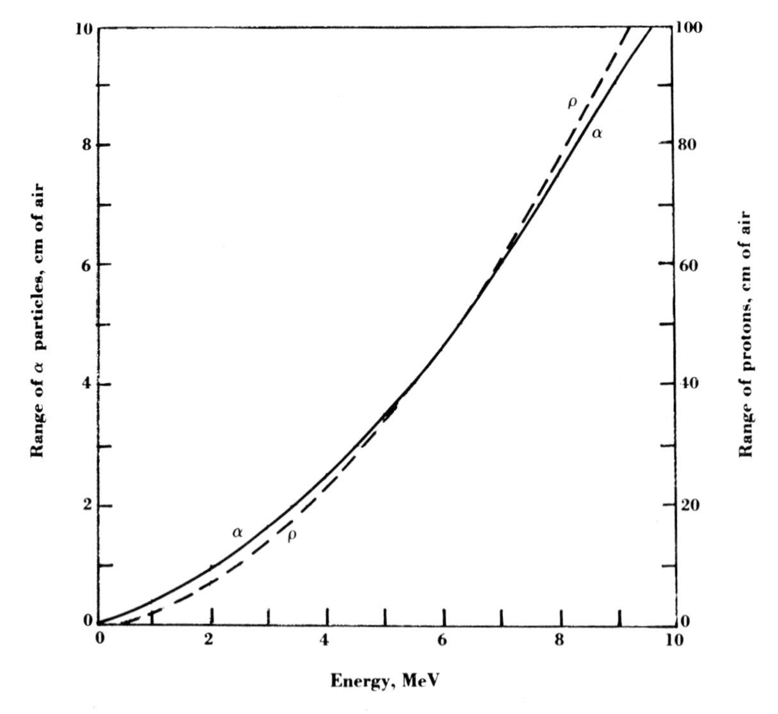

FIG. 9.12 Range–energy relations for α particles and protons (ρ) in standard air. [From "Introduction to Modern Physics," 6th Ed. by F. K. Richtmeyer, E. H. Kennard, and J. N. Cooper. Copyright 1969 by McGraw-Hill Book Company. Used with permission of McGraw-Hill Book Company. Based on data from Segre (1953).]

TABLE 9.1

Energy loss in protein per 100 Å (eV)

	α particle	
1 MeV	4 MeV	10 MeV
2520	1300	662
	Proton	
320	117	57

and Pollard (1962). The data give the energy in eV given to 100 Å of protein by the particles indicated.

It is seen from Eq. (9.35) that, since the Z of an α particle is twice that of the proton and, for the same kinetic energy, V^2 of the proton is 4 times that of the α particle, the loss rate (ignoring the logarithmic dependence in Eq. (9.35)) is expected to be a factor of 8 greater for α particles than for protons. This is essentially what is shown by the data in Table 9.1 and Fig. 9.12 for energies around 1 MeV; but, because of the logarithm term, the ratio rises to 11–12 at higher energies.

Because the stopping power $-dE/dx$ is a complicated function of the energy the range must be calculated from Eq. (9.36) by numerical integration. The energy loss rate is seen in Eq. (9.35) to increase with decreasing particle velocity V. Therefore, the number of interactions of the charged particle increases as it slows down, and near the end of its path the number of interactions, or "damage," to the medium is greatest. This is shown in Fig. 9.13, where the amount of ionization as a function of penetration distance into an aluminum absorber is shown for very high-energy deuterons (ionized deuterium) from an accelerator. Although such high energies are not used in medical application it is necessary in such an experiment to obtain accurate data. The upper curve illustrates the rate of ionization in the medium and, hence, the rate of energy loss of the deuteron. As discussed above, the damage rate is seen to be greatest near the end of the path of the particle, and the maximum rate of energy loss is known as the *Bragg peak*.† The Bragg peak is important in design of modern instruments for tumor irradiation. It means that a particle of suitable energy and charge may penetrate tissue with low damage and be most destructive at a prespecified depth. Designs are in progress for both high-energy ion accelerators and meson-producing

† The lower curve of Fig. 9.13 illustrates a phenomenon known as *straggling*. Because of the varying statistical history of energy and angle of the collisions of each particle they do not all have the same range.

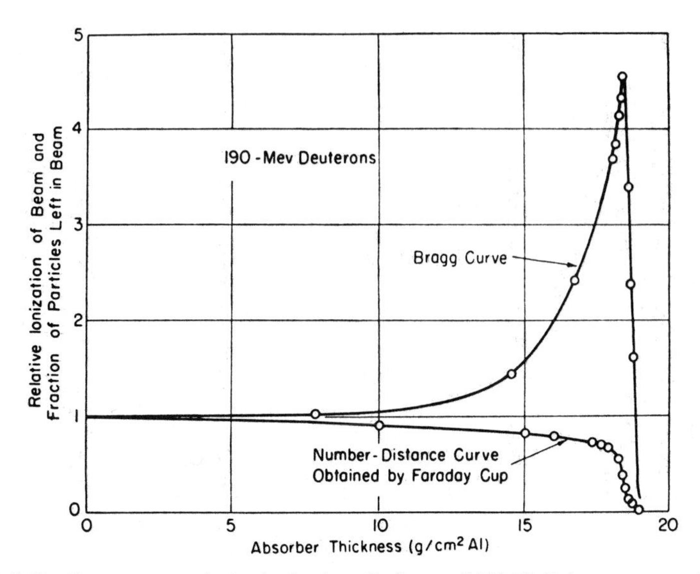

FIG. 9.13 Upper curve, relative ionization of a beam of 190-MeV deuterons as a function of depth in Al illustrating the Bragg peak. Lower curve, the fraction of incident deuterons that penetrate the Al as a function of depth. The loss is due to the straggling. (The absorber thickness in cm is obtained by dividing the value of the abscissa in this figure by the density of Al; $\rho = 2.7$ g/cm^3.) [From Hine and Brownell (1956); based on data by Tobias *et al.* (1952).]

accelerators to be used in the selective irradiation of tumors based on this principle.

It is desirable to be able to scale the stopping power and range between different target materials. Note that the N of Eq. (9.35) is the number of electrons per cm^3. This can be put into more conventional terms in the following way. Since each cm^3 of a material has $\rho N_0/A$ atoms, it has $N = \rho N_0 Z/A$ electrons per cm^3, where ρ is the density, N_0 is Avogadro's number, A is the gram atomic weight and Z the number of electrons per atom. There is also a weak Z dependence through J in the logarithmic term in Eq. (9.35). There is no direct scaling factor because ρ, Z, and A all come into it. However, as the atomic number A increases, Z increases, and since the ratio of Z/A is about unity, N of Eq. (9.35) is proportional to the density of the absorbing material. Therefore the easiest rule for relative stopping power of materials is the ratio of their densities. The relative ranges in absorbing materials would then be the reciprocal ratio of their densities.

DISTRIBUTION OF EXCITATION AND IONIZATION

We may use the formulation discussed above to determine more detailed information on the distribution of the excitation energies. Let KE = Q

for convenience, and differentiate the kinetic energy expression, Eq. (9.25), with respect to b, substituting $b\,db$ from Eq. (9.26) and b^4 from Eq. (9.25). In this way we obtain the number of collisions in distance dx giving rise to excitation energies in the interval dQ about Q:

$$d(dn/dx) = (2\pi Z^2 e^4 N/mV^2)(dQ/Q^2) \tag{9.37}$$

Integrating Eq. (9.37) between a lower energy Q_1 and a higher energy Q_2 gives the number of electrons n_{1-2} with energies in this range:

$$\frac{dn_{1-2}}{dx} = \frac{2\pi Z^2 e^4 N}{mV^2}\left(\frac{1}{Q_1} - \frac{1}{Q_2}\right) \tag{9.38}$$

This relationship enables us to determine the number of energetic electrons in an arbitrary energy range Q_1 to Q_2. A natural upper limit on Q_2 is the highest kinetic energy that can be imparted to an electron, $Q_{max} = 2MV^2$ given by Eq. (9.32). A natural lower limit is $Q_{lim} = J$, the smallest atomic excitation, discussed in that section.

It is not clear whether or not excitations cause as significant biological damage as do ionizations. Lea (1955) states that they are not significant and yet Setlow and Pollard (1962) include them, and Elkind and Whitmore (1967) show data for the killing of cell cultures by ultraviolet irradiation that can cause only excitations. A possible explanation of this controversy is contained in a footnote in Elkind and Whitmore (1967, p. 238). They cite evidence by Han *et al.* (1964b) that while mammalian cells can repair sublethal x-ray damage they cannot repair sublethal ultraviolet damage. Trosko *et al.* (1965) have presented evidence that indicates that mammalian cells cannot excise the thymine dimers that are produced by ultraviolet irradiation, which may explain the different character of ultraviolet damage compared with x-rays.

ENERGY LOSS OF ELECTRONS

The rate of energy loss of electrons is described by a more complicated expression than Eq. (9.35), which we will not derive. In the calculation of energy loss of a charged particle we assumed that only the electrons in the material were deflected and that the path of the charged particle was unaltered because of its large mass compared to that of the electron. This is clearly not the case if the energetic particle itself is an electron. There is a second phenomenon that must be included in the calculation of electron–electron collisions, called quantum mechanical *exchange*. This arises because electrons are indistinguishable and in a collision between two of them an observer cannot tell which electron was the incident one. Thirdly, electrons are easily accelerated to high velocities and hence relativistic effects must be included. Bethe (1933; see also Livingston and Bethe, 1937) has calculated the energy loss rate $-dE/dx$ for electrons including all of these effects, and gives the result:

$$-\frac{dE}{dx}=\frac{2\pi e^4 N}{mV^2}\left\{\ln\left(\frac{V^2 mT}{2\bar{J}^2(1-\beta^2)}\right)-\ln 2[2(1-\beta^2)^{1/2}-1+\beta]+1-\beta^2\right\} \tag{9.39}$$

where $T = mc^2(1/(1-\beta^2)^{1/2} - 1)$ is the relativistic kinetic energy of the electron, $\bar{J}$ is the average ionization potential of the absorber, and $\beta = V/c$ is the ratio of the velocity of the electron to the velocity of light.

Although electron collisions are the main cause of energy loss of heavy charged particles, electrons themselves are generally not used in radiation therapy because of their limited range.

ENERGY LOSS OF PHOTONS

The loss of energy of photons such as x rays and γ rays occurs by three important processes: (1) Compton scattering, (2) photoelectric absorption, and (3) pair production.

In *Compton scattering* the photon collides with an electron as shown in Fig. 9.14. If the photon energy is large compared with the electron binding energy in the atom the electron can be considered to be initially at rest. In the collision energy is transferred from the photon to the electron. Following the collision the electron has a finite momentum and the photon has a lower frequency, since the energy of a photon is $E = h\nu$, where ν is frequency and h is Planck's constant.†

In *photoelectric absorption* the photon enters an atom and gives its energy to an electron. The electron then leaves the atom with a kinetic energy equal to that of the photon less the original binding energy of the electron to the atom. With the subsequent readjustment of the electrons in the atom, further energy is radiated in the form of atomic x rays and Auger electrons, as discussed earlier. These x rays generally have sufficient energy to leave the absorption material to which the particular atom belongs, although they frequently undergo Compton scattering on the way out. The Auger electrons lose their energy as charged particles passing through matter and thereby cause local radiation damage. The probability of an incident photon

† $h = 6.6256 \times 10^{-34}$ J/sec. For example, a visible photon with a wavelength of $\lambda = 0.5\ \mu$m (green light) has a frequency of $\nu = 6 \times 10^{14}$ Hz and a quantum energy of approximately 4×10^{-19} J.

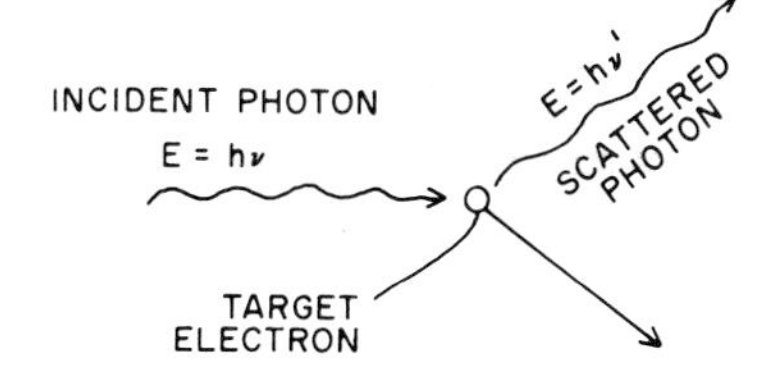

FIG. 9.14 Compton scattering of a photon from an electron at rest.

losing energy through photoelectric absorption is greatest for low-energy photons, particularly just above thresholds for x-ray ionization of the irradiated material. Therefore, collimated x rays or low-energy γ rays can be used to cause radiation damage to a tumor.

The *pair production* theory is one of the most striking successes of modern physics. In 1928 Dirac used the Einstein relation $E = mc^2$ to obtain a quantum mechanical description of a charged particle in an electromagnetic field. He showed that his quantum mechanical wave equation required that the total energy of the particle be written in the form

$$E = (m_0{}^2c^4 + p^2c^2)^{1/2}$$

where m_0 is the rest mass of the particle, p is its momentum, and c is the velocity of light, and where *both positive and negative roots are necessary for consistency*. The positive root permits the total energy of an electron to have any value from its rest energy m_0c^2 to $+\infty$. The negative root permits any value from $-m_0c^2$ to $-\infty$. Systems tend to approach minimum energy, in this case $-\infty$. Since electrons do not seem to be doing this, Dirac proposed that all of the negative energy electron states are already occupied in our part of the universe and there is no room for additional negative energy electrons. This sea of negative energy states lies $-m_0c^2$ and below whereas the electrons with zero kinetic energy occupy energy states $+m_0c^2$, as sketched in Fig. 9.15. According to Dirac's expression, stable particles

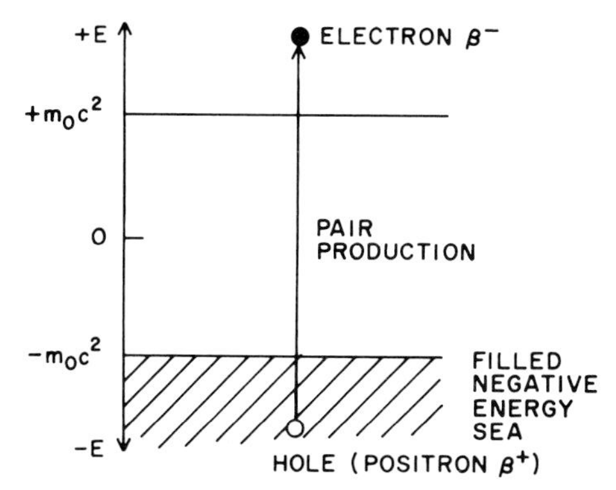

FIG. 9.15 Energy level diagram of pair production.

cannot exist in the "forbidden" energy gap between $+m_0c^2$ and $-m_0c^2$. Electrons can make the transition from negative to the positive energy states, but the reverse cannot occur. The magnitude of the energy gap is $2m_0c^2$ or 1.022 MeV. A γ ray with energy greater than this can impart sufficient energy to an electron in a negative energy state, which then appears in the upper part of Fig. 9.15 as a positive energy electron.† The absence of an electron in the sea of negative energy states is a "hole" that behaves as a positive charge since it represents the lack of a negative charge. Furthermore, since it represents the lack of negative energy it behaves as if it had positive energy. This "hole" is the *positron*, β^+, a particle equal in mass to the electron but with a positive charge. This process of the conversion of a γ ray into a positron and an electron is called *pair production* and, of course, can occur only for γ rays with energy in excess of 1.022 MeV. Such pair production can occur only near an atomic nucleus, in order to satisfy both energy and momentum conservation, and the pair production rate is proportional to the density of nuclei in the medium through which the γ rays pass. When a positron slows down it mutually annihilates with an electron to give *two* γ rays of energy 0.5 MeV in opposite directions.

We have now concluded our brief tour of photon absorption mechanisms. The sum of these three effects causes an attenuation of the incident photon beam. The intensity I of any transmitted beam is given by

$$I = I_0 e^{-\mu x} \tag{9.40}$$

where I_0 is the incident intensity, x is the distance, and μ is called the *linear attenuation (or absorption) coefficient*. Another useful term is the *mass attenuation coefficient*, μ_m, which is the linear attenuation coefficient divided by the density of the absorber μ/ρ. An *atomic attenuation coefficient*, μ_a is the product of the mass attenuation coefficient and the atomic weight A divided by Avogadro's number N_0, and ρ is the density of the medium

$$\mu_a = (\mu_m A/N_0) = (\mu A/\rho N_0) \tag{9.41}$$

A molecular attenuation coefficient can be calculated knowing the chemical compound and adding the atomic attenuation coefficients of the constituents.

A summary of the three major contributions to the linear attenuation coefficient is shown in Fig. 9.16. The three processes have one important common feature—they all involve the transfer of photon energy to charged

† If we rewrite the Dirac expression for positive energy states in the form

$$E = m_0c^2[1 + (p/m_0c)^2]^{1/2}$$

and expand binomially for $p \ll m_0c^2$ we find the total energy to be $E = m_0c^2 + p^2/2m_0 + \cdots$. The first term is the well-known Einstein energy of the rest mass and the second term is the Newtonian kinetic energy.

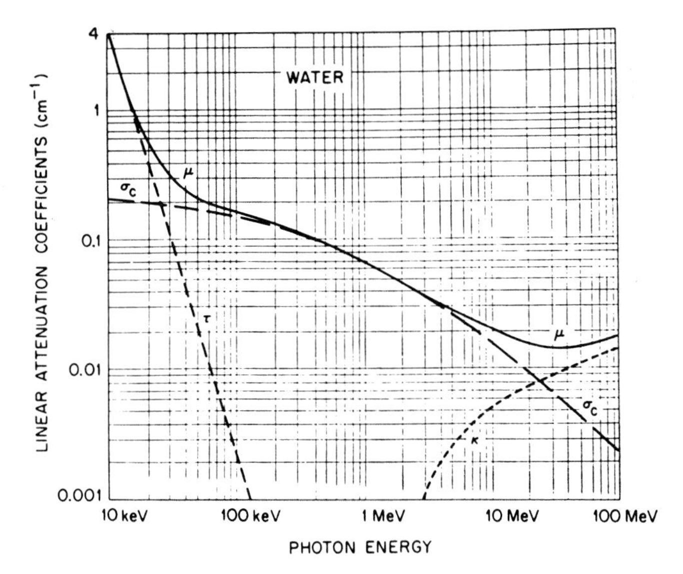

FIG. 9.16 Behavior of the linear attenuation coefficient with energy for photons interacting with water. τ, photoelectric effect; σ_c, Compton scattering; κ, pair production. [From Grodstein (1957).]

particles. It is the charged particles that are detected for confirmation of the theories and, from a medical physics viewpoint, it is the charged particles that cause radiation damage to cells irradiated with x rays or γ rays.

ROENTGENS, REP, RAD, REM, AND RBE

In early experiments on the effects of x rays in air the amount of ionization was measured simply by putting an electric field across the exposed air and recording the current of ions. From this type of measurement the *roentgen* unit of radiation, R, was established. This involves a quantity of free electric charge in a mass of air. The unit of exposure, 1 R, is the exposure of x rays or γ rays obtained in a mass of 1 cm^3 of dry air in standard state, i.e., 1.293×10^{-3} g, when the ions produced have a charge of 1 esu† of either sign. Since

† Most of the older literature on cell damage and radiotherapy is in cgs units and will be followed in this chapter to facilitate reading of the references. Conversion relations to mks units are:

1 joule = 10^7 ergs
1 coulomb = 2.998×10^9 esu
e (electronic charge) = 4.803×10^{-10} esu = 1.602×10^{-19} coulomb
1 roentgen = 1 esu/cm^3 air (STP) = 2.58×10^{-4} coulomb/kg air (STP)
1 rad = 100 ergs/gm = 10^{-2} joules/kg
1 eV = 1.602×10^{-12} ergs = 1.602×10^{-19} joules

the charge on an electron is 4.80×10^{-10} esu, 1 R corresponds to the formation of 2.08×10^{9} ion pairs per cm^3 of air or 1.61×10^{12} ion pairs per g. Because the ionization potential varies for different substances, an average must be agreed upon to compare irradiation doses. Rather than specifying an ionization potential, which used to be taken as an average of 33.9 eV for air, the roentgen has been redefined as that amount of x or γ radiation that deposits precisely 86.9 erg in 1 g of dry air at standard temperature and pressure.

It is seen in Fig. 9.16, however, that the rate of absorption of a 100-keV x ray is about 10 times larger than that from a 100-MeV γ ray in air. For medical-biological applications it is more appropriate to introduce a quantity that represents absorbed dose in body tissue. In older literature this was called the *roentgen equivalent physical* (rep), and was taken as the x or γ radiation that liberated 97 erg of energy per gram of body tissue. The modern term is called the *rad* and is the absorbed dose of any high energy radiation which is accompanied by the liberation of 100 erg/g (0.01 J/kg) of energy in the absorbing material. There is still a problem in comparison because, for example, the same number of rads of x rays and neutrons may produce strikingly different biological effects. This difference is taken into account by the term *relative biological effectiveness* (RBE), which is defined as the ratio of the absorbed dose of x or γ rays of about 200 keV energy to the absorbed dose of some other radiation that produces the same biological effects or *roentgen equivalent man* (rem), where

$$\text{Dose in rem} = (\text{RBE}) \times (\text{dose in rad})$$

TARGET THEORY

The theory of "kill" or "survival" requires an operational definition of these terms. Since most quantitative experiments are done on cell cultures, cessation of reproduction means death, or at least nonviability. There are other types of biological damage, such as mutations, but this is a rare event compared to death. Also there are recovery events in which some cells apparently recover or are restored to viability with time. These factors lead to the rather complex kinetic formulations (Dienes, 1966) of radiation damage. We will treat only one case here, whose formulation involves probability theory. When this case is understood, other cases in the literature will be easily understandable.

Cell culture studies are important because of the dilution possibilities. After irradiation of a colony of high concentration of viable cells ($10^5 - 10^6/cm^3$) it can be diluted so that the surviving fraction can be measured over three or four orders of magnitude with errors within 10%.

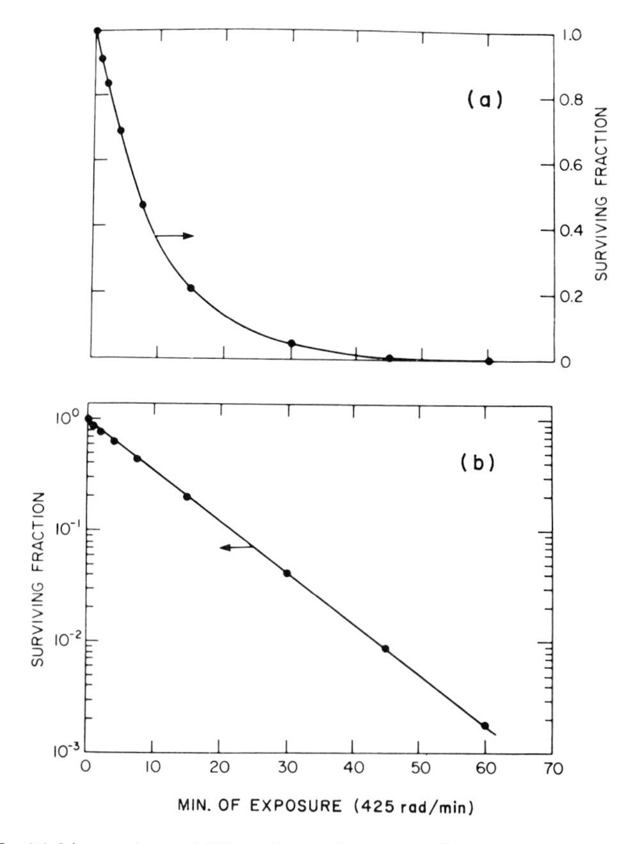

FIG. 9.17 (a) Linear plot and (b) semilogarithmic plot of the x-ray survival of a strain of haploid yeast cells. [From Elkind and Whitmore (1967); based on data by Wood (1953).]

In contrast, animal experiments are limited to the range of about 5% to 95% survival. Thus, in cell studies theory can be compared over a logarithmic range while only a linear range is possible in animal studies. An illustration of this is shown in Fig. 9.17. This figure shows that the logarithm of the survival rate is a linear function of the radiation dose. This is a common but not the only type of curve. Some cells exhibit a shoulder on the curve before the linear slope. Others exhibit an "s" or sigmoid shape, which we will not consider here.

Suppose we consider that part of the cell is sensitive to irradiation and designate this sensitive volume as V. If the radiation creates I inactivating events per unit volume of the cell randomly distributed then the average number of "hits" on the sensitive part of the cell is IV. For a fixed dosage rate I is proportional to exposure time. If we now ask the probability $P(n)$

of exactly n hits occurring in the sensitive volume V of a single cell, the answer is given by the Poisson probability distribution (Appendix C) as

$$P(n) = e^{-LV}(IV)^n/n! \tag{9.42}$$

The probability that a cell is not hit is $n = 0$ and, inserting this into the distribution and noting that $0! = 1$, we obtain

$$P(0) = e^{-IV} \tag{9.43}$$

Hence, if N_0 cells are irradiated, the number N receiving no hit after a dose I is

$$N = N_0 e^{-IV} \tag{9.44}$$

or

$$\ln(N/N_0) = IV \tag{9.44'}$$

which leads to a logarithmic survival curve similar to that shown in Fig. 9.17.

We can now extend this probability approach to see what happens if we eliminate two constraints, and say that (1) more than one hit on a sensitive volume is required and, (2) there is more than one sensitive volume per cell required to be inactivated. Consider the first situation, called the *multihit case*. If n hits are required to inactivate the volume V, the probability of survival, i.e., those that have received not more than $n - 1$ hits, is the sum of the probabilities

$$P(0) + P(1) + P(2) + \cdots P(n-1)$$

or

$$N = N_0 e^{-IV}\left(1 + IV + \frac{(IV)^2}{2} + \cdots + \frac{(IV)^{n-1}}{(n-1)!}\right)$$

which may be written as a survival fraction

$$N/N_0 = e^{-IV} \sum_{k=0}^{k=n-1} (IV)^k/k! \tag{9.45}$$

It is also convenient to define a complementary kill fraction N^*/N_0 as the fraction of those which after a dose I have received at least n hits.

$$N^*/N_0 = 1 - (N/N_0)$$

which, by substitution of Eq. (9.45), is

$$N^*/N_0 = 1 - e^{-IV} \sum_{k=0}^{k=n-1} (IV)^k/k! \tag{9.46}$$

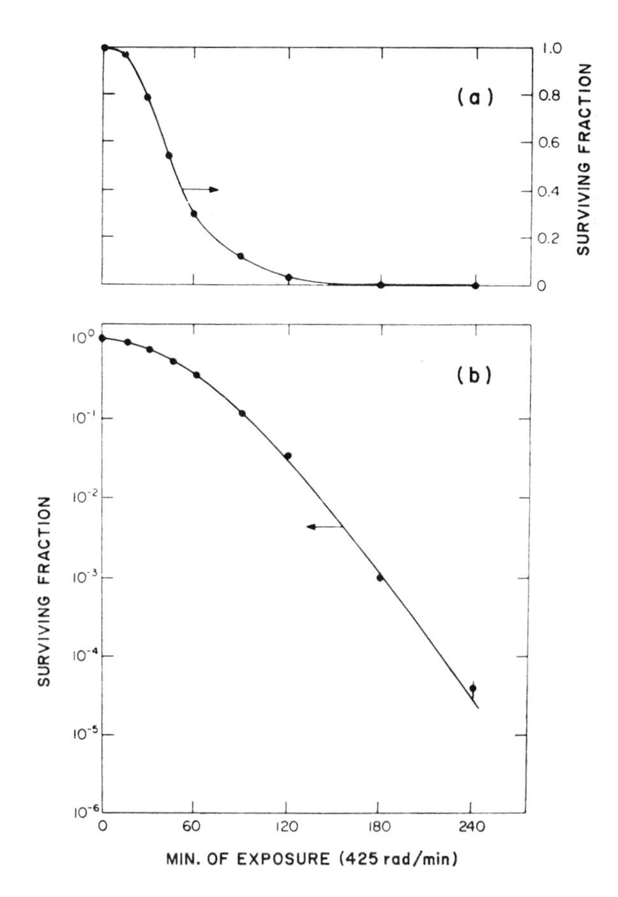

FIG. 9.18 (a) Linear plot and (b) semilogarithmic plot of the x-ray survival of a diploid strain of yeast cells. [From Elkind and Whitmore (1967); based on data by Wood (1953).]

It is seen in Eq. (9.45) that when I becomes large the exponential term $\exp(-IV)$ dominates, so that, although the survival curve does not begin logarithmically, it becomes so at large doses. This behavior leads to an initial shoulder on the survival curve. An example of such a shoulder is shown in Fig. 9.18, although the interpretation is not purely multihit.

Consider now the case of multitargets. Start with two. For example, two genes in a cell, both of which must be hit to inactivate it. The probability that one is not hit is e^{-IV}, and therefore the probability that one is hit at least once is $1 - e^{-IV}$. The probability that both are hit at least once is the product of the two independent events, or

$$\text{probability that both are hit} = (1 - e^{-IV})^2$$

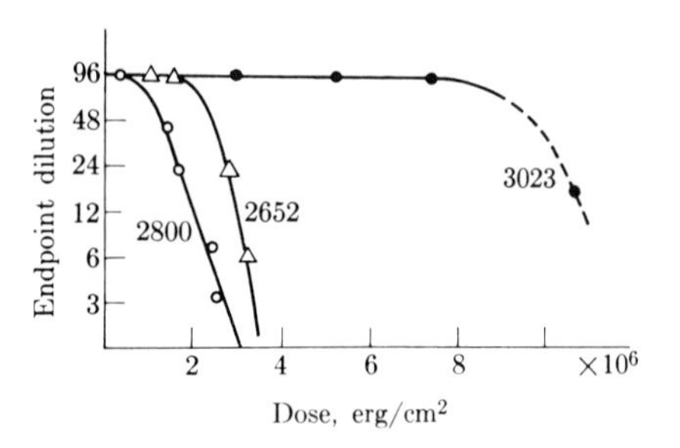

FIG. 9.19 The inactivation of hemagglutinin of influenza virus by ultraviolet light of three different wavelengths. The curves drawn through the experimental points are based on the following numbers of targets: 2800 Å, 5 targets; 2652 Å, 40 targets; 3023 Å, 100 targets. The number of targets varies with wavelength because of the differing sensitivities of parts of the virus to different wavelengths. [From Tamm and Fluke (1950).]

where it has been assumed that both sensitive volumes are equal. In the cell culture experiments it is the survivors that are counted and the probability of survival is one minus the probability that the two hits occur, or

$$\text{surviving fraction} = N/N_0 = 1 - (1 - e^{-IV})^2$$

We can generalize this to p targets within a cell by writing

$$N/N_0 = 1 - (1 - e^{-IV})^p \tag{9.47}$$

The high-dose slope of $\ln(N/N_0)$ versus I of Eq. (9.47) can be obtained by numerical substitution, and curve fitting to experimental data can be performed to obtain significant parameters. By inspection one can see, however, that for low doses very few cells have received p hits and the survival fraction should be approximately constant, with the value $N/N_0 = 1$. At high doses almost every cell that has not received the required p hits to kill it has received $p - 1$ hits, so that one more hit will kill it. Thus at high doses this multihit case reduces to the single hit case and yields a constant slope. An example of this type of behavior is seen in Fig. 9.19.

As an extension of this formulation we may write a generalized equation for the multitarget and multihit surviving fractions as

$$N/N_0 = 1 - (N^*/N_0)^p$$

which, by substitution of Eq. (9.46), is

$$N/N_0 = 1 - \left[1 - e^{-IV} \sum_{k=0}^{k=n-1} (IV)^k/k!\right]^p \tag{9.48}$$

Tables of the Poisson probability function (Thorndike, 1926) exist and the usual procedure is to plot curves with various p's and n's and to try to fit

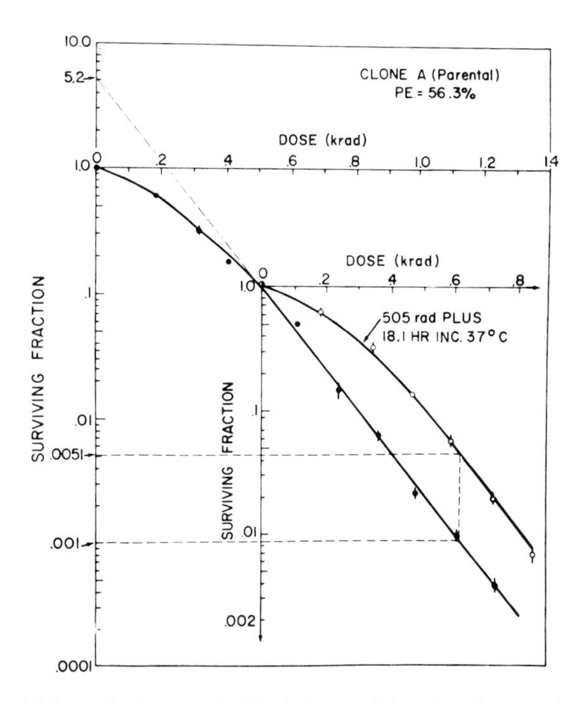

FIG. 9.20 Initial survival curve (solid circles) and fractionation survival curve (open circles) for "clone A" cultured Chinese hamster cells. The fractionation survival curve was determined 18.1 hr after 505 rad and the curve is normalized to the survival corresponding to 505 rad. [From Elkind and Sutton (1960).]

expressions of the form of Eq. (9.48) to the experimental data. Although sometimes simple matching is revealing, the actual situation in many systems of interest is highly complex. This is because some cells that have not received sufficient hits to be killed can recover with time. A damaged cell that recovers is said to have received *sublethal* damage.

Experiments have been performed that show that there is an age dependence of survival of cells. In a given culture of cells of mixed age it cannot be ascertained which are the survivors. However, there is general agreement that in the killing part of the dose curve all surviving cells have received some damage.

An example of a recovery curve is seen in Fig. 9.20. In this experiment the irradiation was interrupted for 18 hr and then begun again. It is seen that a new shoulder develops, which indicates that some cells again require multihits to be killed. These multihits accumulate and the former killing slope is again achieved. If there had been no recovery or cell division within 18 hrs the second radiation kill slope would have superimposed on the first slope.

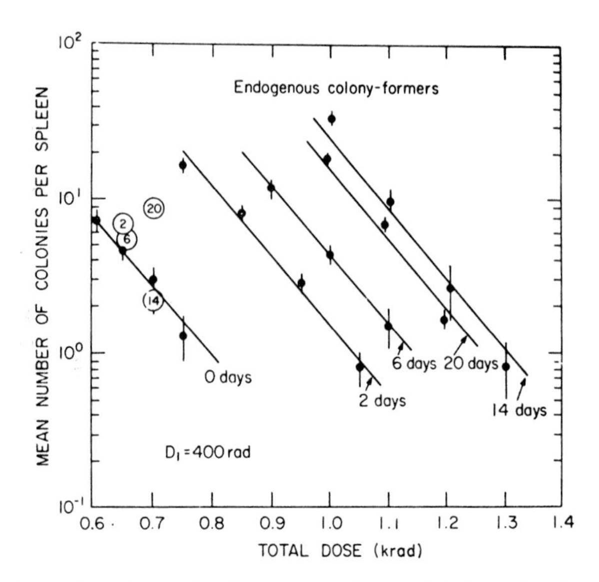

FIG. 9.21 Dose dependence of endogenuous spleen nodule formation in male mice as a function of time after an initial dose of 400-rad Co^{60} radiation. Large open circles refer to the number of colonies obtained after a single exposure. [From Till and McCulloch (1964).]

An example of a series of these curves for different recovery times of endogenous spleen nodule formation cells of mice treated with γ rays from Co^{60} is shown in Fig. 9.21. It is seen that due to recovery plus division of surviving cells the fractionation curves are displaced in the direction of higher total doses.

LINEAR ENERGY TRANSFER (LET)

From our earlier discussion of the mechanisms of energy loss of charged particles and photons, it is obvious that a charged particle has a short range, and therefore its energy is dissipated in a small region. Energy loss rate is also called *linear energy transfer* or LET. It is therefore expected that x rays and γ rays have low LET while α particles will have high LET. Linear energy transfer includes both electron excitation and ionization of atoms in the absorbing material and is expressed as thousands of electron volts of energy per micron of path length (keV/μm) of the particular absorber. Some values of LET in water of a number of types of radiation are shown in Table 9.2.

Neutron damage has been mentioned briefly. Although being a neutral particle it does not interact directly with electrons in the absorbing material, the knock-on that it produces is a heavy ion that has a high LET. Thus,

TABLE 9.2[a]

Radiation	LET (keV/μm) in water
3-MeV x ray	0.3
Co^{60} γ rays (1.2 MeV)	0.3
250-keV x ray	3.0
0.6-keV β from H^3	5.5
Recoil p from fission n	45
5.3-MeV α from Po	110

[a] Data from Casarett (1968).

appropriate interactions from external machines such as one that accelerates deuterons and then permits them to collide with a tritium target can produce neutron beams. Some heavy radionuclides such as californium-252 spontaneously emit fission spectrum neutrons. In some cases the neutrons are accompanied by recoil protons. These emitters can be encapsulated in hollow needles and inserted in tumors to take advantage of the large LET of the neutron and/or recoil proton.

Another particle under serious consideration for use in radiotherapy is the π^- *meson* (pion). This is a subnuclear particle with the charge of an electron but with a mass 273 times greater that decays into first another type of meson and then into an electron. Because of its large mass, however, a large amount of energy is required to create the π^- meson. This is done

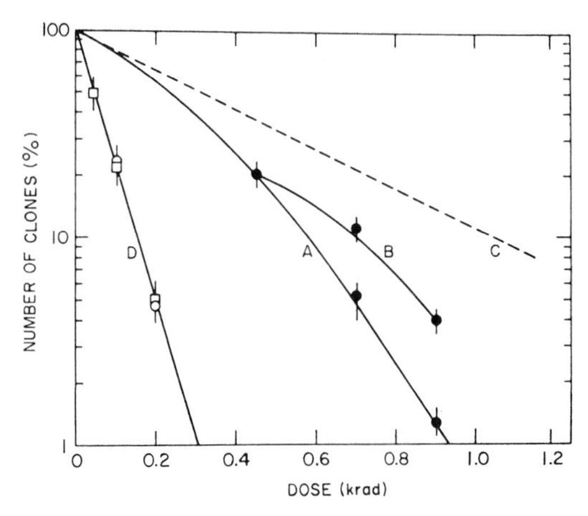

FIG. 9.22 Survival of cultured human kidney cells to x-ray and α-particle irradiation. A, Single 200-kv x-ray doses; B, 450-rad x ray plus 250-rad or 450-rad 12 hr later; C, initial slope of curve A; D, α-particle dose. Circles are single exposures and squares are two equal doses 12 hr apart. [From Barendsen (1962).]

by large accelerators that fire fast protons onto stationary protons. Thus, a large cost is involved and experiments in clinical use have not yet begun.

The killing rate of high-LET α particles is expected to be higher than that of low-LET x rays because many hits can be imparted to a single cell by one α particle and a cell so struck would not be likely to recover. This difference between damage from α particles and x rays is shown in Fig. 9.22. The α particle curve shows no shoulder and hence no recovery from the damage while the x-ray dose curve is similar to that of Fig. 9.20. This figure also illustrates the need for the RBE (relative biological effectiveness) term defined earlier.

MEAN LETHAL DOSE AND SENSITIVE VOLUME

The previous discussion on target theory showed that at high dosages I a logarithmic survival curve results of the form

$$N = N_0 e^{-IV} \tag{9.44}$$

where N_0 is the original number of cells and N is the number receiving no hit. From data within this logarithmic region estimates can be made of the size of the sensitive volume V that, because of the form of Eq. (9.44), must be proportional to the logarithmic decay constant.

An average energy required in a biological medium for an irradiating photon to form one ion cluster is about 110 eV (see Johns and Cunningham, 1977, Table XVIII-3), and we will assume that the formation of one ion cluster in a sensitive volume V will destroy a cell. Let ρ be the density of the cell and recall that 1 rad deposits 100 erg/g. Then the energy released by I rads in a biological tissue of density ρ is

$$\begin{aligned} \text{Energy released by } I \text{ rads} &= I \times 100 \times \text{erg/g} \\ &= I \times 100 \times \rho \text{ erg/cm}^3 \\ &= I \times 100 \times \rho \text{ eV}/1.6 \times 10^{-12} \text{ cm}^3 \end{aligned}$$

Because we are dealing with such a small volume, we express it in cubic microns $(10^{-6}\text{m})^3$. In these units $1 \text{ cm}^3 = 10^{12}\ \mu\text{m}^3$. The energy released in a sensitive volume V expressed in cubic microns is the product of the above energy and V, or

$$\text{Energy released in sensitive volume } V = \frac{I \times 100\, \rho V 10^{-12}}{1.6 \times 10^{-12}}\, eV$$

where V is in cubic microns.

$$\begin{matrix}\text{Numer of ion clusters per sensitive} \\ \text{volume } V \text{ with 110 eV/ion cluster}\end{matrix} = \frac{I \times 100 \rho V}{1.6 \times 110}$$

At this point the mathematical convenience of an exponential behavior becomes evident. If we have a function of the form of Eq. (9.44) and write it as

$$N = N_0 e^{-I/I_0}$$

since the exponential must be dimensionless we can evaluate the ratio N/N_0 when $I = I_0$:

$$N/N_0 = 1/e = 0.37$$

I_0 is therefore the usual exponential decay constant. Since it represents a mean or average value of an exponential in the logarithmic survival region of radiation cell death, it is called the *mean lethal dose*, i.e., the dose required to reduce a cell population to 37% of its original value.

Returning to our estimate of sensitive volume V, it is clear that if we assume a density of about 1 g/cm^3 and that one ion cluster in the sensitive volume will kill the cell, the mean lethal dose can be taken from a graph such as Fig. 9.17 and V can be estimated.

It is obvious that because of the exponential behavior of cell survival a question such as "How many rads will it take to kill all the cancerous cells in a tumor?" is meaningless. Obviously the mathematics predicts an infinite dose. However, one may readily calculate the number of rads required to have a survival fraction of only one in 10^{12}. With such a low number remaining the body's natural defenses would probably account for the remainder.

HYPOXIA AND SOURCE SELECTION

One very important finding by radiation biologists is the fact that cells with reduced oxygen, *hypoxia*, are more resistant to radiation damage than those with normal oxygen content. The resistance factor varies with both dose and dose rate, and an example of the survival of HeLa cancer cells irradiated with x rays while in atmospheres of oxygen and nitrogen is illustrated in Fig. 9.23. Tumors usually have an insufficiency of blood vessels and the tumerous tissue is therefore hypoxic compared to healthy tissue.

This oxygen effect is most pronounced at low LET values. Although the chemistry is not completely understood [see, for example, reviews by Kiefer (1975) and by Adams (1973)] it is known that it is minimized by the use of high LET radiation.

There are many factors that relate to the selection of a suitable radiation source to treat a given cancerous condition. We have discussed only the physical ones and have not even considered the variety of other effects, such

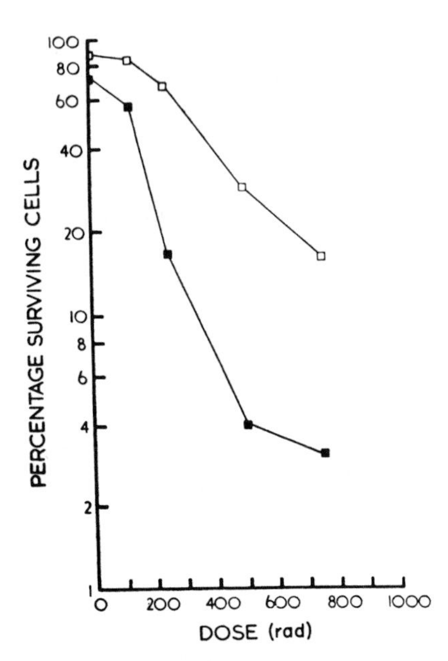

FIG. 9.23 Comparison of survival of HeLa cancer cells irradiated in (lower curve) oxygenated and (upper curve) nitrogenated media, expressed as a percentage of nonirradiated controls. [From Weiss (1960).]

as chemical, kinetic, thermal, etc. Bearing in mind only the physical considerations, we may summarize the preceding sections into important parameters in radiation source selection: (1) a high LET radiation is desirable to overcome the oxygen effect. (2) damage to healthy tissue should be minimized, and (3) dose rate and frequency of doses must be optimally selected. This latter option will be discussed in the next section but we may now add to items (1) and (2).

The desirability of high LET radiation is obvious, but the achievement is expensive. The data shown in Fig. 9.22 for α particles was taken from a thin layer of cells because the range of such particles from radionuclides is only a few microns. Therefore efforts are being directed toward the use of accelerators both for light ions and for neutron interactions and eventually mesons. These machines are both expensive and complicated to operate and are therefore generally not available to clinics. Implantation of radioactive nuclides is regaining popularity and we shall discuss this briefly later.

The minimization of damage to healthy tissue has a mechanical solution. For example, if ionizing radiation is to be directed from an external source to an internal tumor, because of absorption the tissue between the source and the tumor will receive a larger dose than the tumor itself. To correct this the patient is placed in a position in which the tumor is at the center of rotation of the source. Thus, as the source circles the patient both axially and azimuthally the tumor is continuously irradiated while any other part of the body receives only a fraction of the dose, depending upon its radial distance from the tumor.

FRACTIONATION OF RADIATION DOSAGE

The discussion on cell recovery has important implications for radiotherapy. If the cells of a tumor are to be killed by radiation from an external source or from implantation of radioactive needles, what is the optimum dose rate? The radiation is only partly attenuated in the tumor and proceeds to damage healthy body tissues as well. Therefore, the lower the dose rate the less healthy tissue is damaged but, since tumor cells may recover, the greater total dose must be accumulated by the patient. Therefore it is desirable to choose doses and intervals to allow the patient to recover between treatments while continually damaging the tumor.

A calculation of significance would be the *survival decrement*, i.e., the killing slope, per unit dose versus the dose rate. This calculation is similar to that for the creation of an isotope in a reactor when the half-life is short enough so that there is a loss of created isotope while new isotopes are being formed. This model was applied to cell survival by Lajtha and Oliver (1961). Their experiments on *vicia* seedlings led them to assume that the recovery of the surviving cells from a first dose followed an exponential behavior with a relaxation time of $1/\mu$ where μ is a constant to be determined. This assumption was later justified by Oliver (1964). If I_1 is an initial photon dose that damages the surviving cells, the effective dose I_E given to these cells at any time t is

$$I_E = I_1 e^{-\mu t} \tag{9.49}$$

Thus only at $t = 0$ does the effective dose equal the dose given at that time. At any time thereafter, because some cells damaged sublethally are recovering, the effective dose is less than that delivered up to that time. If the rate of delivery of the dose is d rad per unit time the effective dose I_E given in time t will be

$$I_E = d \int_0^t e^{-\mu t}\, dt = (d/\mu)(1 - e^{-\mu t}) \tag{9.50}$$

We can obtain from Eq. (9.50) the effective dose that is used in the survival fraction equations such as Eq. (9.47). A plot of data from an experiment can yield $1/\mu$, the relaxation time, by means of slope analysis. Such an experiment was performed by Berry and Cohen (1962) on leukemia cells in mice for two

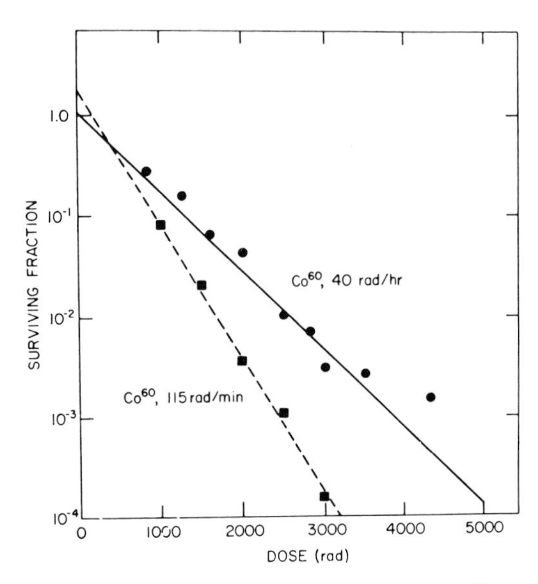

FIG. 9.24 Surviving fraction of mouse leukemia cells at two different dose rates. The dashed and solid lines are calculated using the effective dose rate of Eq. (9.50) with μ, the recovery constant, as the only adjustable parameter. [From Berry and Cohen (1962). Reproduced by kind permission of the British Journal of Radiology.]

different dose rates. The results are shown in Fig. 9.24, in which the solid and dashed lines are based on the Lajtha and Oliver (1961) equation for the single hit model. The agreement is quite good, although Elkind and Whitmore (1967) feel that the degree of agreement is possibly fortuitous.

The problem facing the radiotherapist is now more clearly defined. If data exist for the number of rads of a given type of radiation required to kill a particular cancer cell, the therapist must decide how to administer the required dose to a patient without seriously damaging the patient's well-being. The body tolerance is known and the radiation may be delivered in a long (protracted) dose or in a series of short, high-level (acute) doses. Equation (9.50) indicates that, depending on the dose rate, the effective dose is not the delivered dose, and suitable corrections must be made.

Suppose low-dose radium technique has been used for implantation in a tumor for continuous radiation for a time t. Now, however, we wish to replace this low-dose radium by a high-dose source that will be used periodically N times. How are the times and number of doses to be related to the low-dose time? Note that the total dose required will be different because of the cellular recovery factor. Liversage (1969) has used Eq. (9.50) and an averaging assumption to obtain an approximate equation for such a relationship. Figure 9.25 shows a hypothetical survival–dose curve of the form

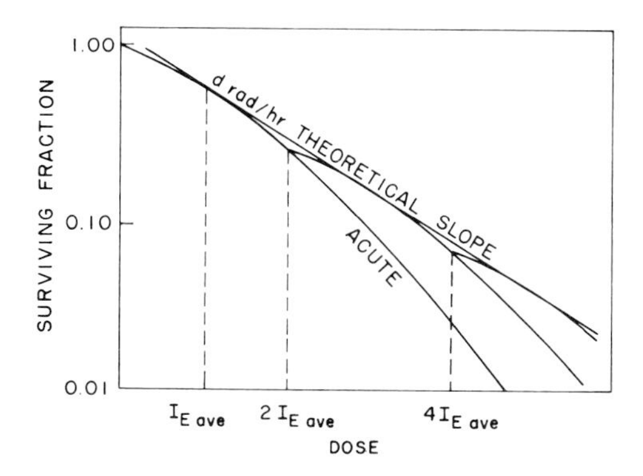

FIG. 9.25 The average slope of the protracted cell survival curve is assumed to be equal to the slope of the acute cell survival curve at point $I_{E\,ave}$ and this is assumed to be equal to the average slope of the acute curve from 0 to $2I_{E\,ave}$. Thus protracted irradiation at d rad/hr is assumed to be as effective as acute irradiation given in fractions of magnitude $2I_{E\,ave}$. [Redrawn from Liversage (1969). Reproduced by kind permission of the British Journal of Radiology.]

of Fig. 9.20. The line tangent to all the curve shoulders is the slope of the protracted cell survival curve. At point $I_{E\,ave}$ it is equal to the slope of the acute dose rate curve. From the appearance of this curve it is reasonable to assume that this slope also equals the average slope of the acute curve from 0 to $2I_{E\,ave}$. In this way the protracted irradiation at d rad/hr is assumed to be as effective as acute irradiation given in fractions of magnitude $2I_{E\,ave}$. By definition, $I_{E\,ave}$ is the average of I_E over the exposure time t. From Eq. (9.50)

$$I_{E\,ave} = (1/t)\int_0^t (d/\mu)(1 - e^{-\mu t})\,dt = (d/\mu)[1 - (1/\mu t)(1 - e^{-\mu t})] \quad (9.51)$$

The approximation developed above is equivalent to the following: if the fractionated dose is given in units of $2I_{E\,ave}$, the mean fractional reduction in the surviving fraction per rad is the same as it is for protracted irradiation given at d rad/hr for t hours. Therefore, if the same total dose is given in both cases, the same biological effects should result. Thus, N acute doses of $2I_{E\,ave}$ rad should produce the same biological effect as protracted irradiation at d rad/hr for t hours for the same total dose. That is,

$$2I_{E\,ave}N = td$$

where N is the number of acute doses. Upon substituting this into Eq. (9.51) Liversage (1969) obtains the relationship between the number of acute doses N and the duration and relaxation time of the equivalent protracted dose:

$$N = \frac{\mu t}{2[1 - (1/\mu t)(1 - e^{-\mu t})]} \quad (9.52)$$

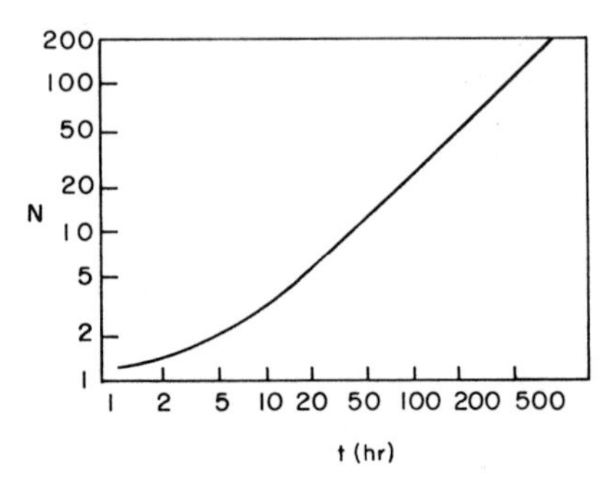

FIG. 9.26 Graph of the number of dose fractions N versus time t of continuous irradiations with $\mu = 0.693/1.5\ \text{hr}^{-1}$ from Eq. (9.52). N is the number of fractions in which acute irradiations should be given in order that the radiation "wasted" due to intracellular recovery between fractions will be equal to that "wasted" during a protracted irradiation lasting t hr. [Redrawn from Liversage (1969). Reproduced by kind permission of the British Journal of Radiology.]

This formula indicates that the radiation lost due to recovery between N fractions of acute irradiations equals the radiation lost due to recovery from a protracted irradiation of length t hours. It is clear that the larger the shoulder on the survival curve the greater will be the wasted irradiation. Liversage tested this formula on seven different cells and found agreement to within 10%. It is interesting to note that this formula is independent of both dose and dose rate d, and depends only on the product μt. Therefore results can be scaled using a universal plot of Eq. (9.52). Experiments have shown that recovery times for mammalian cells are of the order of 1.5 hr. Figure 9.26 is a plot of Eq. (9.52) with $\mu = 0.693/(1.5\ \text{hr})$. When $\mu t \gg 1$, Eq. (9.52) reduces to $N \approx \mu t/2$, which is $\mu \approx \frac{1}{2}\ \text{hr}^{-1}$ in the case shown in Fig. 9.26. As a note of caution it should be recalled that not only is this formula an approximation, but there are some doubts that the fundamental assumption of a single exponential recovery rate is valid.

PHOTON BEAM ATTENUATION

When a beam of x rays or γ rays passes through matter the beam is attenuated by the processes described in an earlier section. Whether a tumor is irradiated by an external beam or from internally planted probes of γ-ray emitters the problem is to determine what fraction of the beam strikes the tumor cells. This problem is an extremely complex one from a theoretical viewpoint and experimental results are generally used. We have seen in Eq. (9.40) that the sum of the attenuation effects on a beam is expressed as an exponential attenuation coefficient μ. If the irradiation originates from a point source the intensity per unit area at a distance r from the source will be

$$I/I_0 = e^{-\mu r}/r^2 \tag{9.53}$$

where I_0 is the total intensity of the source. This relation does not express all of the parameters. For example, there is a multiplicative buildup factor B, which is defined as the ratio of the total absorbed energy due to primary and scattered γ rays to the absorbed energy due to the primary γ rays alone. Experiments must be performed to determine the dose contribution from scattered γ rays in which water is used to simulate the scattering due to body tissues. In these experiments a γ-ray emitter is placed at a fixed distance from a detector and the number of counts recorded in a given time interval. The emitter and detector are then submerged in water and the counts again taken. If this is done at different distances the $1/r^2$ dependence cancels and only the ratio of the delivered dose to the emitted dose is measured for each distance. Such data were taken by Meisberger *et al.* (1968) for several γ-ray

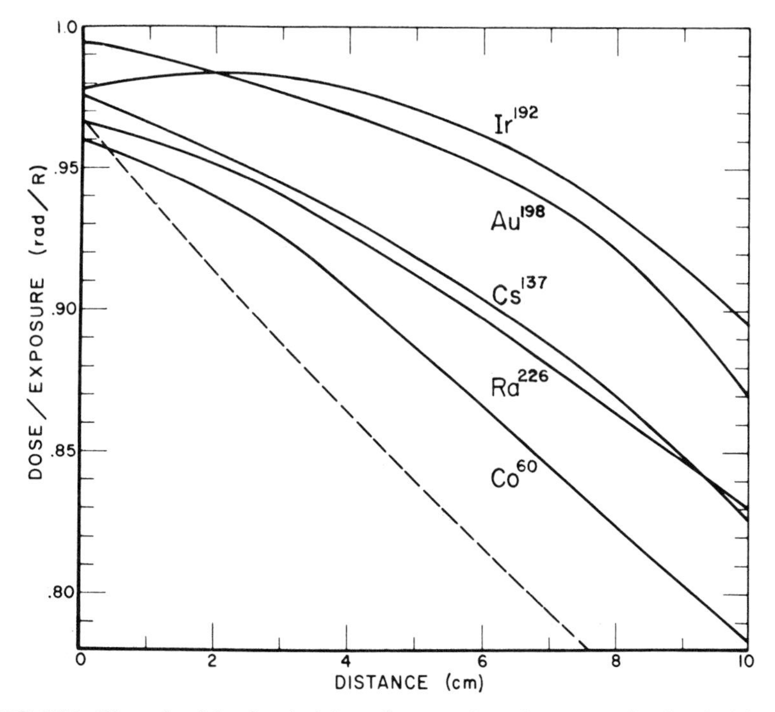

FIG. 9.27 The ratio of the absorbed dose of γ-ray emitters in water to the absorbed dose in air at different distances. The dashed line is calculated from exponential absorption with a constant buildup factor and a narrow beam attenuation coefficient. [From Loevinger (1971). Based on data of Meisberger *et al.* (1968).]

emitters and their data were expressed as coefficients of a third-order polynomial in the separation distance r. Plots of these data are given in Fig. 9.27. Attenuation for other sources such as x rays of different energies have also been obtained and must be considered when calculating doses for radiotherapy. It is seen from Fig. 9.27 that the contribution from scattering is significant. In the case of Ir^{192}, for example, the dose actually increases for the first few centimeters of separation.

ISODOSE

Another area in which physics is needed in radiotherapy is isotope implantation. Such a procedure fell into disuse some time ago because the surgeon, to avoid much personal exposure, would try to set the implants quickly and leave them for a protracted time. More recently the technique of *afterloading* has brought implantation back into favor. In afterloading a series of empty small tubes is inserted by the surgeon with great care and accuracy. Later, by means of threads through the tubes, small seeds of the isotope are drawn into place and left for either protracted or acute doses. The problem in this method is how to place the tubes and the seeds to give a constant or *isodose* to the particular geometry of a tumor. Each seed can be considered a point source at some distance from a given point in the tumor. With the above attenuation data and a $1/r^2$ dependence, how then are the tubes and the seeds in them placed to give the entire tumor nearly the same dose?

Before the advent of computers a series of arrays were calculated for regular geometric shapes and the tumor would be matched to the shape and the geometric parameters adjusted for size. The best known of these computations are those of Paterson and Parker (1934). Most radiotherapy centers are now using computer calculation of possible distributions of isotope seeds to give isodose contours. Still the problem is not a simple one if the number of surgical implantations is to be kept to a reasonable minimum. But with the computer, recovery times and fractionated doses can be also introduced and the best possible arrangement can be calculated for a tumor of arbitrary shape.

RADIOTHERAPY TREATMENT PLANNING

The design of radiation treatment for the killing of cells in the human body involves several factors of which clinical experience is an important one. The physical principles involved have largely been covered in the earlier parts of this chapter. Many important details have been omitted, however, and the interested reader is referred to the book by Johns and Cunningham (1977).

From a survival versus dose curve for the particular cell of interest the total number of rads required to kill all but a minute fraction can be determined from a curve of the form of Fig. 9.18 for the cell. This is the value for a single or acute dose. If the dose is to be fractionated, cells recover between doses and the corrected dose must be evaluated from Eq. (9.50), for which the recovery time constant μ must be known.

When a collimated beam is directed into the body the absorbed dose varies with depth (called depth dose) because of the attenuation. However, as the energy of the beam is increased the peak intensity moves inward. This apparent buildup occurs for the following reason. The electrons set in motion by the radiation have primarily a forward motion. Therefore the number of these electrons will increase with depth until a depth equal to the range of the electrons is reached. Thereafter attenuation causes a decrease of dose (energy deposited per unit time) with distance. With low-energy radiation the electron range is small and the buildup is therefore too small to be observed. An example of this effect was seen for γ rays in Fig. 9.27 for all the emitters illustrated except Ir^{192}. A similar set of curves, which is perhaps better for pedogogical purposes, is shown in Fig. 9.28 for 22-MeV x rays from a betatron, 1.25-MeV (average) γ rays from a Co^{60} source, 200-keV

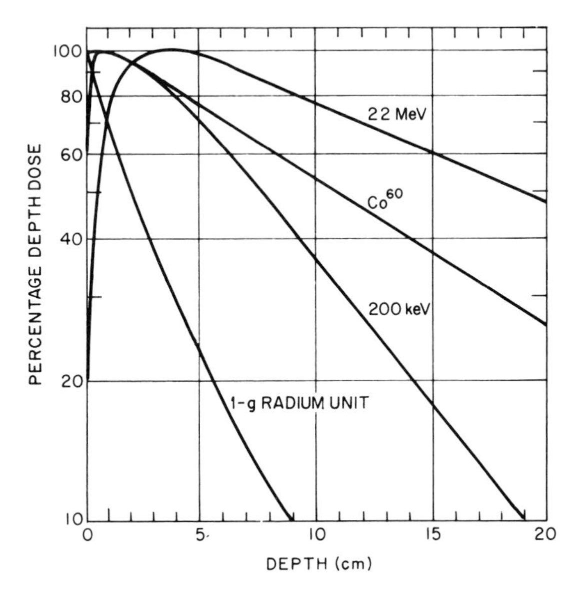

FIG. 9.28 Percentage depth dose versus dose for photons of 22 MeV, 1.25 MeV, 0.2 MeV, and 0.18 MeV. [From Watson *et al.* (1954).]

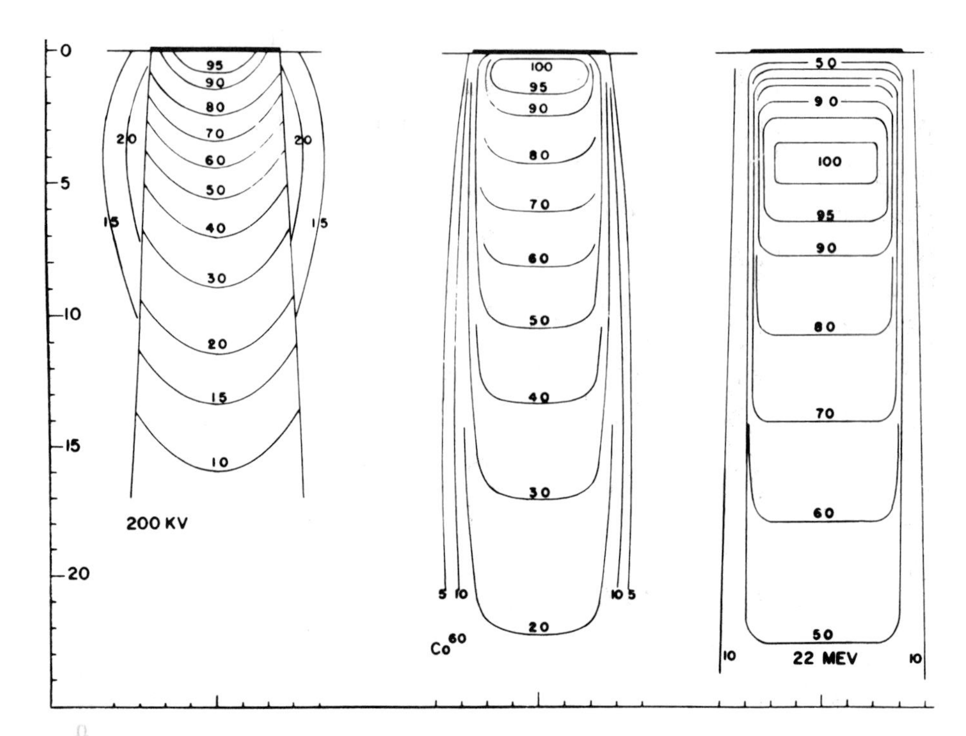

FIG. 9.29 Isodose (isointensity) loci of photons normal to the surface of energies 0.2 MeV, 1.25 MeV, and 22 MeV. The ordinate is depth in water in centimeters. The abscissa is also in centimeters to show beam spread. [From Watson *et al.* (1954).]

x rays and 180-keV γ rays from radium. (Note that these are all energetic photons but the method of production is the source of the designation x ray or γ ray). It should also be noted that the collimation of such beams is not quite parallel and that they diverge somewhat. Here it is seen that the maximum intensity of the 22-MeV photon beam in water is not reached until about 4-cm depth. The profiles, or cross sections of three of the beams (not the radium) are shown in Fig. 9.29. It is seen that the lower energy photons have isodose profiles of greater curvature because, in the greater length of the off-normal paths that parts of the beam traverse, absorption is evident.

The absorbed dose in a medium can be determined by a measurement with a small gas chamber in a dummy or "phantom" equivalent to the geometry of the human, and with reliance on the Bragg–Gray cavity theory and knowledge of the stopping power of the tissue. Suppose a medium is being irradiated with an external source of high-energy photons and a small gas-filled cavity is introduced into the medium at a given position.

The photons impart energy to the electrons in both the gas and the medium. Define the following:

E_m energy in erg/g imparted to medium by excited electrons
E_g energy in erg/g imparted to gas by excited electrons
J_g ion pairs/g of gas produced in cavity by the electron
W average energy in erg/ion pair to ionize gas.

$$S_g{}^m = \frac{E^m}{E_g} = \frac{\text{energy imparted to unit mass of medium by electrons}}{\text{energy imparted to unit mass of gas by electrons}} \tag{9.54}$$

and $S_g{}^m$ is called the *mass stopping power ratio.* Clearly $E_g = WJ_g$ and

$$E_m = S_g{}^m E_g = S_g{}^m W J_g \tag{9.55}$$

This relation is known as the Bragg–Gray formula. The gas in the cavity is usually air for which W at room temperature is approximately

$$W = 34 \text{ eV/ion pair} = 54.4 \times 10^{-12} \text{ erg/ion pair} \tag{9.56}$$

If a charge Q is collected in a volume V in a Bragg–Gray cavity

$$\text{ionization/unit volume} = \frac{Q(\text{esu})}{V(\text{cm}^3)} = \frac{Q}{V} \times 2.082 \times 10^9 \frac{\text{ion pairs}}{\text{cm}^3} \tag{9.57}$$

where 1 esu $= 1/4.8 \times 10^{-10}$ esu/electron $= 2.08 \times 10^9$ electrons and it is assumed that one electronic charge is measured per ion pair. Taking the density of air as 0.0013 g/cm^3 we can write the number of ion pairs per gram as

$$\text{ion pairs per gram of air} = J_{\text{air}} = \frac{Q}{V} \times \frac{2.082 \times 10^9}{0.0013} \frac{\text{ion pairs}}{\text{g}} \tag{9.58}$$

Now substitute Eqs. (9.58) and (9.56) into Eq. (9.55) and obtain the energy absorbed per gram of the medium into which the cavity was inserted:

$$E_m = S^m_{\text{air}} \times 54.4 \times 10^{-12} \frac{\text{erg}}{\text{ion pair}} \times \frac{Q}{V} \times \frac{2.082 \times 10^9}{0.0013} \frac{\text{ion pairs}}{\text{g}}$$

$$= 87.7 \times S^m_{\text{air}} \frac{Q}{V} \frac{\text{erg}}{\text{g}}$$

or, since 1 rad $= 100$ erg/g,

$$E_m = 0.877\, S^m_{\text{air}} \frac{Q}{V} \text{ rad} \tag{9.59}$$

TABLE 9.3[a]

		$\bar{S}^{m}_{air}$		
Source	Energy (MeV)	Water	Muscle	Polystyrene
Au^{198}	0.41	1.149	1.149	1.139
Cs^{137}	0.67	1.145	1.145	1.133
Cu^{60}	1.25	1.135	1.133	1.120

[a] Data from ICRV Handbook 85, U.S. Nat. Bur. St. (1964).

When a suitable S^{m}_{air} value is used the effective dose received in a medium can be obtained from measurement with a small ionization chamber in the same location. This is done by constructing phantoms or dummies. These are usually made of water although some plastics are suitable. Clearly, the dummy material must have nearly the same mass stopping power ratio as the tissue to be irradiated. The determination of S^{m}_{air} for different energies required the work of many investigators over a period of years because a number of factors enter into the term. For example, γ rays can impart energy to electrons by Compton scattering, the photoelectric process, or pair production. Each of these primary excited electrons may in turn excite other electrons and it is only with some difficulty that an average excitation energy and number of electrons can be determined for each photon energy. The development of this theory is beyond the scope of this book. However, these data now exist in tables and the mass stopping power ratio can be looked up for any energy. Some examples are given in Table 9.3, in which $\bar{S}^{m}_{air}$ is given as an average. Note that the values in this table are suitable only for skin dose; as the radiation penetrates it loses energy and the $\bar{S}^{m}_{air}$ for the energy of the photon at the tumor must be used.

Note that in Fig. 9.29 although a tumor at the position of maximum intensity would certainly be irradiated, so would a lot of healthy tissue in the path of the beam. As we have seen in Fig. 9.23, healthy tissue is at a disadvantage and the cells could be killed by a dose smaller than that required to kill the cancerous ones. However, irradiation of a medium at the same time by different sources is additive. Imagine from Fig. 9.29 the irradiation of a tumor in a human neck with four such sources. Two sets of opposing pairs, 20 cm apart, could be arranged at right angles. In this way the center would have a dose rate of 200% of the contact intensity of any one source. Furthermore, the contact, or maximum intensity location, need not be at the skin level but could be several centimeters away. This is the common practice and the distance from the skin is called sfd, skin focus distance. It should be noted that most tumors do not have neat geometrical configurations and the arrangement of multiple sources for additive radiation fields is not a trivial calculation. Computer programs have been developed to assist physicians in these computations.

REFERENCES

Adams, G. E. (1973). Chemical radiosensitization of hypoxic cells, *Brit. Med. Bull.* **29**, 48.

Barendsen, G. W. (1962). Dose-survival curves of human cells in tissue culture irradiated with alpha-, beta-, 20 kV x-ray, and 200 kV x-radiation, *Nature* (*London*) **193**, 1153.

Berry, R. J., and Cohen, A. B. (1962). Some observations on the reproductive capacity of mammalian tumor cells exposed *in vivo* to gamma-radiation at low dose rates, *Brit. J. Radiol.* **35**, 489.

Bethe, H. A. (1933). Quantenmechanik der Ein- und Zwei-Elektronenproblem, *in* "Handbuch der Physik," Vol. 24, Part 1. Springer, Berlin.

Bohr, N. (1915). On the decrease of velocity of swiftly moving electrified particles in passing through matter, *Phil. Mag.* **30**, 581.

Casarett, A. P. (1968). "Radiation Biology." Prentice-Hall, Englewood Cliffs, New Jersey.

Christensen, E. E., Curry, T. S., and Nunnally, J. (1972). "An Introduction to the Physics of Diagnostic Radiology." Lea and Febiger, Philadelphia.

Dewey, D. L. (1963). The x-ray sensitivity of *Serratia Marcescens*, *Radiat. Res.* **19**, 64.

Dienes, G. J. (1966). A kinetic model of biological radiation response, *Radiat. Res.* **28**, 183.

Donato, L. (1973). Basic concepts of radiocardiography, *Seminars Nucl. Med.* **3**, 111.

Elkind, M. M., and Sutton, H. (1960). Radiation of mammalian cells grown in culture, I. Repair of x-ray damage in surviving chinese hamster cells, *Radiat. Res.* **13**, 556.

Elkind, M. M., and Sutton, H. (1959). X-ray damage and recovery in mammalian cells in culture, *Nature* (*London*) **184**, 1293.

Elkind, M. M., and Whitmore, G. F. (1967). "The Radiobiology of Cultured Mammalian Cells." Gordon and Breach, New York.

Errera, M., and Forssberg, A. (1961). "Mechanisms in Radiobiology." Academic Press, New York.

Evans, R. D. (1955). "The Atomic Nucleus." McGraw-Hill, New York.

Fichardt, T. (1973). Determining optimum dose-time-area relationships in curative radiotherapy of cancer, *S.A. Med. J.* **23**, 1070.

Groch, M. W., Gottlieb, S., Mallon, S. M., and Miale, A., Jr. (1976). A new dual-probe system for rapid bedside assessment of left ventricular function, *J. Nucl. Med.* **17**, 930.

Grodstein, G. W. (1957). X-ray Attenuation Coefficients from 10 keV to 100 MeV, Nat. Bur. Std. Circular #583.

Han, A., Miletic, B., and Petrovic, D. (1964a). Survival properties and repair of radiation damage in L-cells after X-irradiation, *Int. J. Radiat. Biol.* **8**, 201.

Han, A., Miletic, B., and Petrovic, D. (1964b). The action of ultra-violet light on repair of X-ray damage in L-cells grown in culture, *ibid.* **8**, 187.

Hine, G. J., and Brownell, G. L. (1956). "Radiation Dosimetry." Academic Press, New York.

Jacquez, J. A. (1968). Tracer kinetics, *in* "Principles of nuclear Medicine" (H. N. Wagner, Jr., ed.). Saunders, Philadelphia, Pennsylvania.

Johns, H. E., and Cunningham, J. R. (1977). "The Physics of Radiology," 3rd ed. Thomas, Springfield, Illinois.

Kiefer, J. (1975). Theoretical aspects and implications of the oxygen effect in radiation research, *Proc. Congr. Radiat. Res., 5th* (O. F. Nygaard, H. I. Adler, and W. K. Sinclair, eds.). Academic Press, New York.

Korst, D. R. (1968). Blood volume and red blood cell survival, *in* "Principles of Nuclear Medicine" (H. N. Wagner, Jr., ed.) Saunders, Philadelphia, Pennsylvania.

Lajtha, L. G., and Oliver, R. (1961). Some radiobiological considerations in radiotherapy, *Brit. J. Radiol.* **34**, 252.

Lea, D. E. (1955). "Actions of Radiations on Living Cells." Cambridge Univ. Press, London and New York.

Lederer, C. H., Hollander, J. H., and Perlman, I. (1967). "Table of Isotopes." Wiley, New York.

Liversage, W. E. (1969). A. General formula for equating protracted and acute regimes of radiation, *Brit. J. Radiol.* **42**, 432.

Livingston, M. S., and Bethe, H. A. (1937). Nuclear physics, *Rev. Mod. Phys.* **9**, 245.

Loevinger, R. (1971). Absorbed dose from interstitial and intracavity sources, *in* "Afterloading in Radiotherapy" (N. Simon, ed). HEW Bureau of Radiological Health, Rockville, Maryland.

Meisberger, L. L., Keller, R. J., and Shalek, R. J. (1968). The effective attenuation in water of the gamma rays of Gold 198, Iridium 192, Cesium 137, Radium 226, and Cobalt 60, *Radiology* **90**, 953.

Oliver, R. (1964). A comparison of the effects of acute and protracted gamma-radiation on the seedlings of vicia faba. Part II. Theoretical calculations, *Int. J. Radiat. Biol.* **5**, 475.

Paterson, R., and Parker, H. M. (1934). A dosage system for gamma ray therapy, *Brit. J. Radiol.* **7**, 592; (1936). A system of dosage for cylindrical distributions of radium, *ibid.* **9**, 487; (1938). A dosage system for interstitial radium therapy, *ibid.* **11**, 252, 313.

Puck, T. T. (1960). The action of radiation on mammalian cells, *Am. Naturalist* **94**, 95–109.

Richards, P. (1965). The technetium-99m generator, *in* "Radioactive Pharmaceuticals." (Proc. of a Symp. at Oak Ridge Inst. of Nucl. Studies, Nov. 1–4, 1965). U.S. Atomic Energy Commission Div. of Tech. Info. (Available from Clearinghouse for Federal, Scientific, and Technical Information, Natl. Bur. Standard, Dept. of Commerce, Springfield, Virginia.)

Richtmyer, F. K., Kennard, E. H., and Cooper, J. N. (1969). "Introduction to Modern Physics." McGraw-Hill, New York.

Robertson, J. S. (1962). Mathematical treatment of uptake and release of indicator substances in relation to flow analysis in tissues and organs, *in* "Handbook of Physiology" (P. Dow, ed.), Sect. 2, Circulation Vol. 1. American Physiological Society, Washington, D.C.

Segré, E. (1953). "Experimental Nuclear Physics," Vol. 1. Wiley, New York.

Setlow, R. B., and Pollard, E. C. (1962). "Molecular Biophysics." Addison-Wesley, Reading, Massachusetts.

Sheppard, C. W. (1962). "Basic Principles of the Tracer Method." Wiley, New York.

Simon, N. (ed.) (1971). Afterloading in radiotherapy, *Proc. Conf. New York City, May.* Bur. of Radiolog. Health, Div. of Radiat. Exposure, Dept. of Health, Education, and Welfare Publ. No. FDA 72-8024, BRH/DMRE 72-4.

Tamm, I., and Fluke, D. J. (1950). The effect of monochromatic ultraviolet radiation on the infectivity and hemagglutinating ability of the influenza virus Type A strain PR-8, *J. Bacteriol.* **59**, 449.

Thorndike, F. (1926). Applications of Poisson's probability summation, *Bell Syst. Tech. J.* **5**, 604.

Till, J. E., and McCullogh, E. A. (1964). Repair processes in irradiated mice hematopoietic tissue, *Ann. N.Y. Acad., Sci.* **114**, 115.

Tobias, C. A., Anger, H. O., and Lawrence, J. H. (1952). Radiological use of high energy deuterons and alpha particles, *Am. J. Roentgenol. Radium Therapy Nuc. Med.* **67**, 1.

Trosko, J. E., Chu, E. H. Y., and Carrier, W. L. (1965). The induction of thymine dimers in ultraviolet-irradiated mammalian cells, *Radiat. Res.* **24**, 667.

Wagner, H. N., Jr. (ed.) (1968). "Principles of Nuclear Medicine." Saunders, Philadelphia, Pennsylvania.

Watson, T. A., Johns, H. E., and Burkell, C. C. (1954). The Saskatchewan 1,000 Curie cobalt 60 unit, *Radiology* **62**, 165..
Weiss, L. (1960). Some effects of hypothermia and hypoxia on the sensitivity of HeLa cells to x-rays, *Int. J. Radiat. Biol.* **2**, 20.
Wood, T. H. (1953). Reproducibility of x-ray survival curves for yeast cells, *Proc. Soc. Exp. Biol. Med.* **44**, 446.
Zimmer, K. G. (1961). "Studies on Quantitative Radiation Biology." Oliver and Boyd, Edinburgh.

CHAPTER **10**

Computerized Tomography: γ-Ray and X-Ray Brain Scanning

INTRODUCTION

A picture of the interior of the body accomplished by a noninvasive method is of inestimable advantage in diagnosis. A commonly used technique is that of x-ray photography, called radiography. This is based on the principle developed in Chapter 9 that a photon's energy is degraded by absorption in passing through media. Reference to Fig. 9.16 shows that in the range of x-ray photons used, 50 to 100 keV, the primary causes of energy loss are Compton scattering and the photoelectric effect. Regardless of the mode of energy loss the intensity I of any transmitted beam is given by Eq. (9.40)

$$I = I_0 e^{-\mu x} \tag{9.40}$$

where I_0 is the initial intensity, x the distance traveled, and μ the linear attenuation or absorption coefficient. The attenuation increases with increasing density of the absorbing medium. A radiograph, by the relative amounts of film darkening, records the differences in absorption of the initial x-ray beam by the different tissues as the x rays pass through the body.

One need not use an x-ray tube to generate the photons—a γ-ray nuclide will do just as well, and strong radioactive sources of high energy γ rays are used for radiography of metal parts that have too great an absorption for conventional x rays. Correspondingly, external γ-ray sources are too insensitive for use on humans. However radionuclides do have an advantage in a technique called *scintigraphy*. By the selection of the proper isotopes and by compounding into appropriate chemicals they will go to a specific part of the body. When there, they will emit γ rays in all directions. By suitable scanning and triangulation the location of the emitting source can be determined.

In both x radiography and scintigraphy additional information is obtained from a stack of pictures, called *tomographs*, that represent a series of adjacent slices through the body or particular organ (*tomo* is the Greek work for slice). Rapid construction of these pictures is called *computerized tomography* (CT), which will be described in this chapter.

ABSORPTION CHARACTERISTICS OF BODY TISSUES

X rays are produced when electrons from a hot filament in a vacuum tube are accelerated by a voltage and allowed to strike a solid. The radiation is produced in two ways: (1) *bremsstrahlung* (braking radiation), which is a broad spectrum of x radiation arising from the deceleration of the electrons in the electric fields of the nuclei, and (2) ionization of the target atoms in which the inner-shell electrons are knocked from their orbits by the impinging electrons and subsequently replaced, accompanied by the release of x-ray photons of precise energies. Variation of the accelerating voltage will alter the spectrum of the bremsstrahlung. The x-ray energies produced by ionization are governed by the binding energies of the orbital electrons of the target atoms, however, since 80–90% of the x rays produced arise from bremsstrahlung, for most medical x rays the selection of target atoms is not an important factor. The target is a metal so that the heat of the impinging electrons will be carried away rapidly. It is clear from Fig. 9.16 that a high-energy x ray passing through a thin specimen with small differences in tissue absorption would not show as much contrast as would a lower-energy x ray, while through a thick specimen a low-energy x ray may be too greatly absorbed during its traverse.

Roentgen discovered x rays in 1895 and within a few months they were being used in hospitals. They were used exclusively for bone fracture studies initially because of the large difference between absorption by bone and other body tissues. Figure 10.1 shows the relative absorption characteristics between body tissues if water is taken as the zero reference. The ordinate scale will be explained later. The density difference on a high-contrast x-ray

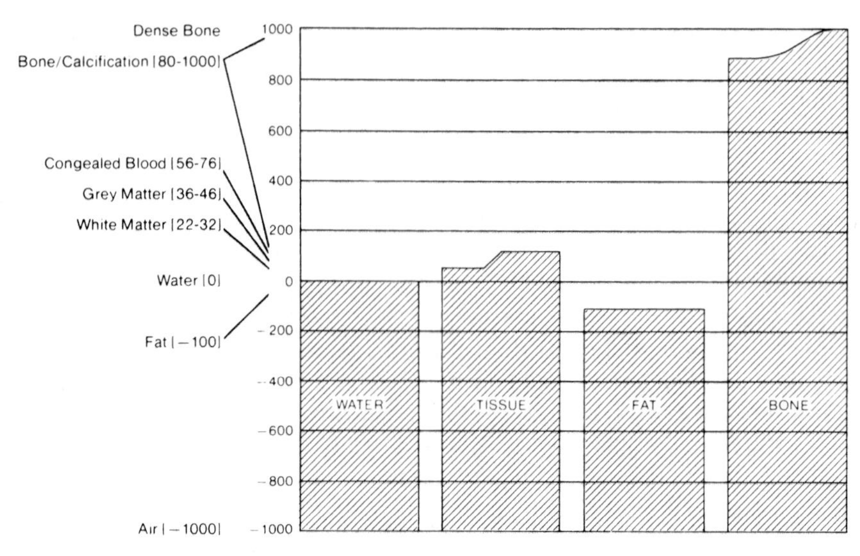

FIG. 10.1 Relative x-ray absorption of body tissues with water–air difference taken as 1000. See text for meaning of ordinate scale. Note that one unit of this scale is called a *brain unit*. Body scanning devices use a water–air difference of 500 with a corresponding body unit twice the size of a brain unit. [Courtesy of EMI Medical, Inc.]

TABLE 10.1

Attenuation of 60-keV photons in various tissues and fluids of the brain[a]

Material	Linear attenuation coefficient μ (cm^{-1})	Precent of photons through 1 cm transmitted relative to water
Fat	0.1846	101.85
Water	0.2029	100.00
Plasma	0.2063	99.58
White matter	0.2088	99.48
Gray matter	0.2106	99.24
Menigioma[b]	0.2125	99.05
Whole blood[c]	0.2129	99.02
Blood clot	0.2209	98.23

[a] Data compiled by Phelps *et al.* (1975b).
[b] Tumor of the membrane.
[c] Hematocrit of 40.

film between bone and other tissue can be 10^2 to 10^3. If, however, one wishes to x-ray a part of the body other than bone, the differences in absorption are considerably smaller, as seen in Table 10.1, and it is difficult to obtain significant proper contrast on a film. The situation is much more difficult in radiography of the interior of the head. Bone surrounds the interior tissues and the photons must pass through it while still displaying small differences in tissue absorption. In addition, in a conventional radiograph the entire tissue absorption is projected onto a planar film. This latter problem is solved by a tomograph.

GEOMETRIC GENERATION OF TOMOGRAPHS

If one wishes to have a radiograph of a thin cross section of some internal body tissue it is obvious that the photons must also pass through other tissue. The object then is either to subtract the nonrelevant part or to cause it to blur into a diffuse background so that only the desired slice is in sharp focus. This has been accomplished by geometric techniques long before computer assistance was available. Some of these techniques will first be described.

Consider the arrangement of an x-ray source and a film in Fig. 10.2. The source is at a distance a from a position called the *focal plane* and is moving

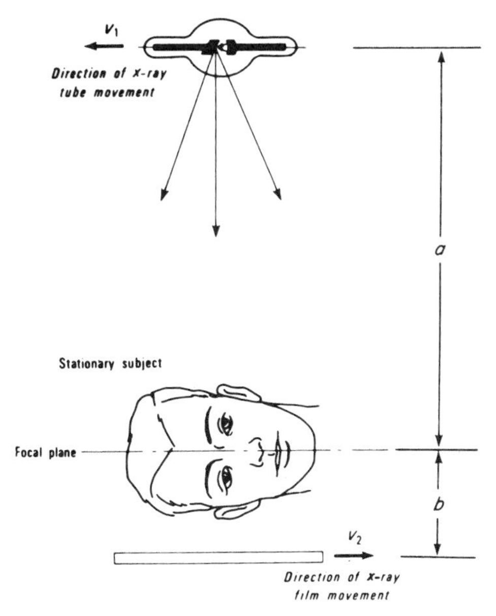

FIG. 10.2 Method for obtaining longitudinal tomograph. Structures lying on or near focal plane are sharp while off-plane structures are blurred. [From Anger (1974).]

with velocity v_1 to the left. The film is at a distance b from the focal plane and is moving with a velocity v_2 to the right. If the relation exists that

$$v_1/a = v_2/b$$

all of the structures lying in the focal plane are sharply imaged because the movement of the projected image of points in the plane on the film is exactly equal to the movement of the film. All other structures above and below this plane are also projected onto the film but are blurred because the movement of their projections is different than that of the film. The result is a sharply focused picture of the focal plane against a blurred background. A different plane may be brought into focus simply by changing the velocity of either the x-ray source or the film. This technique has been used to obtain a longitudinal tomograph.

An axial tomograph can be obtained by circular types of motion with a different geometric arrangement. This is seen in Fig. 10.3. The patient stands

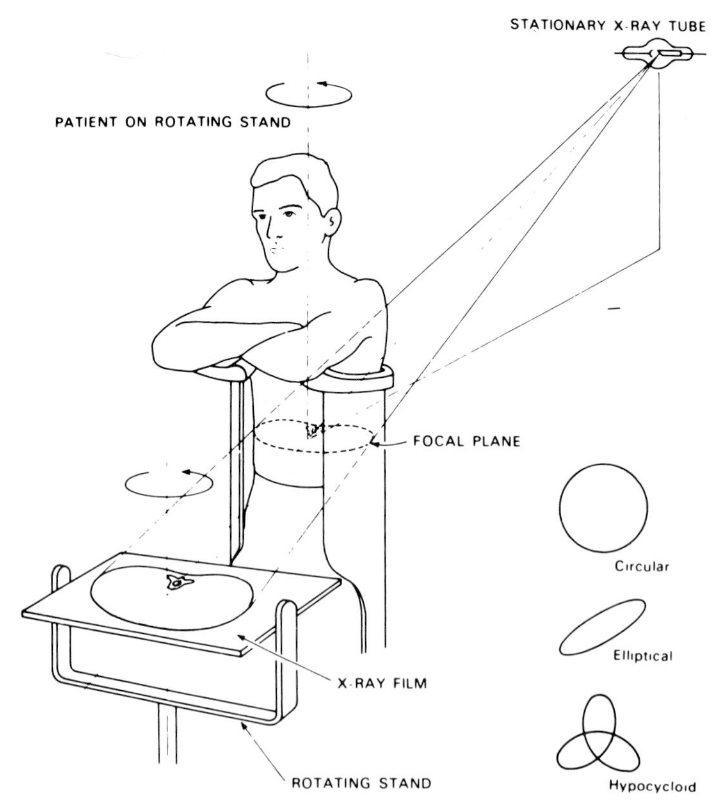

FIG. 10.3 Method of obtaining axial tomograph on film with possible modes of rotation. [From Anger (1974).]

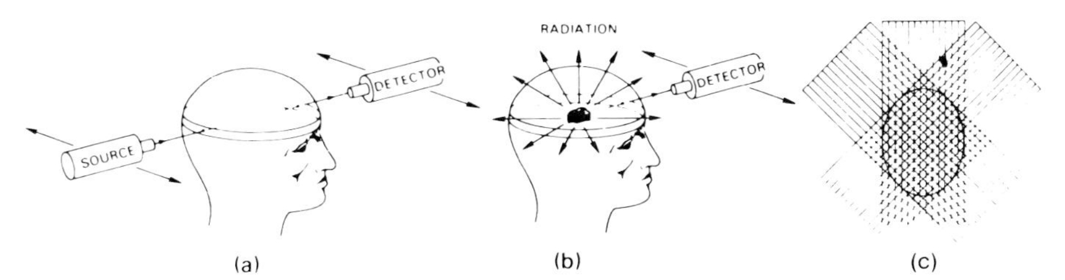

FIG. 10.4 Axial tomography of the head with (a) x-ray external source, and (b) γ-ray internal source, both resulting in a series of rays which, when taken at different angles as in (c), can be reconstructed to locate areas of absorption of x rays or emission of γ rays. [From Brooks and Chiro (1975).]

on a rotating table and in front of him lies the x-ray film in a horizontal plane that rotates in synchrony with him. A fixed x-ray beam is directed in a slanting direction so that the desired focal plane is projected on the film. The rotation need not be circular; elliptical or hypocycloidal modes are better. The arrangement in Fig. 10.3 shows that off-plane points will have a different angular velocity than the film and become blurred.

If a source and a detector are moved at the same velocity linearly across the head the result is a line of varying absorption density. If the assembly is

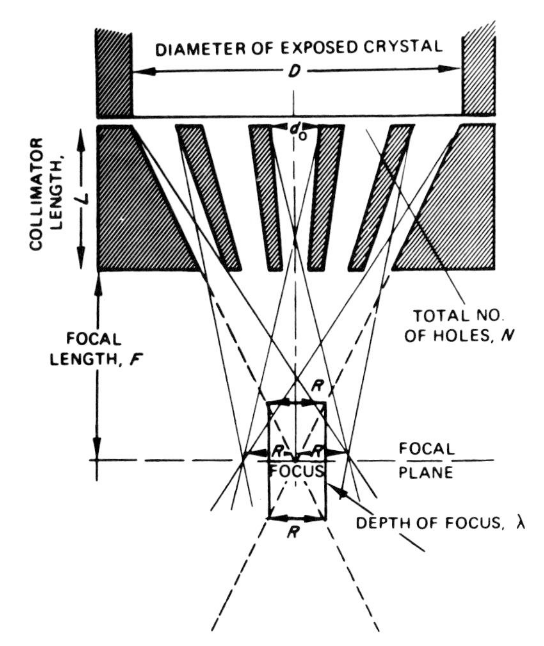

FIG. 10.5 Cross section of collimator and collimator hole geometry, which indicates the extreme beam paths that determine the focal plane, the resolution radius R, and the depth of focus λ. [From Hine and Erickson (1974).]

rotated and again traversed a different line of projected densities will result. This is shown in Fig. 10.4 for three traverses and the result is a star-like pattern in which higher density areas can be located from x-ray absorption or γ-ray emitting areas by triangulation of their coordinates. This technique will be discussed in more detail later.

If a collimated scanner is used it has a focal plane and a depth of field determined by its geometry. A cross section of such a collimator is shown in Fig. 10.5. It is seen that by the design of the size of the holes, their angle as well as the relative angle of the holes in the collimator head, a focal plane is determined. There are many factors that go into the design of a collimator head, such as the spacial resolution required for a given depth of focus. Usually, several collimator heads are supplied with a scanning machine with instructions to the radiographer on design parameters for intended use. Returning to tomography, it is clear that since a scanner with a collimator has a focal plane such as scan will produce a tomograph.

DETECTION OF PHOTONS

The earliest and still most common technique of photon detection is by photographic emulsion. In its simplest form, solutions of silver nitrate and potassium bromide are mixed, from which silver bromide forms and floc-ulates into small grains. A gelatin is mixed with this to keep the silver bromide suspended as small crystals and also to serve as a medium that will swell when wet to allow the developer to reach the grains. On these grains of silver bromide are small "sensitivity spots" believed to be silver sulfide, the sulfur coming from the gelatin. The silver bromide is an ionic crystal and there is some thermal migration of the ions. When a photon strikes an ion in the crystal an electron can be released by the photoelectric effect. This electron, being stripped from its ion with considerable energy, is free to migrate through the crystal until it becomes trapped at a sensitivity spot of silver sulfide. Trapped at this point, it gives the silver sulfide a negative charge, which attracts one of the migrating positive silver ions. This ion combines with the electron and forms a neutral silver atom. Upon repetition of this process a small cluster of silver atoms forms, which becomes a nucleus for further precipitation of silver in the development process. In develop-ment, the emulsion is treated with a solution that converts the silver bromide to silver, and it works fastest on grains that already have silver nuclei, i.e., the ones that have been exposed to light. The precipitating silver is amor-phous and appears black; therefore a developed film, called a "negative," has black regions where photons have struck.

Clearly, such a process is linear to photon exposure only in a limited range of intensities. Instead of going through the intermediate step of film, more sensitive automatic recorders were developed. These are based on the numbers and energies of the photoelectrons as they are created and two types will be briefly described, the semiconductor detector and the scintillation detector.

The Pauli exclusion principle states that only two electrons—and these must be with opposite spins—can occupy the same energy state. Thus, if 10^{21} atoms are condensed into a cubic centimeter of a solid, the electrons that had common orbital energies in the separated atoms must now rearrange themselves into discrete states infinitesimally apart in energy. These are called energy bands. The higher energy orbits into which electrons could be excited also form bands although these are normally unoccupied when the electrons are in their lowest energy states. These unoccupied bands into which electrons could be excited are called conduction bands, while the bands which are formed by the atomic electrons in their orbits are called filled bands. In an insulator the conduction band is several electron volts above the filled band of highest energy. Thus, if an electric field is placed across the crystal, electrons cannot move because there are no unoccupied energy levels. That is, they would have to acquire a new energy to move in the electric field but there are no new energy states available. In a metal, however, the unoccupied conduction band overlaps the filled band, so when and electric field is placed across the metal the electrons may assume a new energy and move in the field. In semiconductors the conduction band is above the filled band, but only slightly. Thus increasing thermal energy can excite electrons into the conduction band from the filled band and there is increasing electrical conductivity with increasing temperature. In addition to the electrical conduction caused by electrons excited into the conduction band, each electron that leaves the filled band leaves behind a "hole" where it had been. Since this hole is the absence of an electron, it behaves as if it had a positive charge and moves through the filled band under the influence of an applied electric field, although in the opposite direction to that of the electron. When an incoming photon imparts sufficient energy to an electron to excite it from the filled band to the conduction band an electron–hole pair is created and two electronic charges are available to contribute to conduction. Thus the arrival of a photon in a semiconductor detector that has an electric field across it will cause a sudden flow of current through the crystal. Usually one high-energy photon, by means of Compton scattering and the photoelectric effect, will excite hundreds of electrons so the resulting pulse of current is readily measured.

A scintillation detector is based on the same solid state model but a somewhat different process. The crystal used is an insulator, usually sodium iodide, NaI. The incoming photon excites electrons across the rather large energy band gap into the conduction band where they can migrate until they fall back to the filled band. Upon falling back to the filled band they release the energy of the band gap as a photon, similar to an excited electron in a free atom returning to the valence shell. Such a high-energy photon is difficult to monitor and it is desired to reduce this energy to the visible light region. This is done by introducing impurity atoms, which will form traps whose energy lies in the band gap between the top of the filled band and the bottom of the conduction band. The impurity is chosen, thallium is used in NaI, so that the trap energy is about half-way down to the filled band. In the case of Tl in NaI this energy drop is about 3 eV, which corresponds to a photon energy with a wavelength of 4030 Å. To recapitulate the process, the incoming photon creates Compton or photoelectrons, which in turn create other excited electrons, in the filled band. All of these have sufficient energy to go from the filled band to the conduction band. Here they migrate about until they fall into a thallium trap, which is at an energy of 3 eV below their energy in the conduction band. All of this takes place in a fraction of a microsecond. Upon falling into the trap they have decreased their potential energy by 3 eV, and this energy is released as a photon of 4030-Å wavelength. The impurity concentration is designed so that most of the electrons will fall into a trap before they fall back to the filled band.

The NaI crystal is in a light-proof container to which is attached a photomultiplier tube. This tube is designed so that the 4030-Å photon strikes an inner surface and releases a photoelectron. This electron is accelerated by an applied voltage to another plate, where it collides with sufficient energy to knock out several electrons. These electrons are then accelerated by a high voltage to strike another plate, releasing still more electrons. The usual photomultiplier tube has ten of these stages. Thus, if each stage emits five electrons for each one striking it, the amplification is more than one million.

Usually, NaI crystals employed in scintigraphy are large, about one-half-inch thick and several inches in diameter. If each γ ray entering the crystal has the same energy and excites the same number of electrons, and if the light released all strikes the photomultiplier, the pulse of current will be the same for each. If a γ ray has a different energy, the difference in the current pulse will be proportional to the energy difference. However, even for γ rays of identical energy, not all current pulses have the same magnitude. This is because there are statistical variations in the number of 4030-Å photons produced by each γ ray, variations within the photomultiplier tube as well as variations in the amount of light reaching the photomultiplier, e.g., some light strikes the container walls and is scattered. Also there are other small losses due to the details of the atomic nature of the process. What results is a distribution of current pulse sizes, called pulse heights. This distribution is the origin of the energy resolution of the detector and is one of the inherent limitations of distinguishing between two γ rays that have been through slightly different absorber paths.

In actual usage the detector or the photon beam is being moved and it is desired to record the beam energy along a path. Such a recording is an intensity versus distance line made up of discrete points taken close together. Rather than measure current directly, modern instruments usually have the photomultiplier or the semiconductor in parallel with a capacitor. The capacitor is charged a known amount, and the current produced in the detector partially discharges it. This amount of discharge is then recorded as the beam intensity at that spatial point. The detector (or beam) then moves to the next position and the process is repeated. This is continued until a line of predetermined length and spatial separation between points is traversed.

THE BLOOD–BRAIN BARRIER

It was pointed out in the Introduction that the most difficult area of the body from which to obtain internal information lies within the skull. It is difficult with x rays because one is trying to assess low tissue density changes against a high background of x-ray absorption by the bone.

Scintigraphy by a γ-ray emitting substance is quite different in that differences in attenuation are not the measured quantity. In addition, for successful γ-ray scintigraphy, another difficulty had to be overcome because of a phenomenon known as the *blood–brain barrier*. Such a barrier was first proposed by Goldman in 1913, for when he injected a dye into the blood stream and allowed it to equilibrate in the body none was found in either the central nervous system or the brain.

The barrier is selectively permeable because, obviously, nutrients and oxygen penetrate as well as ethyl alcohol, aspirin, and many other drugs. Detailed studies have revealed distinct differences between the capillaries of the brain and those of other body tissues. A cross-sectional schematic is shown in Fig. 10.6. Figure 10.6a is that of a nonneural capillary in which are shown the general routes of exchange blood plasma inside the capillary and the extracellular fluid outside. Small molecules of molecular weight less than 20–40 $\times$ 10^4 can diffuse through either the intercellular cleft or the membrane bridging the fenestrations. Molecules of any size can be transported across the capillary cell by pinocytosis (drinking by cells), which is the method by which cells absorb liquids. Solutes could also be exchanged directly through cell membranes and cytoplasm but these routes are inefficient when compared to the extracellular ones.

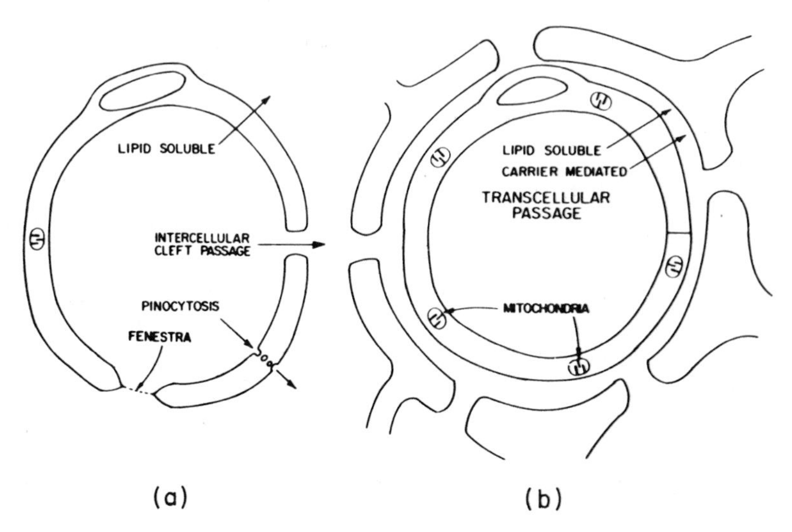

FIG. 10.6 Cross-sectional diagrams of capillaries: (a), nonneural capillaries; (b), brain capillaries. [From Oldendorf (1975).]

Figure 10.6b shows a diagram of a brain capillary. The extracellular routes are closed. The intercellular clefts are sealed shut by the formation of tight junctions between adjacent cells. The fenestra are missing and the pinocytotic cells are 15–20 times less numerous. Therefore, with none of the bypass mechanisms available, all molecules going from the blood into brain cells must pass through the cells of the capillary walls. The membrane wall is lipid, or fatty, and the molecule being delivered is in blood, or an aqueous solution. In order to penetrate the brain a particular molecule must therefore be transported by an aqueous solution, penetrate a lipid membrane, traverse the cytoplasm of the cell, and penetrate the lipid membrane on the other side of the cell. One general characteristic, clearly, is that the molecule must have some solubility in both water and lipids. Extensive studies by Oldendorf (1974, 1975) and others have shown that the relative solubility of molecules in water and oil in many cases determines the rate of uptake by the brain.

The barrier is only effective for healthy, normal tissue. In diseased or injured tissue the barrier is weakened. Thus some chemicals will be taken up from the blood stream by tumors or lesions in the brain. Bakay (1971) has summarized some possibilities for this behavior.

(1) An increase in both normal and abnormal blood vessels in brain tumors.

(2) Possible abnormal permeability of blood vessel walls in tumors.

(3) Enlarged extracellular spaces in tumors.

(4) Increased cellular metabolism in tumors.

(5) Reactivity of abnormally large amounts of fluid in diseased brain tissue.

(6) Increased rate of liquid absorption in diseased or tumor tissue.

Several of these possibilities may be operative at the same time. Regardless of the mechanism, however, it is possible to introduce a radioactive isotope in a chemical compound intravenously and have it selectively absorbed by damaged or tumorous brain tissue.

The compound selected must be sufficiently small to have a negligible biological effect. The radioactive isotope must be chosen so that its half-life is not much longer than the measurement time. It should emit γ rays of sufficient energy to be detectable by an external scanner. Also it should either emit no particle or, if it does, the particle should be of low energy, because it is absorbed by, and may damage, the body. It was shown in Fig. 9.3 that the metastable isotope of technicium ${}_{43}Tc^{99m}$ produced from a ${}_{42}Mo^{99}$-generator emits a 140-keV γ ray with a half-life of 6 hr. This is made into the compound sodium pertechnetate ($NaTcO_4$), which is injected intravenously with an activity of 15×10^{-3} curies into a patient prior to a brain scan.

AXIAL TOMOGRAPHY BY γ-RAY EMISSION

When a γ-ray emitting isotope is injected into the body in a chemical compound which will concentrate in a brain tumor, the tumor becomes a source of γ rays radiating equally in all directions. It was seen in Fig. 10.4 that by triangulation the tumor can be located within a plane; in this case the source is inside (Fig. 10.4b). If other sweeps of the detector are made at closely spaced distances along the axis of the head, a series of slices are obtained, and within each the tumor can be located. In this way the three-dimensional coordinates of the tumor are obtainable.

A device for obtaining such data rapidly was constructed by Kuhl and Edwards (1970) and is shown schematically in Fig. 10.7. Four scanning detectors move on a gantry and obtain the indicated projections in one sweep. NaI scintillation crystals with 2-in diameter, are used in the detectors with specially designed collimators which have the same resolution at 1 in. (brain surface) as at 4 in. (brain center), since there is an equal need for good resolution at the surface as at the center. The scanner speed can be adjusted to 0.5, 1, or 2 cm/sec and 85 picture elements (called *pixels*) are taken per line. The counting data are recorded digitally and the number of the pixels in a tomograph grid or matrix is $85 \times 85 = 7{,}225$.

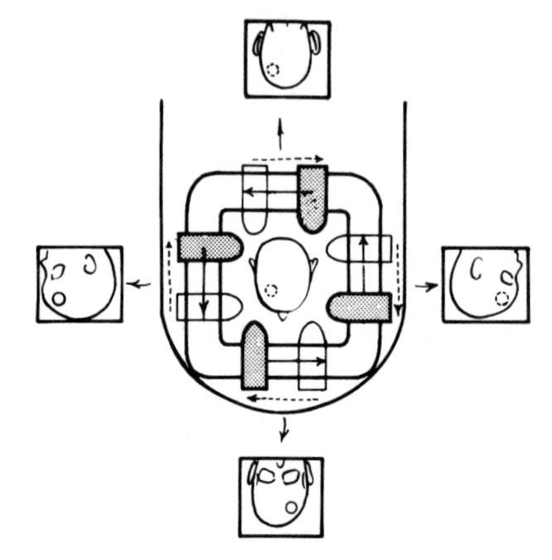

FIG. 10.7 Rectilinear scanning of a γ-ray emitter in the brain showing the four views obtained by the four moving detectors in one position of the gantry. [From Kuhl and Edwards (1970).]

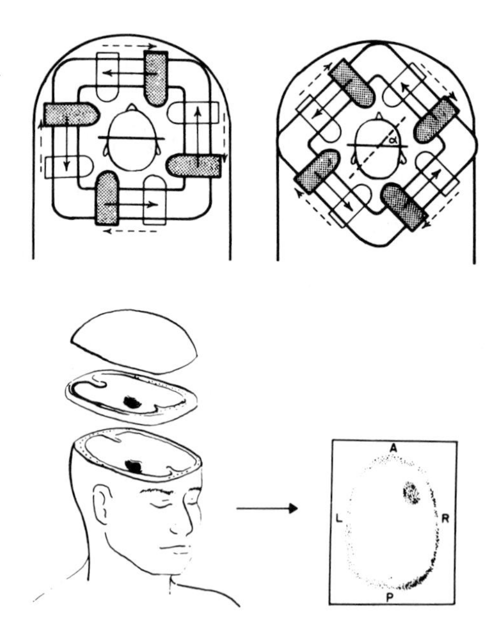

FIG. 10.8 Transverse-section scanning by the system shown in Fig. 10.7 is a sequence of transverse detector motion taken at each of a series of angles so that the entire circumference of the head is surveyed. Upon reconstruction of the data the resulting tomograph is obtained. [From Kuhl and Edwards (1970).]

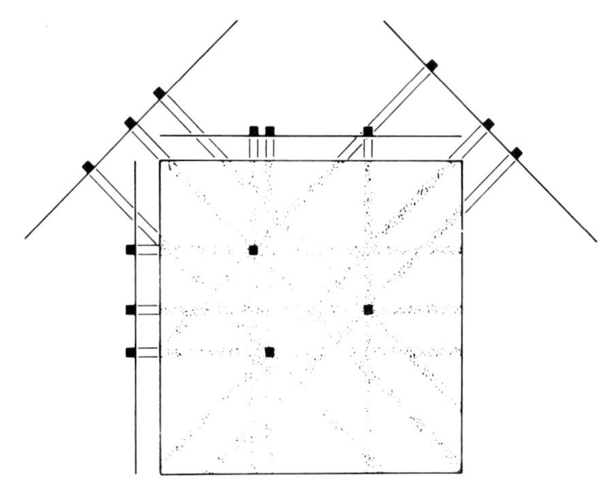

FIG. 10.9 The star pattern that results from uncorrected reconstruction of three emitting sources from four scan angles. [From Muehllehner and Wetzel (1971).]

It is not enough simply to locate the emitting area; one is interested in its shape. In order to accomplish this, scans must be taken at a series of azimuthal angles. Figure 10.8 shows how the gantry is rotated for another set of scans. This is generally done by rotation in units of 7.5° for twelve scans and the detectors will then have scanned 360°.

When these scans are superimposed there will be high density regions of the emitter locations surrounded by decreasing densities or "star patterns" caused by the other projections. Star patterns generated by three emitting sources onto four scan angles are shown in Fig. 10.9. If $360/7.5 = 48$ scan lines were drawn on this figure, although the centers of the stars could be located, the background would be uninterpretable.

To see how the matrix would appear when formed by the digitized scanning information let us consider a line source of emitters of equal strength and four scans at 45° from each other. The line source can be five equal strength emitters, which would each contribute 20 counts to the detector as it passed. Figure 10.10 shows the location on the diagonal of the tomograph of the five emitters and four scan lines around it at 0°, 45°, 90° and 135°. The numbers on the scan lines indicate what each scan has recorded as having seen. The problem then is to take the data from the scan lines and contruct a matrix that represents the emitters and their strengths. Start with a blank matrix and add to each element the values in the 0° scan line to obtain Fig. 10.11. Next add the 45° values to each of the points in the matrix "seen" by points on the scan line, and Fig. 10.12 is obtained. Doing the same for the 90° and 135° scan lines, each contributes 20 to every element of the matrix. Finally, divide each element of the matrix by the number of scan

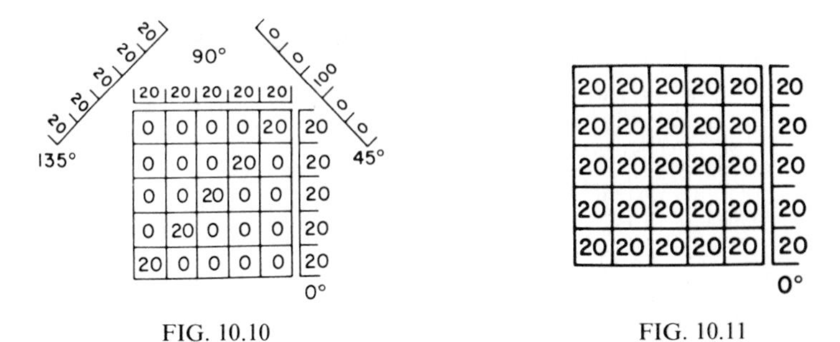

FIG. 10.10

FIG. 10.11

FIG. 10.10 Hypothetical scans at four angles with data that would be obtained from a diagonal array of five emitting sources each of strength 20. [From Kuhl *et al.* (1972).]

FIG. 10.11 The first step in simple reconstruction from scans of Fig. 10.10 is to assign values seen by first scan to each square in an originally blank matrix. [From Kuhl *et al.* (1972).]

lines used, i.e., four, and the result is Fig. 10.13. This is the simple reconstructed two-dimensional matrix, which is a back projection of data on one-dimensional scan lines. The line of emitting elements is present but there is a high background, so the contrast is not as great as in the situation of Fig. 10.10, which was scanned. As scans from more angles are contributed, the background near the emitters will become greater at a more rapid rate than that farther away. A mathematical technique with an associated computer program has been devised by Kuhl *et al.* (1972) to correct for this background star pattern and is explained in their original paper on this

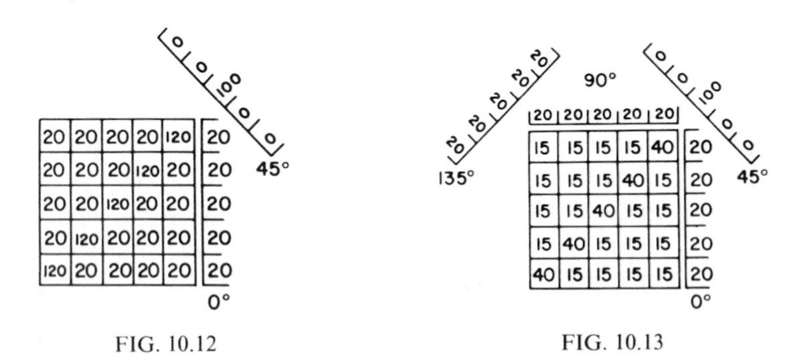

FIG. 10.12

FIG. 10.13

FIG. 10.12 The second step in simple reconstruction is to add values of the second scan to each square in the matrix from which the scan has indicated emission. [From Kuhl *et al.* (1972).]

FIG. 10.13 The third and fourth steps in simple reconstruction are addition to each matrix element of the values seen by each of the other two scans and division by the number of scans. [From Kuhl *et al.* (1972).]

subject. In a later section we shall discuss a method of averaging away the background in the x-ray scanning device.

X-RAY COMPUTERIZED TOMOGRAPHY (CT)

The description given above of the effect of photons on film grains and the subsequent development process indicate that it is a rather primitive way of counting photons. It is highly nonlinear in density change and the eye is notoriously bad in distinguishing values of shadings. For these reasons x-ray crystallographers have been using NaI detectors instead of film for over a quarter of a century; the precision of photon counting is much higher than film and the data are recorded digitally for later examination.

Starting in 1967, G. N. Hounsfield of EMI Laboratories in England began to calculate the possible precision of a computer reconstruction of several scans of the brain with an x-ray source and an NaI crystal detector. The calculations resulted in an estimate that with such a system the absorption coefficient of each element of the slice could be evaluated to an accuracy of $\frac{1}{2}\%$, which reference to Table 10.1 shows to be the diagnostic requirement. Because tumors have a high x-ray absorption coefficient relative to the brain while cholesterol deposits have a low one, these pathological conditions were, in principle, detectable.

A prototype machine was constructed and tested in 1971. The clarity of detail that was revealed exceeded expectations. The importance of this new technique was immediately recognized and in 1972 Hounsfield won the McRobert Award.

The machine was improved upon with better software and miniaturized computer circuits and put into production. Although it is expensive, about half a million dollars, it has revolutionized brain and optic nerve examination by giving physicians information that heretofore was simply unavailable.

Although developed originally for brain tomography, a modified version is now available for axial tomography of the rest of the body. We will confine the discussion of the principles to the version originally developed for the brain. Also, while other manufacturers have brought out competing machines, the EMI Scanner was the first, and most of the subsequent papers have been written about its operation. We will therefore discuss only the EMI Scanner.

CALCULATION OF ATTENUATION COEFFICIENT

In the discussion leading to Eqs. (8.13) and (8.15) intensity was defined as power transmitted in the direction of a wave. In the case of x-ray photons, absorption takes place both by Compton scattering and by the creation of photoelectrons. In the latter case the x-ray photon is lost from the beam so that the power (energy per unit time) arriving at the detector is reduced by this loss. In Compton scattering, however, the photon may continue in the same general direction but with a reduced energy. Thus, Compton scattering by an absorber will reduce a monochromatic x-ray beam to one of lower photon energy, while it remains essentially monochromatic. On the other hand, if different parts of a monochromatic beam go through either different absorbers or different thicknesses, a polychromatic beam will be transmitted. The EMI Scanner for the head has an arrangement in which a 120-keV x-ray beam (which can also be operated at 140 keV) passes first through a water-filled compartment 27 cm long so that the emerging beam for data taking has been reduced to a 73-keV monochromatic beam.

Equation (9.40) shows that the fraction of intensity remaining in a photon beam I/I_0, in passing a distance x through an absorber is written as

$$I/I_0 = e^{-\mu x} \tag{9.40}$$

where μ is a function of the density of the absorber and its atomic weight. Also we have seen in Fig. 9.16 that the linear attenuation coefficient μ is a function of the photon energy.

Let us first compute the fraction of the photon beam remaining after passing through two absorbers of the same thickness x but with different absorption coefficients μ_1 and μ_2. By Eq. (9.40) the fraction of the beam I_1/I_0 remaining after passing through thickness x of absorber 1 is

$$I_1/I_0 = e^{-\mu_1 x} \tag{10.1}$$

The beam intensity entering absorber 2 is then I_1 and the beam emerging is I_2 and the fraction of the emerging beam remaining is I_2/I_1. Reapplying Eq. (9.40) results in

$$I_2/I_1 = e^{-\mu_2 x}$$

Solving for I_1 gives

$$I_1 = I_2 e^{+\mu_2 x}$$

and substituting this into Eq. (10.1) yields

$$I_2/I_0 = e^{-(\mu_1 + \mu_2)x} \tag{10.2}$$

which is an expression for the fraction of the original intensity remaining after having passed through two absorbers of equal thickness x but with different absorption coefficients. Often the ratios I_1/I_0 and I_2/I_0 are given the symbol T for transmittance. We could continue to apply Eq. (9.40) as the photon beam passed through other absorbers, and if they all had the same thickness but different absorption coefficients Eq. (10.2) would be written for n absorbers as

$$T_n = \exp\left(-\sum_{i=1}^{n} \mu_i x\right) \tag{10.3}$$

However, we must introduce the fact that the attentuation coefficient is a function of the energy of the photon, or $\mu(E)$. In a polychromatic beam of intensity I_0 the number of photons in the energy interval dE is dI_0/dE. Therefore the intensity of the beam transmitted through n regions of equal length x is the sum of the products of each group of photons and its exponential attentuation. This is expressed by the integral from zero energy photons up to the maximum energy photon, or

$$T_n = \int_0^{E_{\max}} (dI_0/dE) \exp\left[-\sum_i \mu_i(E)x\right] dE \tag{10.4}$$

Because of the relation

$$e^{(a+b)} = e^a e^b$$

we may add and subtract the term $\exp(\sum \mu_w x)$ in the exponential of Eq. (10.4) to obtain

$$T_n = \int_0^{E_{\max}} (dI_0/dE) \exp\left[-\sum_i \mu_w x\right] \exp\left\{-\sum_i [\mu_i(E) - \mu_w(E)]x\right\} dE \tag{10.5}$$

where μ_w is the absorption coefficient for water. Note that in the EMI device there is a fixed length of water X and this length is nx. Therefore we may write

$$\sum_i \mu_w x = \mu_w X$$

and Eq. (10.5) becomes

$$T_n = \int_0^{E_{\max}} (dI_0/dE) \exp[-\mu_w X] \exp\{-\sum [\mu_i(E) - \mu_w]x\} dE \tag{10.6}$$

where μ_w is written as being independent of energy because the beam passes through it before it encounters the distribution of absorbers in the brain.

The first two terms under the integral represent the effect of the water path on the incident beam I_0, and what emerges is the intensity I_w transmitted through the water. The number of photons in each energy increment dE of this beam is dI_w/dE and Eq. (10.6) may be rewritten as

$$T_n = \int_0^{E_{\max}} (dI_w/dE) \exp\{-\sum [\mu_i(E) - \mu_w]x\} dE \tag{10.7}$$

Referring to Fig. 10.1, whose scale will be explained shortly, it is seen that tissue and fat are very close to water in attenuation, and therefore the difference $(\mu_i - \mu_w)$ in Eq. (10.7) can be said to be independent of energy, to a reasonable approximation. With this assumption the exponential can be taken out of the integral and the remaining integration is simply the intensity of the beam emerging from the water path:

$$\int_0^{E_{max}} (dI_w/dE)\, dE = \int_0^{E_{max}} dI_w = I_w$$

Therefore we may write

$$T_n = I_w \exp[-\sum (\mu_i - \mu_w)x] \tag{10.8}$$

Computer algorithms may then be set up to rapidly solve for $(\mu_i - \mu_w)$ from Eq. (10.8). This equation is based on the assumption that the attenuation coefficient of tissue is very little different from that of water. Reference to Fig. 10.1 shows that this assumption is not valid for bone. Presumably the EMI Scanner has a correction factor built into the computer software to account for typical amounts of bone encountered in a scan of the brain. This suggestion was tested by McCullough *et al.* (1974, 1976) who showed that the quality of the image deteriorated in scans of the lower portion of the brain where there is not only less bone but its distribution is uneven.

Most of the tissues within the brain have attenuation coefficients higher than that of water by only 3–5%. In order to enlarge the scale, EMI has defined a number, called Hounsfield Brain Unit, which is ten times the percentage difference between the μ_i of the tissue and that of water divided by the attenuation coefficient of water

$$\text{Brain Unit} = 10[(\mu_i - \mu_w)/\mu_w] \times 100 \tag{10.9}$$

so that the Brain Unit becomes an "attenuation value." For example, a tissue with a 4% higher attenuation coefficient than water has $\mu_i = 1.04\mu_w$. Inserting this into Eq. (10.9) results in a Brain Unit for that tissue of 40. Also, from this relation

$$\text{Brain Unit} = 0.001\mu_w$$

and is equivalent to an attenuation coefficient of 0.1% greater than that of water. By the definition of Eq. (10.9) water and air are 1000 units apart† and if water is taken as zero, any attenuation coefficient less than water is negative. This scale of 1000 units is the ordinate of the EMI Brain scale of Fig. 10.1.

† The EMI body scanner uses a water–air scale difference of 500 and therefore the coefficient on the right-hand side of Eq. (10.9) is 5 rather than 10. This scale is called the Body Unit.

IMAGE RECONSTRUCTION

The geometry of rays from a source passing through an object to a detector is illustrated in Fig. 10.14. If the object is described by the x, y coordinate system, a given ray is specified by the angle ϕ that it makes with the y axis and the distance x' that it is from the origin. A coordinate y' is the distance along the ray. If $f(x, y)$ represents the linear absorption coefficient of the coordinate position (x, y) of the object, the integral of all of these absorption coefficients along the path of the ray give its *ray sum*

$$p(x', \phi) = \int f(x, y)\, dy'$$

It is clear that p is proportional to the total attenuation of the ray

$$p = -\ln(I/I_0) \tag{10.10}$$

where I is the transmitted intensity and I_0 the incident intensity. Because the final display is in a matrix or grid form there are actually a series of small square cells at each coordinate point as indicated in Fig. 10.15. Four possible scans of the object as 45° angles from each other, labeled A, B, C, and D are shown in Fig. 10.16, in which the path of one ray is indicated.

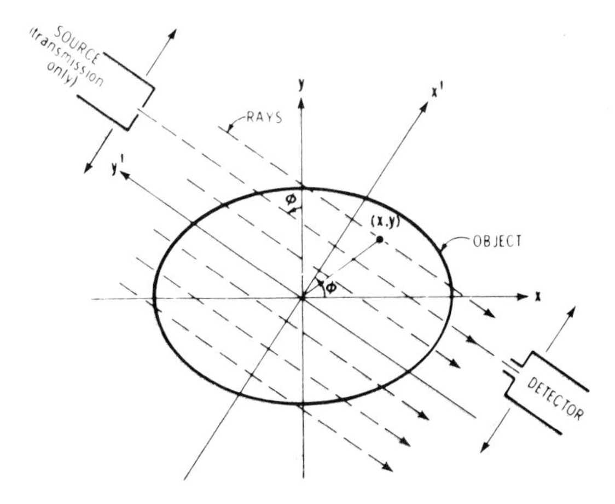

FIG. 10.14 Coordinate system for x-ray transmission tomography. The object being surveyed is in the x, y coordinate system. A ray is specified by its angle ϕ from the y axis and its distance x' from the origin, and the y' coordinate denotes distance along the ray. [From Brooks and DiChiro (1975).]

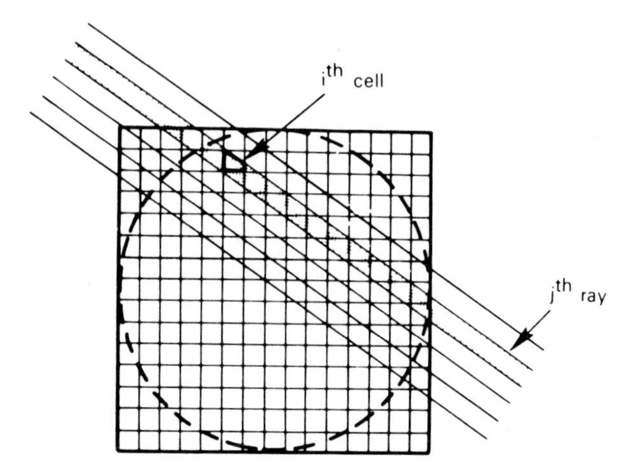

FIG. 10.15 Ray geometry for iterative reconstruction. The object, bounded by the dashed circle, is reconstructed on an $n \times n$ array of cells. The rays are actually strips of finite width. [From Brooks and DiChiro (1975).]

Reconstruction of an object from ray sums is a problem common to many areas of physics and a variety of ways have been developed for dealing with it. We will show one of the simpler ones, called the algebraic reconstruction technique (ART) because it was used in the first version of the EMI Scanner, although more sophisticated methods are used in later models.

Suppose four scans were taken 45° apart, as in Fig. 10.16, and that the data of Eq. (10.10) resulted in the ray sums indicated in Fig. 10.17. Although the individual contributions to the ray sums are in the four boxes of Fig. 10.17, these are not known initially.

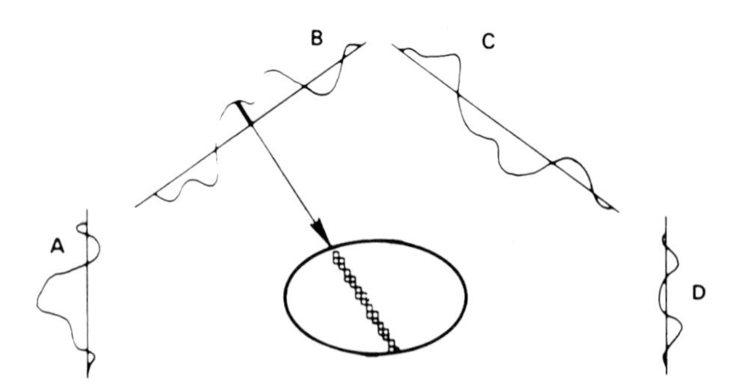

FIG. 10.16 Scan data from four different angles in a hypothetical situation. One datum point is shown to arise form the absorption of a ray that has passed through a sequence of the matrix or cells indicated in Fig. 10.15. [From Brooks and DiChiro (1975).]

FIG. 10.17 Ray sums at scans 45° apart from a 2 × 2 grid in an absorbing object.

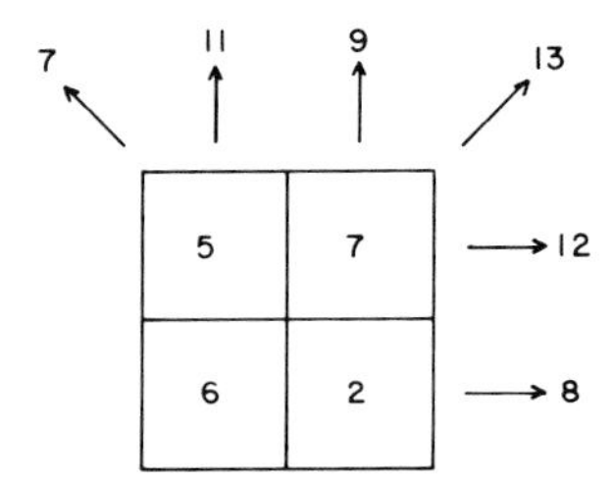

One starts with an empty grid, adds the sums vertically, (see Fig. 10.18a), subtracts these from the ray sums, and places half of the total in each vertical box (Fig. 10.18b). The resulting numbers are then added horizontally as in Fig. 10.18b. The horizontal sums are then subtracted from the horizontal ray sums and half of each result is added algebraically to the corresponding horizontal boxes (Fig. 10.18c). The diagonals are then added as in Fig. 10.18c. These diagonal sums are then subtracted from the diagonal ray sums and half is put into each corresponding diagonal box (Fig. 10.18d). In this simple case the original image is reconstructed exactly. Many more iterations are required for complicated figures and it has been shown that successive iterations should be separated by large angles to ensure that succeeding corrections are independent of each other so that errors do not accumulate. Although the EMI Scanner takes ray sums every 3°, the iterations are not done in the order in which the scans are taken.

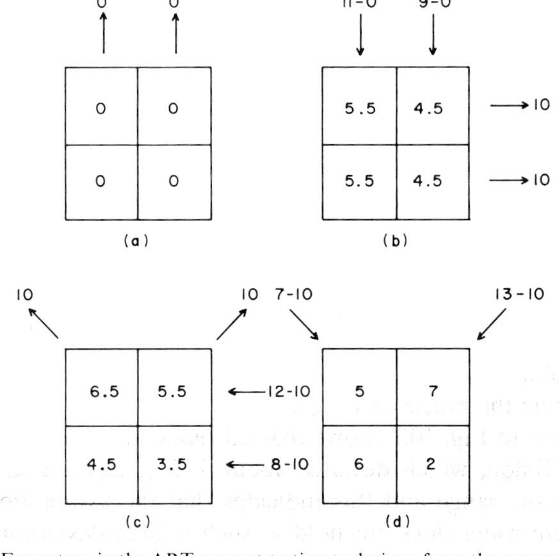

FIG. 10.18 Four steps in the ART reconstruction technique from the ray sums of Fig. 10.17.

PRECISION OF EMI MEASURED ATTENUATION

In obtaining Eq. (10.8) an approximation was used: for small deviations of absorption coefficient μ_i from that of water the difference $\mu_i - \mu_w$ could be considered independent of energy. Since such an approximation is an inherent part of the software of the EMI Scanner, with what precision and accuracy can different tissue absorption coefficients be determined? McCullough *et al.* (1976) used the EMI Scanner to obtain the Body Units for different plastics of standard length, as well as water, and compared them with independent determinations of the attenuation coefficients of the materials for 73-keV x rays.

If one uses the known attenuation coefficient of water of 0.190 cm^{-1} for this energy, Eq. (10.9) can be solved for μ_i (using Body Units, i.e., a water-air difference of 500, see footnote p. 300).

$$\mu_i = 0.38 \times 10^{-3} \text{ (Body Unit)} + 0.190 \tag{10.11}$$

TABLE 10.2

Material	Body Unit	μ (cm^{-1}) (from Body Unit)	μ (cm^{-1}) (73-keV x rays)
Polyethylene	-50.9 ± 1.8	0.171 ± 0.001	0.171
Polystyrene	-14.3 ± 0.3	0.185 ± 0.001	0.186
Water	1.2 ± 0.2	0.190 ± 0.0005	0.190
Nylon	48.6 ± 2.1	0.208 ± 0.001	0.208
Lexan	52.8 ± 2.5	0.210 ± 0.001	0.210
Plexiglas	62.4 ± 2.3	0.213 ± 0.001	0.214
Bakelite	131.8 ± 3.8	0.240 ± 0.001	0.236
Teflon	442.1 ± 21.6	0.358 ± 0.008	0.374

Their measured values of Body Units for the series of materials are listed in the second column of Table 10.2. The μ_i are calculated from Eq. (10.11) in the third column where the deviation is rounded off to the first integer in several cases, and therefore the true deviation is often less than indicated. The fourth column is the independent evaluation of μ_i. It is seen that up through Bakelite the precision of μ_i determinations by the EMI Scanner is 0.5%. Reference to Fig. 10.1 shows that all tissue falls within this range of Body Units. Teflon, which deviates about 2–3%, has a Body Unit well outside the tissue range and this indicates that the assumption for small deviations from water does not hold at such a large deviation. The tissue precision of 0.5% is considerably superior to that achievable by film.

SCANNING METHODS

A fundamental concern in the collection of data is that a maximum number of the photons that pass through the tissue arrive at the detector because of harmful effects of excess exposure. We have seen that in the emission of internal γ rays only a planar slice is recorded of the 4π solid angle continual emission. In the design of a scanner for an external source of photons more adjustable parameters are at the designer's disposal. For example, by appropriate shielding the beam width can be quite narrow, so that only the plane of tissue being recorded is exposed. In the design of the EMI Scanner the beam is wide enough so that two adjacent banks of detectors can record in one scan, resulting in two adjacent tomographs, each a slice 1.3 cm thick. Although this does not reduce the exposure it shortens the time of production of a series of adjacent tomographs.

It is of interest to consider, as did the designers of the EMI Scanner, two possible methods of collecting the data. These are illustrated in Fig. 10.19. In arrangement (a) the x-ray tube and the detectors are attached to a gantry indicated by the oval-shaped outline. This gantry moves in one direction during the scan, stopping spatially either 160 or 320 times to collect data,

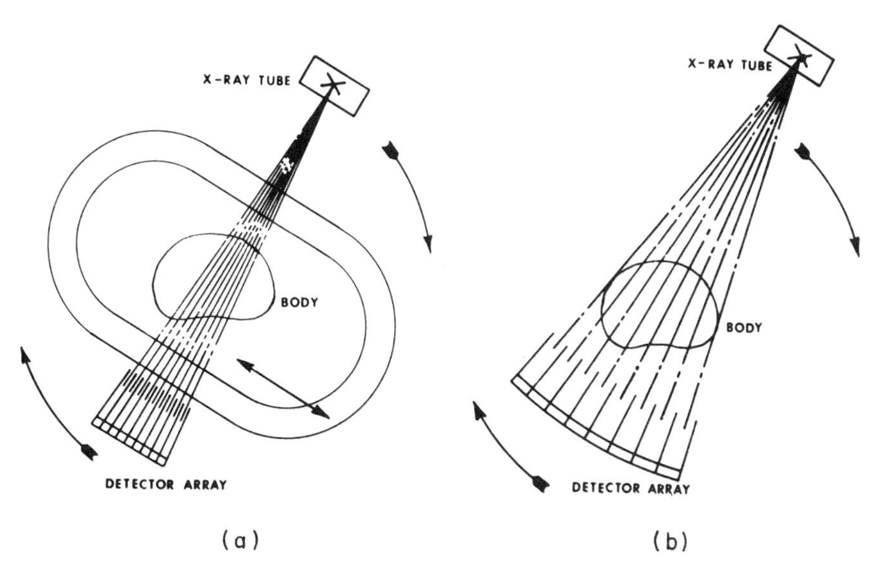

FIG. 10.19 Two possible design arrangements for x-ray tube and detectors to scan at different angles. (a) x-ray tube and detectors mounted on (oval) gantry that has translational motion for one scan. Gantry then rotates to a new position for another scan. (b) x-ray tube and detectors mounted on a gantry (not shown) that rotates only. [From G. N. Hounsfield, *Amer. J. Roentgenol.* **127**, 3 (1976).]

and then returns quickly. It then rotates 3° and repeats the process until the preselected 180° or 240° rotation is completed. The x-ray beam is arranged in a fan so that each part reaches a detector and none passes through the body without being detected. At the end of each traverse, when the beam no longer passes through the body, the detectors can be recalibrated so that electronic stability is not a serious problem. Both the length of the traverse and the spatial interval between readings can be varied. The scale of the picture can thus be varied, magnifying small areas if desired.

In arrangement (b) the gantry (not shown) has rotational motion only, stopping to record between each step of rotation. This system can operate

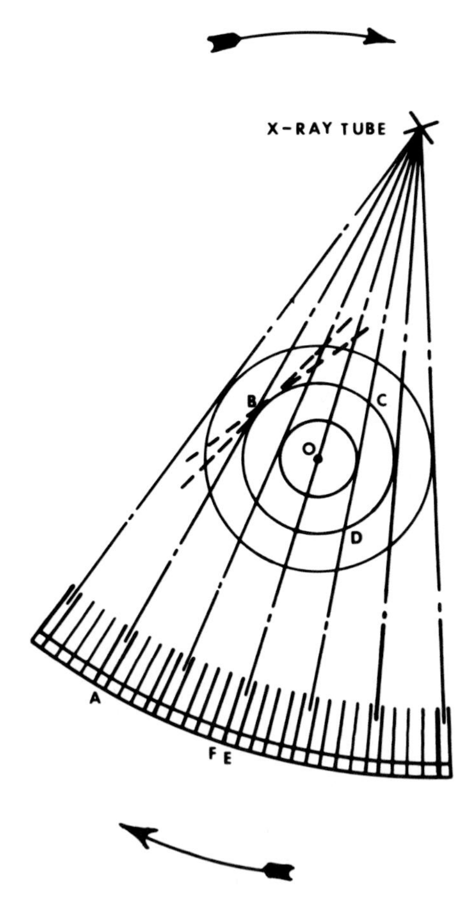

FIG. 10.20 Diagram of Fig. 10.19b indicating that, since different detectors look exclusively at material a fixed distance from the axis of rotation, any imbalance in the detectors will produce light or dark rings on the picture. [From G. N. Hounsfield, *Amer. J. Roentgenol.* **127**, 3 (1976).]

faster since no mechanical return is required, and the complete body is exposed at one time instead of being traversed. The designer's estimate is that it could take the necessary data three times faster than arrangement (a). The disadvantage is that the beam width, and therefore the detector spacing and direction, must be designed for the largest patient and the resolution cannot be varied for greater detail of small areas. During operation the detectors cannot be recalibrated and the center detectors see exclusively the center of the body, so detector error is not averaged over the whole scan. Figure 10.20 indicates the type of error that would accumulate in this mode of operation. For example, detector A will see mostly tissue lying close to the circle *BCD*, and a small error in the gain of the detector relative to the others would translate into the resulting picture as a bright or dark ring. The worst situation is at the center *O*, which is seen by only detector E. Should a minute variation exist between the gain of detector E compared with its neighbor F a spot would occur in the processed picture.

When the advantages and disadvantages of these two designs are considered, accuracy is clearly to be preferred to speed and the design of Fig. 10.19a was selected for the EMI Scanner.

THE EMI SCANNER

The position of the patient in the EMI Scanner is shown in Fig. 10.21. The gantry, which is within the compartment shown in Fig. 10.21, moves as in Fig. 10.19a. During the scan the intensity of the x-ray beam after it emerges from the water path is continuously monitored by a reference detector to correct for any x-ray tube fluctuations, and the system is recalibrated for any drift at the end of each traverse. The number of reconstruction computations made in each of the transverses can be either 160, for coarse examination, or 320 for detailed examination, although there are many more transmission measurements. The final display after computation will be a matrix of either 25,600 pixels (160×160) or 102,400 pixels (320×320), which result from 320,000 transmission measurements.

The computations are completed within a few seconds after the scan is completed and are stored on a magnetic disk. This disk, which contains the digital information, can be used in several ways. A printout of the numerical values can be made to obtain the precise information that the machine has computed, or the disk can be inserted into a video display system, which can be photographed. While in the video display system, sections of the display may be magnified to fill the screen and the intensity contrast is adjustable. This latter feature is advantageous because the sensitivity of the instrument to small changes in ray sums is greater than that of the human

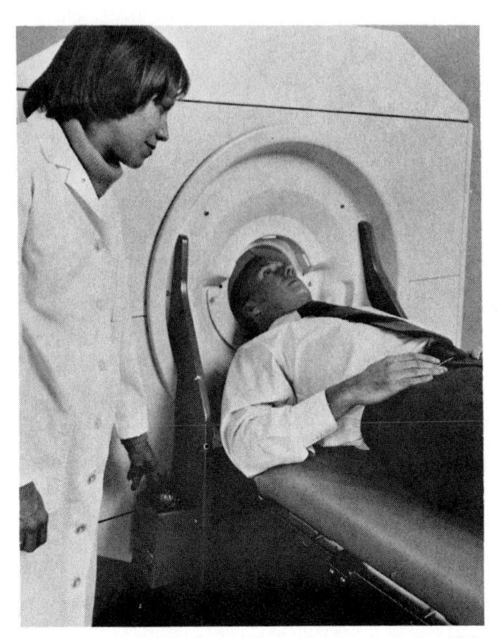

FIG. 10.21 Position of patient in EMI CT for brain scanning. [Courtesy Dr. R. Sassetti, EMI Medical, Inc.]

eye. Therefore, the contrast of a region with apparent low contrast at one setting of the instrument on the video may be enhanced by a simple knob adjustment to a contrast that is more favorable to the range of visual acuity. The image on the cathode ray display can be varied in three ways. (1) A portion of the total picture can be expanded to fill the entire screen. (2) Window width controls the range of the density, or gray, scale (see Fig. 8.20) (3) The window level adjustment sets the midpoint of the density range. As an example, consider Fig. 10.1. If the window width is 400 and window level 0 all the densities below -200 appear black and those above $+200$ appear white. If the window level is shifted to $+100$ then densities below -100 are black and those above $+300$ are white. If instead of shifting the window level the window width is reduced to 200 then the density range from -100 to $+100$ is expanded and more detail is visible for objects within that density range (see Banna, 1976). A series of four different variations of the same video are shown in Fig. 10.22.

The ability of the instrument to take consecutive adjacent tomographs 1.3-cm thick enables a physician to see the axial shape of the pathological area as well as its transverse or cross-sectional shape. Figure 10.23 shows

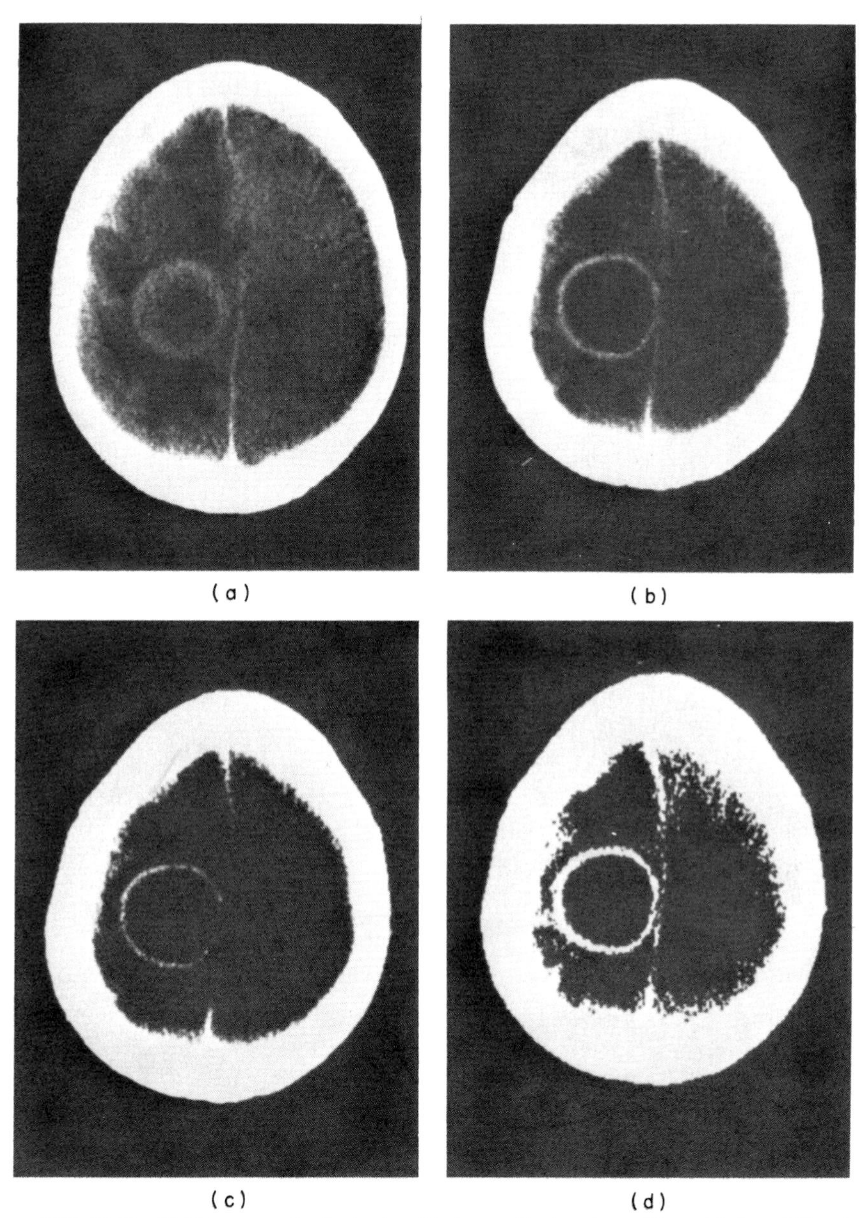

FIG. 10.22 EMI tomograph of a brain with tumor displayed on video with four different variations (a) Window level $L = +38$, window width $W = 75$; (b) $L = +34$, $W = 50$; (c) $L = +30$, $W = 20$; (d) $L = +24$, $W = 1$. [Courtesy Dr. R. Sassetti, EMI Medical Inc.]

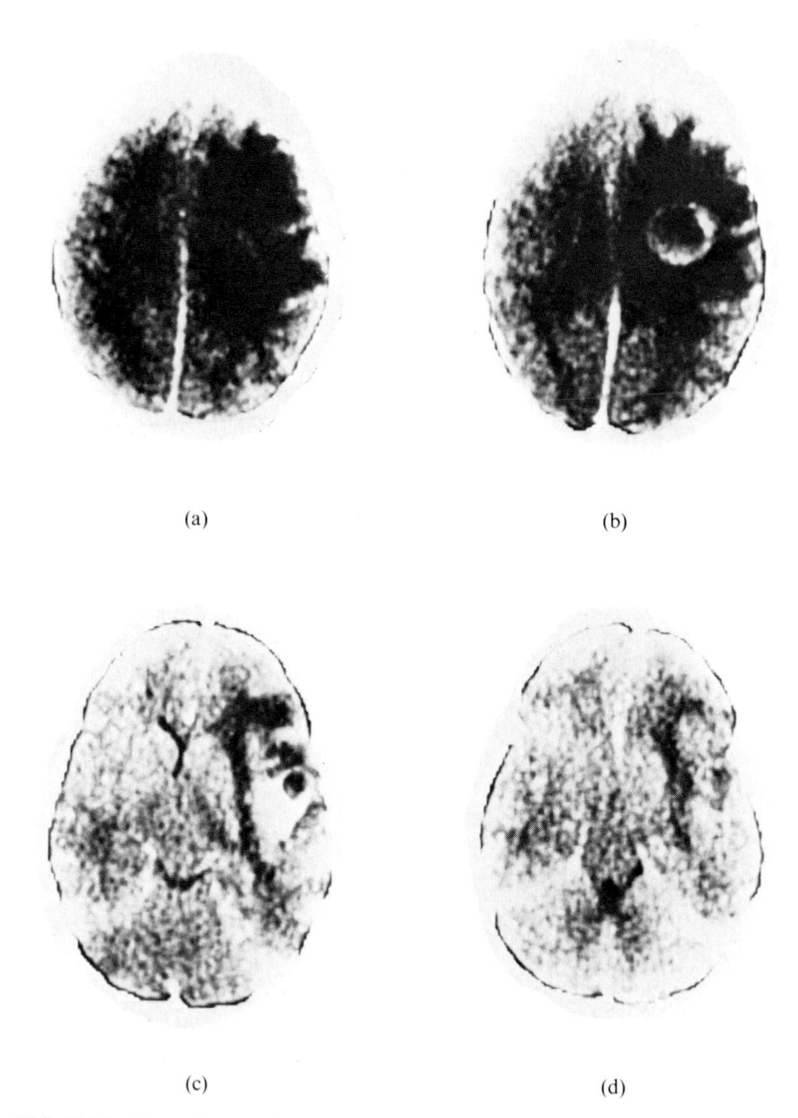

FIG. 10.23 Four adjacent 13-mm wide tomographs of brain with tumor taken with EMI Scanner and photographed from video display. [Courtesy Dr. R. Sassetti, EMI Medical, Inc.]

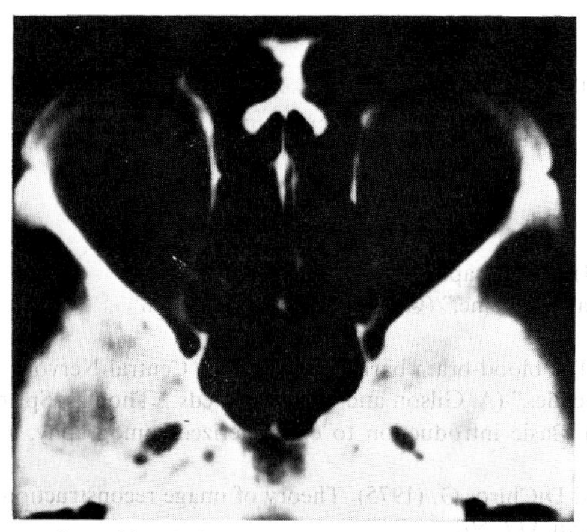

FIG. 10.24 Cross section of head through the optic nerves taken with EMI Scanner. The optic nerves and ophthalmic arteries are clearly visible. [Courtesy Dr. R. Sassetti, EMI Medical, Inc.]

four adjacent tomographs of a region of the brain at the same intensity of display.

One area that has always been extremely difficult to diagnose by conventional x-ray film techniques has been the optic nerve region behind the eyeball. With a tomographic slice in the proper axial position this is now visually accessible, as seen in Fig. 10.24. The resolution of the CT is so good that even the small blood vessels are visible.

Because contrast is lost in reproducing the video output on a printed page as well as in the limitations of the eye to perceive such contrasts in pictures that do not have the adjustable feature of the CT, Figs. 10.22–10.24 do not do justice to the available information that can be seen with the instrument itself.

What are the possibilities for improvement of the spatial resolution in x-ray computerized tomography? If one accepts that the radiation dose currently given a patient is the upper limit of safety then very little improvement can be expected. Statistical analysis has shown (Chesler *et al.* 1977) that in reconstruction tomography the total number of photons N required to obtain an image with a given level of statistical noise varies inversely with the third power of the resolution Δx in the cross-sectional plane, i.e., $N \propto 1/(\Delta x)^3$. Thus, an increase of the resolution by a factor of two requires an increase of photons by a factor of eight. It should be noted that an increase in the number of pixels without increasing the number of photons sacrifices noise level for resolution. The pixel density offered by the EMI Scanner is an optimization of the most reliable data within a pixel with respect to an acceptable statistical noise level (Hounsfield, 1976b).

REFERENCES

Anger, H. O. (1974). Tomography and other depth-discrimination techniques, *in* "Instrumentation in Nuclear Medicine," (G. J. Hine and J. A. Sorenson, eds.), Vol. 2. Academic Press, New York.

Bakay, L. (1971). The blood-brain barrier concept, *in* "Central Nervous System, Investigation with Radionuclides" (A. Gilson and W. Smoak, eds.), Thomas, Springfield, Illinois.

Banna, M. (1976). Basic introduction to computerized tomography, *J. Can. Assoc. Radial.* **27**, 143.

Brooks, R. A., and DiChiro, G. (1975). Theory of image reconstruction in computed tomography, *Radiology* **117**, 561.

Brownell, G. L., Correia, J. A., and Hoop, B., Jr. (1974). Scintillation scanning of the brain, *in* "Annual Review Biophysics and Bioengineering" (L. J. Mullins, ed.), Vol. 3. Annual Reviews, Palo Alto, California.

Chesler, D. A., Aronow, S., Correlle, J. E. *et al.* (1977). Statistical properties and simulation studies of transverse section algorithms, *in* "Reconstruction Tomography in Diagnostic Radiology and Nuclear Medicine" (M. Ter-Pogossian *et al*, eds.). University Park Press, Baltimore, Maryland.

Cho, Z. H. (1974). General views on 3-D image reconstruction and computerized transverse axial tomography, *IEEE Trans. Nucl. Sci.* **NS-21**, 44.

Gado, M. H., Phelps, M. E., and Coleman, R. E. (1975). An extravascular component of contrast enhancement in cranial computed tomography, *Radiology* **117**, 589.

Hine, G. J., and Erickson, J. J. (1974). Advances in scintigraphic instruments, *in* "Instrumentation in Nuclear Medicine" (G. J. Hine and J. A. Sorenson, eds.), Vol. 2. Academic Press, New York.

Hounsfield, G. N. (1976a). Historical notes on computerized axial tomography, *J. Can. Assoc. Radiolog.* **27**, 135.

Hounsfield G. N. (1976b). Picture quality of computed tomography, *Amer. J. Roentgenol.* **127**, 3.

Kuhl, D. E., and Edwards, R. Q. (1970). The Mark III Scanner: A compact device for multiple-view and section scanning of the brain, *Radiology* **96**, 563.

Kuhl, D. E., Edwards, R. Q., Ricci, A. R., and Reivich, M. (1972). Quantitative section scanning using orthogonal tangent correction, *J. Nucl. Med.* **14**, 196.

McCullough, E. C., Baker, H. L., Jr., Hauser, O. W., and Reese, D. F. (1974). An evaluation of the quantitative and radiation features of a scanning x-ray transverse axial tomograph: the EMI Scanner, *Radiology* **111**, 709.

McCullough, E. C. *et al.* (1976). Performance evaluation and quality assurance of computed tomography scanners, with illustrations from EMI, ACTA, and Delta scanners, *Radiology* **120**, 173.

McDavid, W. D., Waggener, R. G., Payne, W. H., and Dennis, M. J. (1977). Correction for spectral artifacts in cross-sectional reconstruction from x-rays, *Med. Phys.* **4**, 54.

Muehllehner, G., and Wetzel, R. A. (1971). Section imaging by computer calculation, *J. Nucl. Med.***12**, 76.

Oldendorf, W. H. (1974). Lipid solubility and drug penetration of the blood brain barrier, *Proc. Soc. Exp. Biol. Med.* **147**, 813.

Oldendorf, W. H. (1975). Molecular criteria for blood-brain barrier penetration, *in* "Non-invasive Brain Imaging" (H. J. DeBlanc, Jr. and J. A. Sorenson, eds.). Society for Nuclear Medicine, New York.

Phelps, M. E., Gado, M. H., and Hoffman, E. J. (1975a). Correlation of effective atomic number an electron density with attenuation coefficients measured with polychromatic x-rays, *Radiology* **117**, 585.

Phelps, M. E., Hoffman, E. J., Gado, M., and Ter-Pogossian, M. M. (1975b). Computerized transaxial transmission reconstruction tomography, *in* "Non-invasive Brain Imaging" (H. J. DeBlanc, Jr. and J. A. Sorenson, eds.). Society for Nuclear Medicine, New York.

CHAPTER **11**

Cryobiology: Cell Freezing and Cryosurgery

INTRODUCTION

In this chapter we consider the effects of an artificial, controlled environment on a cell, namely, freezing. It is a technique that has increasing application in the storage of cells, tissues, and organs for use later in another living organism, and is also used in surgical procedures to selectively kill groups of cells. It is therefore important to understand the principles involved so that proper use of this technique can be made and so that its use may be extended beyond that of the present. There is also a philosophical aspect to the understanding of the principles of cryobiology. Many people believe, or want to believe, that a deceased human may be frozen, preserved, and thawed out in the future when medical science has a cure for their ailment. By the end of this chapter the reader should not only be convinced that such a procedure is unlikely but should also understand why. However, before considering the effects of freezing on groups of cells we must first examine the effect on a single cell.

The freezing and thawing of living cells or more complex organisms has been a subject of much speculation and research. A good deal of empirical data has been amassed for specific problems, such as the storage and transport of bull spermatozoa. In this case, as in others, it has been found that the survival rate is dependent on three primary physical factors: (1) cooling rate, (2) storage time and temperature, and (3) the warming rate. There are other factors that are more intrinsic to the material, such as the medium in which the cells are suspended, the medium in which they are grown, and their rate of growth at the time of freezing.

The physics of the processes is beginning to be understood and we will primarily follow the quantitative work of Mazur.

GENERAL FINDINGS

The interior of the cell is a mixture of proteins, ions, and water. When cells are in an aqueous suspension the external water freezes and upon a further lowering of the temperature the ionic solution of the cytoplasm inside the cell, which has a freezing point of −6°C to −10°C, becomes supercooled until a temperature low enough is reached, at which the intracellular water freezes. Evidence indicates that the intracellular water does not freeze spontaneously but is nucleated by the external ice. This is seen in Table 11.1, which shows that the freezing temperature within the cell is lowered if external ice is absent, a condition obtained experimentally by suspending the cells in silicone oil, paraffin, etc. In addition, Rasmussen *et al.* (1975) have shown that as long as the external water is not frozen, i.e., it is supercooled, there is no evidence of freezing of yeast cells or E. Coli bacteria; this was done to −40°C. These interior freezing temperatures are determinable on the cold stage of a microscope because the cell turns opaque upon freezing.

The freezing of water generally requires nucleating agents, i.e., either an ice crystal or crystalline substances relatively insoluble in water that are crystallographically similar to water. Fletcher (1962) has reported that, in the absence of nucleating agents, droplets of water in the size range of 1 μm to 1 cm in diameter can be supercooled to −39°C and −30°C, respectively.

TABLE 11.1

	External ice present (°C)	External ice absent (°C)	Reference
Guinea pig testis	−6 to −10	−18 to −20	Smith and Smiles (1953)
Sea urchin eggs	−8	< −15	Asahina (1961)
Insect larvae	−10	−20 to −30	Salt (1953, 1963)
Muscle fibers	−10	< −15	Chambers and Hale (1932)
Various plant cells	−7	−20	Levitt (1966)

Molecules and ions in solution are apparently not nucleating agents. For example, if a water sample that normally supercools to −20°C has its freezing point lowered by solutes to −2°C it will supercool to −22°C.

Since the freezing of cells seems to require external ice, it is concluded the ice must be the nucleating agent. If it is assumed that the cell membrane must remain intact for viability upon thawing, the contact of the internal water with the ice must be through the cell pores. This effect has been demonstrated by piercing the cell wall with an ice-filled glass capillary. It has also been shown that if a supercooled muscle fiber is touched with an ice-filled glass capillary, external ice grows along the fiber but produces no intracellular freezing until it reaches the cut end (Chambers and Hale, 1932).

SURFACE TENSION

Since surface tension alters the shape of the water–ice surface in a capillary pore, it must be considered in freezing of water in a capillary. However, we must first acquire a general understanding of surface tension and the Young relation.

Molecules in a system are, in general, attracted to one another by intermolecular forces, e.g., electrostatic, van der Waals, etc. These forces are a function of distance—the closer the molecules, the greater the forces. Therefore, a molecule in the interior of a liquid is surrounded by like molecules and has a certain potential energy. A molecule on the surface of a liquid has close attraction to the liquid but not to the gas or vapor, and there is a resultant attraction inward. If each molecule on the surface has a resultant attraction inward then the lowest energy configuration of the surface is spherical. If this spherical drop of liquid is put onto a solid surface there may be some attraction of the liquid molecules to the solid. If the attraction is negligibly small the liquid will tend to remain spherical although distorted by gravitational force, whereas if the attraction is greater to the solid the liquid will spread out. This latter phenomenon is known as wetting the surface.

If the surface of the liquid sphere spreads out the surface area increases. In order to increase the surface area it has been necessary to bring molecules out from the interior to the surface. Because less molecules surround a surface molecule, work is required to remove the molecule from the interior to the surface. This work is done against the attractive forces of the molecules adjacent to the one being brought from the interior. The *free surface energy* is defined as the work required to increase the surface area by 1 cm^2. The surface has a continuing tendency to contract, i.e., if the solid were removed the drop would return to a spherical shape. This contraction tendency is

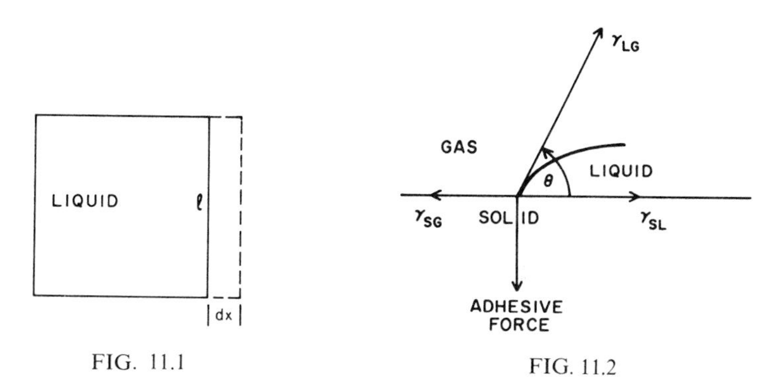

FIG. 11.1

FIG. 11.2

called *surface tension*, which obviously is the same at every point of the surface and in every direction along the surface. Surface tension is defined as the force per unit length acting at right angles to any line on the surface. Its symbol is usually γ. The work of extending the surface area by 1 cm^2 is the force along 1 cm times a 1-cm distance through which the force is moved, i.e., $(\gamma) \times$ (area) (see Fig. 11.1). An incremental amount of work dW, may be written as

$$dW = \gamma l\, dx = \gamma\, dA = E_F\, dA. \tag{11.1}$$

Thus E_F, the surface free energy (energy/unit area), and surface tension γ (force/unit length) are both numerically and dimensionally equal, and the terms are often used interchangeably. Note that the surface free energy E_F in the above process is actually the Gibbs free energy G. (see Appendix A, Eq. (A.5).)

If we consider the case of a liquid in contact with a solid and its saturated vapor (a condition under which this type of measurement is made) we can introduce a term θ called contact angle, as in Fig. 11.2.

The general definition of "wetting" is that $\theta < 90$ degrees. In Fig. 11.2 there are three different surface tensions or free energies involved: (1) γ_{SG}, free energy of the solid–gas interface; (2) γ_{LG}, free energy of the liquid–gas interface; and (3) γ_{SL}, free energy of the solid–liquid interface.

The relationship between contact angle and the surface tensions can be seen in the figure. Since γ_{GS}, γ_{SL}, and γ_{GL} represent forces, and the forces are in equilibrium at the point where all three are in contact, the following relation must exist because the sum of the forces is zero at the contact point:

$$\gamma_{LG} \cos\theta + \gamma_{SL} - \gamma_{SG} = 0$$

or

$$\cos\theta = (\gamma_{SG} - \gamma_{SL})/\gamma_{LG} \tag{11.2}$$

This equation is known as *Young's relation* and in today's literature the determination of surface free energies by measurement of the contact angle is known as the "*sessile drop*" method.

FREEZING OF WATER IN A CELL PORE

Now let us consider the effect of surface tension on the freezing point of water in a capillary, which is the idealization of a cell pore. Assume an ideal cylindrical pore as in Fig. 11.3, in which the ice has a spherical surface of radius r whose tangent angle at the membrane wall is θ, the radius of the pore being a.

If the interfacial tensions are defined at equilibrium as γ_{SL} for ice–water, γ_{LC} for water–capillary, and γ_{SC} for ice–capillary, then by analogy with Eq. (11.2) we may write

$$\cos\theta = (\gamma_{SC} - \gamma_{LC})/\gamma_{SL} \tag{11.3}$$

The ice–water interface will be spherical to minimize the surface area, and its radius is defined in Fig. 11.3 as

$$r = a/\cos\theta \tag{11.4}$$

The change of radius in ice in a small capillary will change its freezing temperature. The relation of the radius to the freezing temperature is derived by Davies and Rideal (1961) in the following way. If dn moles of ice are transferred at constant temperature, volume, and surface area from the interior of a large mass of ice with a planar surface to the interior of the ice in the capillary, following the definition of Eq. (A.13′), the change in the Gibbs free energy is

$$\partial G_S = (\mu_{SC} - \mu_S)\,\partial n \tag{11.5}$$

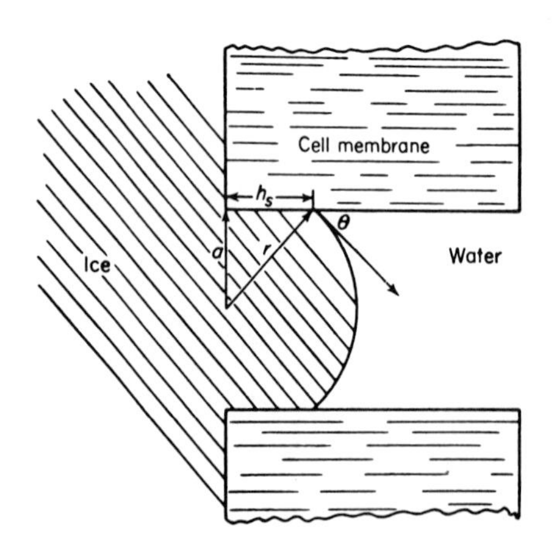

FIG. 11.3

where μ_{SC} is the chemical potential of ice in the capillary and μ_S that of ice in the exterior planar crystal. ∂G, from Fig. 11.3, is equal to the work done to expand the ice–capillary interface. If h_S is the length of the ice–capillary interface, its area is $2\pi a h_S$ and therefore

$$\partial G_S = \gamma_{SC}\, dA_{SC} = \gamma_{SC}\, d(2\pi a h_S) = 2\pi a \gamma_{SC}\, dh_S \tag{11.6}$$

The increase in the volume of ice in the capillary is

$$dV_S = d(\pi a^2 h_S) = \pi a^2\, dh_S \tag{11.7}$$

and if v_S is the molar volume of ice the above increase in volume can also be expressed as

$$dV_S = v_S\, dn \tag{11.8}$$

Substituting Eqs. (11.6)–(11.8) into Eq. (11.5) yields the relation

$$(\mu_{SC} - \mu_S)(a/v_S) = 2\gamma_{SC} \tag{11.9}$$

If we follow the same procedure for the transfer of dn moles of water from exterior water with a planar surface to the interior of the water inside the capillary we obtain a similar relation:

$$(\mu_{LC} - \mu_L)(a/v_L) = 2\gamma_{LC} \tag{11.10}$$

where μ_{LC} and μ_L are the chemical potentials of water in the capillary and of bulk water, respectively, and v_L is the molar volume of water.

When ice and water in the capillary are at the freezing point they are in equilibrium and their chemical potentials are equal:

$$\mu_{LC} = \mu_{SC} \tag{11.11}$$

If we assume $v_S \simeq v_L$, then by substitution of Eq. (11.9) and (11.10) into Eq. (11.11) we obtain

$$\mu_L - \mu_S = (2v_L/a)(\gamma_{SC} - \gamma_{LC}) \tag{11.12}$$

Substituting Eq. (11.3) into Eq. (11.12) yields

$$\mu_L - \mu_S = (2v_L/a)\gamma_{SL} \cos\theta \tag{11.13}$$

From Appendix A, Eq. (A.16) can be written as

$$\mu_L = \mu^0 + RT \ln P_L \qquad \text{and} \qquad \mu_S = \mu^0 + RT \ln P_S \tag{A.16$'$}$$

where P_L and P_S are the equilibrium vapor pressures over planar surfaces of water and ice, respectively. Substitution of these relations into Eq. (11.13) yields

$$RT_C \ln(P_L/P_S) = 2v_L \gamma_{SL} \cos\theta/a$$

and thus the freezing point of water in the capillary T_C can be expressed as

$$T_C = 2v_L \gamma_{SL} \cos\theta / aR \ln(P_L/P_S) \qquad (11.14)$$

This equation states that if $\cos\theta$ is positive, i.e., $0 < \theta < 90$ degrees, the vapor pressure of the liquid P_L must be greater than the vapor pressure of the ice P_S for T_C to be positive. Note that in the geometry of Fig. 11.3 this is the condition in which the water wets the walls of the capillary. Substitution of Eq. (11.4) into (11.14) gives

$$T_C = 2v_L \gamma_{SL} / Rr \ln(P_L/P_S) \qquad (11.15)$$

We may write the integrated form of the Clapeyron–Clausius equation (A.55) with the temperatures T_L and T_S of the liquid and solid water, respectively, as limits:

$$\ln(P_L/P_S) = (\Delta H/R)(T_L - T_S)/T_L T_S \qquad (A.55')$$

where ΔH is the heat of fusion of ice.

Outside the capillary the freezing point of water is 0°C and therefore $T_S = 273$°K. The temperature of the supercooled liquid, which is T_L in this equation, is its freezing temperature in the capillary, defined as T_C in Eq. (11.14). Thus by making these changes in notation, substitution of Eq. (A.55′) into (11.14) gives an expression for the lowering of the freezing point of water in a capillary as

$$\Delta T \equiv T_S - T_C = (2v_L \gamma_{SL} T_S \cos\theta)/(a\,\Delta H)$$

and, by substituting Eq. (11.4) we obtain

$$\Delta T = (2v_L \gamma_{SL} T_S)/(r\,\Delta H) \qquad (11.16)$$

Equation (11.16) is identical with the Kelvin equation, which relates the vapor pressure, and thereby the melting point, over a spherical solid surface to the radius of curvature r.

The γ_{SL} of Eq. (11.16) has been estimated from experiments of freezing in capillaries, and, although r is difficult to measure, γ_{SL} is given by Fletcher (1962) as being between 10 and 25 erg/cm^2. If we assume that the liquid water perfectly wets the capillary pore, i.e., $\theta = 0$ degrees ($\cos\theta = 1$) we obtain from Eq. (11.16)

$$\Delta T \simeq 300/a \qquad \text{where } a \text{ is in angstroms}$$

If -10°C is taken as the freezing temperature this would give a pore radius of 30 Å, somewhat larger than the 4 to 8 Å estimated from diffusion data. Probably the most serious assumption is that $\theta = 0$ degrees. Jackson and Chalmers (1958) used a similar formulation in their consideration of the buckling of northern roads when the mud freezes and pointed out that θ

may well be greater than 0 degrees, if one considers the strain energy of the interface between the solid substrate and the ice. Thus there is no requirement that $\theta = 0$ degrees. If, for example, $\theta = 75$ degrees, $\cos\theta = 0.25$ and the pore radius for $-10°C$ supercooling would be about 7 Å, a value close to other measurements (Banin and Anderson, 1975).

It is seen from the above arguments that although an uncertainty of some factors still exists, the mechanism of freezing within a cell has a satisfactory explanation in that it is nucleated by the external ice passing through the pores. It is now possible to understand the data of Table 11.1, which shows that if external ice is eliminated greater supercooling is achieved.

CELL DEHYDRATION KINETICS

We have seen by the data and the preceding discussion that when a cell is supercooled it is not in equilibrium and hence will freeze when a suitable nucleating agent touches the fluid inside. Another way of achieving equilibrium is for water to leave the cell. It may do this because the vapor pressure of the supercooled fluid is higher than that of the external ice. When water leaves the cell a higher solute concentration is present in the cell, i.e., the same solutes are within the cell but the amount of water has decreased, and the freezing point is lowered.

We can see by Eq. (A.57) that the freezing point of a solution is lower the higher the concentration of solute. With the higher concentration of solute the vapor pressure is reduced and there is a lowered tendency for water to leave. The time dependence of the outflow of water is therefore concentration dependent and a differential equation will result. There are a variety of factors that enter into this situation, but to keep the concepts simple we will consider the important parameters to be: (1) the rate of cooling B; (2) the permeability of the cell to water k; (3) the temperature coefficient of this permeability; (4) the initial volume of water V_i; and (5) the number of osmoles of solute in the cell n_2.

Note on osmole. When pure water is separated from a solution by a membrane permeable only to water the force causing the flow of water through the membrane, i.e., the osmotic pressure, is proportional to the number of particles in solution. 1 gram-equivalent of a dissolved nondiffusible substance is equal to 1 osmol. If the substance ionizes into two ions, such as NaCl, then one half of a gram-molecular weight is 1 osmol. In other words, each Avagadro's number of nondiffusibility entities in solution constitute 1 osmol. Since an osmole is such a large quantity, the term milliosmole (10^{-3} osmol) is more commonly used. Note that since the number of particles per gram is inversely proportional to the molecular weight, a gram of

NaCl of molecular weight 59 will have a vastly greater osmotic pressure than a gram of a protein molecule of typical molecular weight 300,000. That is, since 300,000/59 = 5102, for equal weights the osmotic pressure of NaCl is $2 \times 5102 \simeq 10^4$ times greater than that of the protein.

Let us now assume that the protoplasm of the interior of a cell is an ideal solution, in which case Raoult's law of partial pressures applies, viz., if P^0, P_i, and X_i are the vapor pressure of pure water, the vapor pressure of water in the protoplasm, and the mole fraction of intracellular water, respectively, one may write

$$P_i = P^0 X_i \tag{11.17}$$

Taking logarithms of both sides yields

$$\ln P_i = \ln P^0 + \ln X_i$$

and differentiating with respect to temperature yields

$$d \ln P_i/dT = (d \ln P^0/dT) + (d \ln X_i/dT) \tag{11.18}$$

If we substitute the form of the Clapeyron–Clausius equation given by Eq. (A.52) for $\ln P^0$ into Eq. (11.18) we obtain

$$d \ln P_i/dT = (\Delta H_v/RT^2) + (d \ln X_i/dT) \tag{11.19}$$

where ΔH_v is the molar heat of vaporization.

Eq. (A.52) can also be written for the ice in the external medium as

$$d \ln P_e/dT = \Delta H_s/RT^2 \tag{11.20}$$

where ΔH_s is the molar heat of sublimation and P_e is the vapor pressure of pure ice. Or, if there are solutes present externally, it is assumed that the solution remains in equilibrium with ice, and hence its vapor pressure will also be P_e.

If we subtract Eq. (11.19) from Eq. (11.20) and note that $\Delta H_s - \Delta H_v = \Delta H_f$, the molar heat of fusion, we obtain

$$d \ln(P_e/P_i)/dT = (\Delta H_f/RT^2) - (d \ln X_i/dT) \tag{11.21}$$

We may substitute for X_i, the mole fraction of water in the protoplasm, if we note the following relation. If n_1 and n_2 are the moles of water and solute, respectively, then by definition

$$X_i = n_1/(n_1 + n_2)$$

If we write $\overline{V}_1$ as the partial molar volume of water (Eq. (A.9′)) multiply the above numerator and denominator by it, and define $V = n_1\overline{V}_1$ as the volume of water in the cell, we obtain

$$X_i = n_1\overline{V}_1/(n_1\overline{V}_1 + n_2\overline{V}_1) = V/V + n_2\overline{V}_1) \tag{11.22}$$

It is now assumed for simplicity that the partial molar volume of water $\overline{V}_1$ may be replaced by the molar volume V_1^0 and Mazur (1963a) cites data that show this to be a valid assumption. Equation (11.22) may be substituted into Eq. (11.21) and, upon taking the logarithm and differentiating, the following expression is obtained:

$$d \ln(P_e/P_i)/dT = \Delta H_f/RT^2 - [n_2 V_1^0/(V - n_2 V_1^0)V](dV/dT) \quad (11.23)$$

This equation gives a relation of ratio of the vapor pressure of supercooled water inside the cell to that of water at thermodynamic equilibrium outside the cell as a function of the volume of the intercellular water and its change with temperature.

The rate of the water volume change of a cell is proportional to the osmotic pressure, the area, and the permeability of the cell wall to water, or

$$dV/dt = kA\pi \quad (11.24)$$

where the permeability constant (called k in this section to avoid confusion with other symbols) is given in units of μ^3 of water per μ^2 of surface area per minute per atmosphere difference in osmotic pressure between the inside and the outside of the cell, A is the surface area of the cell, and π is the osmotic pressure. We will use a modified form of Eq. (A.40) for the osmotic pressure:

$$\pi = (RT/V_1^0) \ln(P^0/P) \quad (A.40')$$

where V_1^0 has been substituted for $\overline{V}_1$ and P^0 and P are the vapor pressures of pure water and water in solution, respectively. Substituting Eq. (A.40') into Eq. (11.24) we obtain

$$dV/dt = (kART/V_1^0) \ln(P_e/P_i) \quad (11.25)$$

where a substitution of the P_e for P^0 and P_i for P has been made to apply to the present situation.

It is not expected, however, that k is temperature independent. Indeed, quite the contrary is the case. Since the transport of a molecule across a barrier is a thermally activated process one expects a Boltzmann exponential temperature dependence and experiments have shown this to be true. Figure 11.4 is a plot of the temperature dependence of k of various mammalian blood cells to water and the data are fitted by the relation

$$\ln k = a + bT \quad (11.26)$$

Since b is the slope, if we take k_0 as the permeability at T_0 we may write

$$\ln k_0 = a + bT_0 \quad (11.27)$$

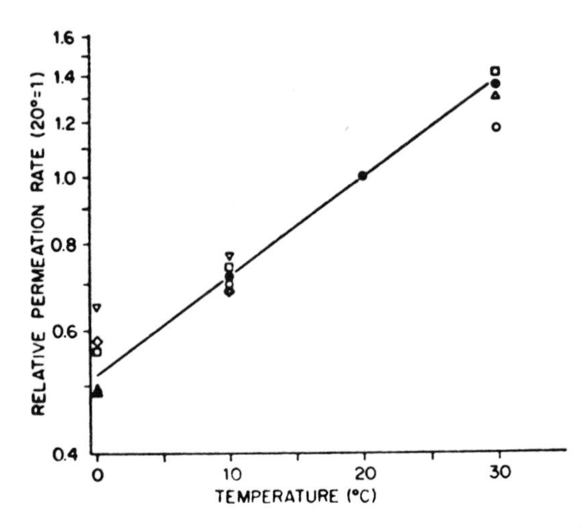

FIG. 11.4 The effect of temperature on the relative permeability of various mammalian red blood cells to water. The relative permeation was calculated as 1/time for hemolysis. ●, rat; ○, man; ◇, rabbit; □, guinea pig; Δ, dog; ∇, ox. [From Mazur (1963a), based on data of Jacobs *et al.* (1935).]

and, upon subtracting Eq. (11.27) from Eq. (11.26) we have

$$\ln (k/k_0) = b(T - T_0) \tag{11.28}$$

Expressed exponentially, Eq. (11.28) may be written as

$$k = k_0 e^{b(T - T_0)} \tag{11.29}$$

The simplest mathematical rate of cooling is linear, i.e.,

$$dT/dt = B \tag{11.30}$$

where B is a constant in degrees per unit time, and only this case will be considered. Time can be eliminated by combining Eq. (11.30) with Eq. (11.25) to obtain

$$dV/dT = (kART/BV_1{}^0)\ln(P_e/P_i) \tag{11.31}$$

Eq. (11.29) can be substituted for k in Eq. (11.31) to obtain

$$\ln(P_e/P_i) = (dV/dT)(BV_1{}^0/k_0 e^{b(T - T_0)}ART)$$

which is substituted into Eq. (11.23), obtaining finally

$$\frac{d}{dT}\left(\frac{V_1{}^0 B}{ARk_0}\frac{1}{T}e^{-b(T - T_0)}\frac{dV}{dT}\right) = \frac{\Delta H_f}{RT^2} - \frac{n_2 V_1{}^0}{(V + n_2 V_1{}^0)V}\frac{dV}{dT}$$

If one assumes V and T are the only variables and performs the differentiation, the following differential equation is obtained

$$Te^{b(T_0-T)}\frac{d^2V}{dT^2} - \left[(bT+1)e^{b(T_0-T)} - \frac{ARk_0 n_2}{B(V+n_2V_1{}^0)}\frac{T^2}{V}\right]\frac{dV}{dT} = \frac{\Delta H_f A k_0}{BV_1{}^0} \tag{11.32}$$

This equation relates the volume of the intracellular water to temperature, the cooling rate, and the four parameters characteristic of the cell—A, k_0, b, and n_2.

This equation is not soluble analytically, so computer solutions were obtained. Initial values which were taken (a) at $dV/dT = 0$ when $T \geq T_f$, the freezing point of protoplasm; and (b) at temperatures $T \geq T_f$, for $V = V_i$, the initial volume of the intracellular water. Typical solutions are presented in Fig. 11.5. They show the percentage of initial water remaining in a cell at various temperatures for different cooling rates.

Figure 11.6 shows the percentage of water remaining in different size cells at two different cooling rates. In these figures the left-hand curve is that

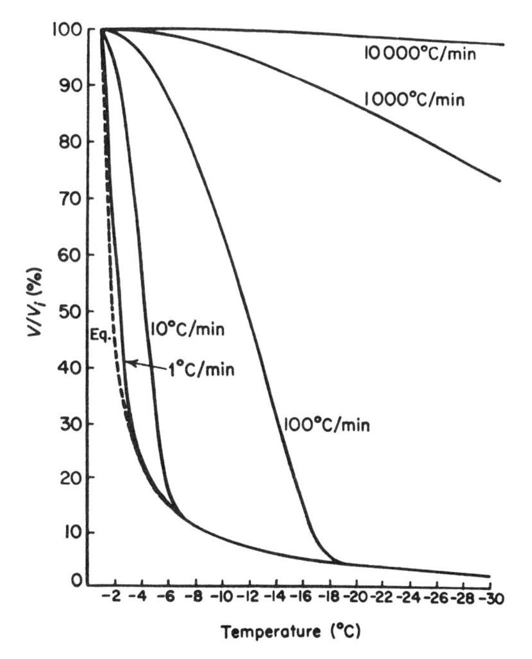

FIG. 11.5 Calculated percentages of supercooled intracellular water remaining at various temperatures in 5.84-μm diameter spherical cells cooled at several rates. [From Mazur (1963a).]

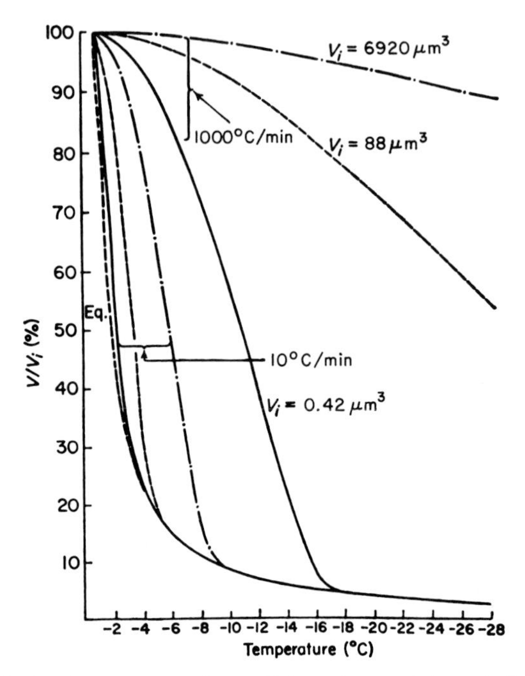

FIG. 11.6 Calculated percentages of intracellular water remaining at indicated temperatures in cells of various sizes undergoing cooling at 10°C/min or 1000°C/min. The value of k is the same as in Fig. 11.5. [From Mazur (1963a).]

obtained for cooling at an infinitely slow rate, i.e., $B = 0$. This is the condition in which there is ample time for diffusion of water out of the cell at each incrementally lower temperature.

The constants used in the numerical solutions are given by Mazur (1963a), but our purposes are simply those of comparison. It is seen in these figures that the higher the cooling rate the more water remains in the cell at a given temperature or, for a given fraction of remaining water, the greater it is supercooled. If the permeability is higher, the cells lose water faster and the curves will move to the left; conversely, if the permeability is lowered, the curves move to the right since more water remains behind for a given rate of cooling.

Since the volume of a sphere increases as r^3 and its surface by r^2, larger cells have a smaller surface to volume ratio than do smaller cells. Thus, for a given rate of cooling, larger cells retain more water than do smaller cells. This effect is shown clearly in Fig. 11.6, in which there are two groups of curves, one group for a cooling rate of 10°/min and the other for 1000°/min. Within both of these groups are three curves, with initial volumes of 0.42, 88, and 6920 μm^3. These were selected for cells with diameters of 1, 6, and 25 μm and with corresponding surface-to-volume ratios of 7.2, 1.2, and 0.28. It is seen that there is a dramatic change in the water retention for a constant permeability and cooling rate.

CHEMICAL DAMAGE

The slowest cooling rate, i.e., the one that approximates the equilibrium curve, is, from the above considerations, the best to eliminate water from the cell and thereby prevent intracellular freezing. However, the elimination of water to lower the freezing point has the resulting effect of increasing the concentration of solutes within the cell. This can result in chemical damage to the cell, and it is suspected that NaCl is the prime offender.

Not much is known about internal chemical damage at the present time. At best it can be stated as a general principle that if chemical damage is taking place then it too has an exponential rate with temperature, i.e., the higher the temperature the faster the damage. From this standpoint, if the cell is to be dehydrated then the faster this occurs, the better. Thus, for the prevention of ice damage, the slower the cooling rate the better while the opposite holds for the prevention of chemical damage. Figure 11.7 summarizes these effects. The heavy curve *COB* represents survival from the combination of the effects.

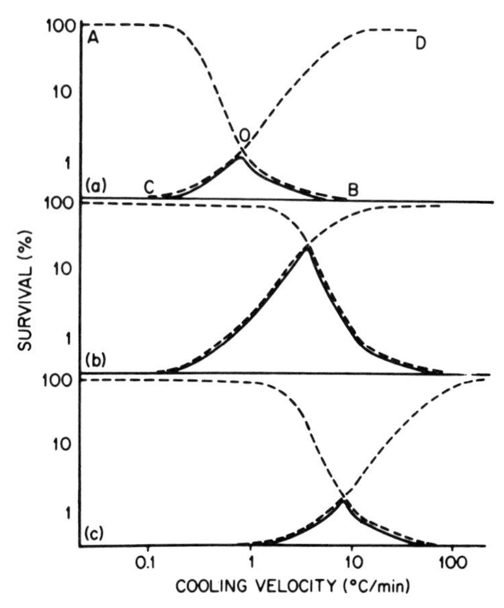

FIG. 11.7 Hypothetical relation between cooling velocity and survival. Curve *AOB* represents the effect of intracellular freezing by itself. Curve *COD* represents the effect of solute concentration by itself. Curve *COB* represents the combined effects. (a), (b), and (c) represent three different hypothetical combinations of cooling rates and chemical damage rates. [From Mazur (1966).]

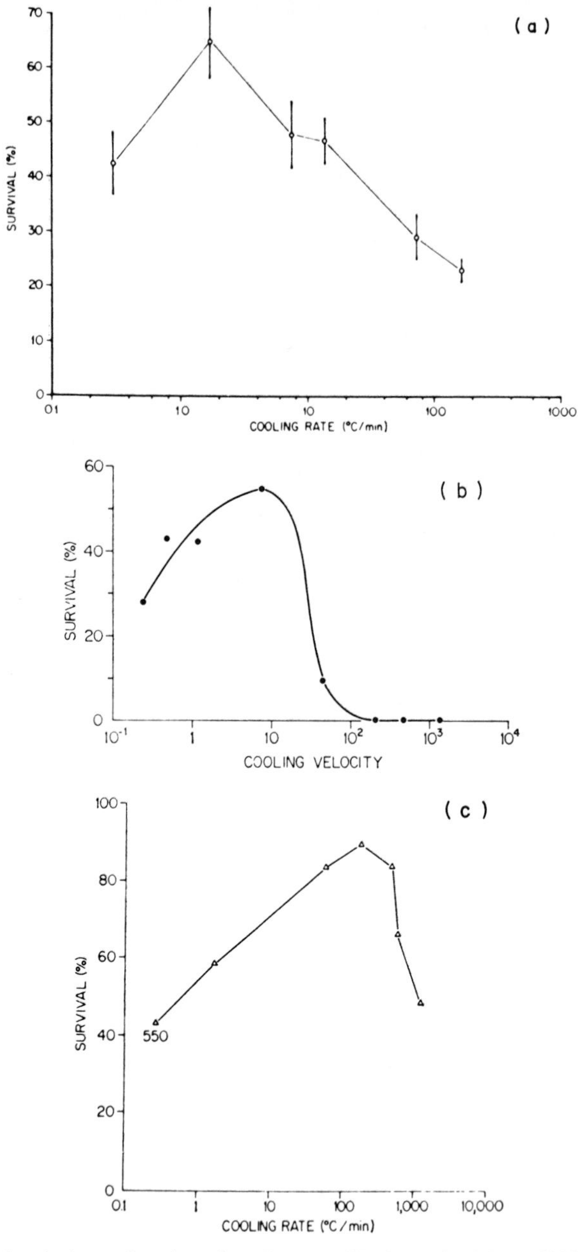

FIG. 11.8 Survival as a function of cooling rate. [Redrawn by Mazur (1977).] Cells were frozen at controlled rates to about −75°C, transferred to liquid nitrogen (−196°C) and then thawed rapidly at 500 to 1000°C/min. (a) Mouse marrow stem cells suspended in solutions of glycerol in saline. [From Leibo *et al.* (1970).] (b) Yeast cells suspended in water. [From Mazur and Schmidt (1968).] (c) Human red blood cells suspended in solutions of glycerol in saline. [From Miller and Mazur (1976).]

Different constants of different types of cells can lead to the variety of curves indicated by Figs. 11.7a, 11.7b, and 11.7c, where shifts in the relative positions of *AB* and *CD* can lead to higher survivals at the optimum rate. The maxima *O* observed in the curves are found in experiments. Figure 11.8 illustrates the percentage survival of different types of cells as a function of cooling rate. Not only is the height of the maximum different for the different types of cells but the cooling rate required to obtain the maximum is quite different for each. These experiments confirm the model illustrated in Fig. 11.7.

It is seen that a 100% survival rate is unattainable in the cases shown. In a complex organism, which contains different types of cells, each with its own constants for dehydration and chemical constituents with their individual damage rates, the average survival rate for all cells would be less than that which could be attained for the best single cell. Therefore a general rule may now be stated: the more complex the organism, the lower is its chances of survival under even the best possible controlled conditions of freezing and thawing. This is a result of the general principles of diffusion and chemical reaction rates.

In the single cell case if the maximum water loss for survival is known one may use Eq. (11.32) to predict the minimum cooling rate necessary to prevent freezing. For example, the constants for red blood cells used in Eq. (11.32) indicate that the minimum cooling rate is about 2500 to 5000°C/min. Experiments confirming this are discussed below.

THAWING RATE

If the cooling is done sufficiently slowly down to −10 to −20°C to cause significant dehydration, then further cooling to any lower temperature can be done as rapidly as desired without further damage. However, the subsequent warming rate is a significant factor in survival for the same reasons as those during cooling. The chemical reaction rate can damage cells while water is being reabsorbed, and this reaction rate depends on time, temperature, and concentration. Obviously, the water diffusion rate across the membrane is a controlling factor in how rapidly dilution of the intracell solutes may be achieved during warming. Slow warming *per se* is not necessarily deleterious, but time spent at certain temperatures for given cells can be damaging. In fact, there is recent evidence that the damage that can occur during different rates of thawing depends on the rate at which the specimen has been frozen (Miller and Mazur, 1976).

An example of death during warming is shown in Fig. 11.9. In this figure, *S. cerevisiae* has been cooled rapidly to below −150°C. It was then rapidly

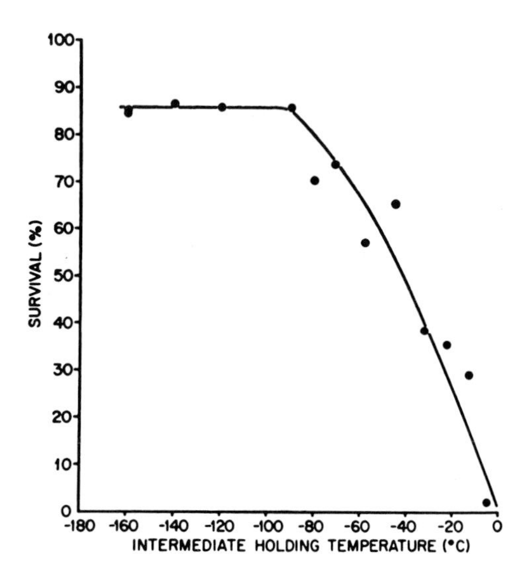

FIG. 11.9 Survival of *S. cerevisiae* cooled rapidly to below $-150°C$, warmed rapidly to the indicated temperatures, held there for 30 min, and then thawed rapidly. [From Mazur (1966). Based on data by Goetz and Goetz (1938).]

warmed to the temperatures indicated by the data points and held for 30 min, then rapidly warmed to room temperature. It is seen that holding above $-80°C$ causes death—the higher the temperature, the greater the percentage of death. The general conclusion from this figure is that the more rapidly the warming, the greater the survival.

STORAGE DEATH

If internal ice crystals form during cooling it is a very rapid process. The individual crystals do not have time to form in an equilibrium configuration and there is both nonequilibrium strain energy and surface energy present. With time, ice crystals tend to fuse together to form larger crystals. This phenomenon is known as *regelation* and was studied by many well-known scientists of the last century, e.g., Faraday, Tyndall, and Lord Kelvin. Only in the last two decades, however, has a satisfactory model been developed. This model arose out of the general study of the phenomenon of sintering. i.e., the fusing of particles by heating.

Although the problem is quite complex, the elementary principles elucidated by Kingery (1960) can be summarized in the following way.

If we examine the contact of two spherical particles as in Fig. 11.10 at two different times we see that the welding process is the growth of the neck

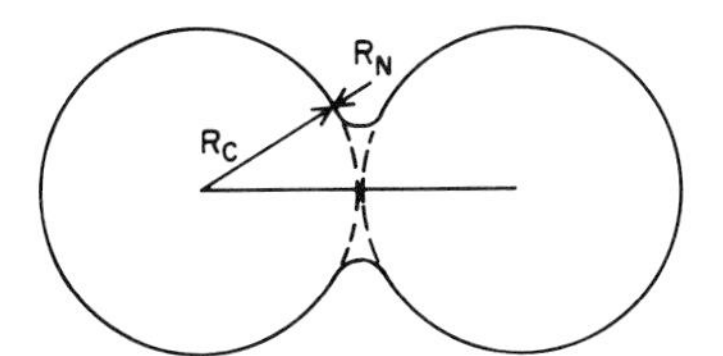

FIG. 11.10 Model for initial stages of sintering of small ice spheres where R_N and R_C are the radii of curvature of the necks and crystals (spheres), respectively. The dashed lines are the configurations at time = 0. [Redrawn from Kingery and Berg (1955).]

between them. This neck growth occurs because material flows from the crystals to the neck under the influence of a chemical potential difference. This difference arises from the difference in radius of curvature between the neck R_N and the crystal R_C. We have seen from the law of Laplace (Eq. (1.13)) that the pressure varies with the radius of curvature as

$$P = \gamma/R$$

where γ is the surface tension. For the pressure difference that arises from the two radii we may write

$$\Delta P = P_N - P_C = \gamma[(1/R_N) - (1/R_C)]$$

From our definition of chemical potential in terms of molar volume $\overline{V}_1$, and by Eqs. (A.6′), (A.9′), (A.13′), and (A.15) we may write

$$\mu - \mu_0 = \Delta P \overline{V}_1$$

Using this, we obtain the relation

$$\mu_N - \mu_C = \overline{V}_1 \gamma[(1/R_N) - (1/R_C)] \tag{11.33}$$

Since $\Delta G = \mu_N - \mu_C$ tends to a minimum, the direction of minimization of the Gibbs free energy is to increase the radius of the neck. Kingery (1960) has shown that the transport of material occurs by surface diffusion of H_2O from the large radius region to that of the small radius. Since this is a surface diffusion rather than a bulk diffusion, it occurs at a much lower activation energy and can take place well below freezing. However, it is still a thermally activated process and proceeds exponentially with temperature. Thus, although it takes place in a conventional deep freeze unit, with modern liquid nitrogen storage facilities there is no regelation. The growth of large ice crystals from smaller ones during storage could damage cell membranes and therefore there is a time–temperature relation for storage death—the lower the storage temperature, the greater the survival for a given time. However, high intracell solute concentrations at intermediate temperatures can reverse the relationship. An example of storage death at different temperatures is shown in Fig. 11.11, where the survival of cells as a function of

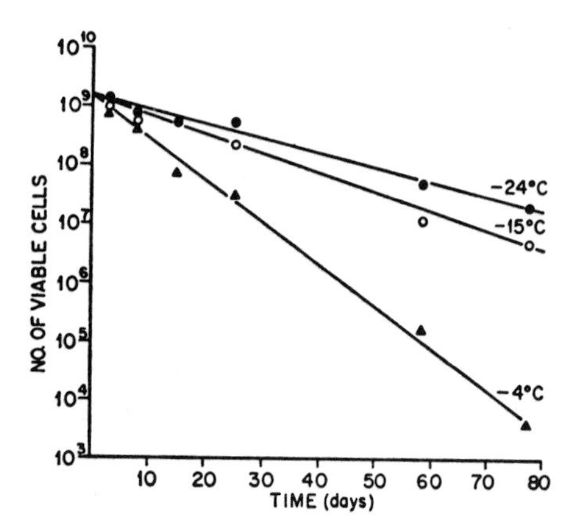

FIG. 11.11 Survival of cells of *S. cerevisiae* as a function of storage time at three different storage temperatures. [From Mazur (1966). Based on data of Stille (1950).]

time at three different storage temperatures (−4°C, −15°C, and −24°C) is plotted. It should be noted that even at −24°C more than 90% die in two months time. However, no storage death has been observed at −196°C, at which temperature chemical activity in cells does not take place.

Considering the death rates that occur upon freezing, storage, and thawing, it is remarkable that any cells survive. However, many can, and experiments have indicated optimum conditions. Furthermore, there are indications that better control of these conditions as well as chemical changes in the freezing solution can lead to improvements.

FREEZING OF RED BLOOD CELLS

The freezing and storage of human tissue is an extremely important process in medicine. The most extensive studies of the basic phenomena involved have concerned the freezing of blood and some data will now be shown that indicate that the principles developed in the previous sections are valid. However, it is rare for a biological system to behave quite so simply and it will be evident that there are other factors involved.

It has been shown by Lovelock (1953) that a critical temperature region for hemolysis (destruction of red blood cells) is −4°C to −40°C. In a series of experiments he subjected red blood cells to freezing and thawing in different temperature ranges. The data of Table 11.2 indicate that the time spent in thawing can be even more injurious than the freezing.

Lovelock (1953) further investigated the effect of changing the external concentration of NaCl and found that red blood cells frozen in a higher concentration of the saline solution could be kept equally well at −45°C as

TABLE 11.2

Time (sec) spent between $-3°C$ and $-40°C$		
Freezing	Thawing	% hemolysis (cell death)
0.3	1.5	7
0.3	50	90

at $-180°C$, but at $-39°C$ total hemolysis occurred within 1 min. Clearly there is a temperature–time–concentration relationship. Red blood cells are not normally affected by sudden chilling or by centrifuging. However, in the presence of NaCl solutions greater than 0.8 *M* they become unstable and hemolyze if suddenly cooled or centrifuged at an acceleration greater than 20*g*.

If the cells are put in a saline solution of slightly higher concentration than that of their interior they tend to become shrunken, presumably because of osmotic diffusion (see Section A.8 of Appendix A) of water from the inside of the cell to the higher concentration on the outside. However, this does not occur if the cells are put in a solution stronger than 0.8 *M*. Apparently the membrane becomes more permeable to the sodium ions and a new osmotic equilibrium is reached. However, this new permeability weakens the cell wall, as mentioned above, and the cell is less resistant to the stresses of thermal shock. Clearly, not only is there a time–temperature–concentration relation for the diffusion of water through the cell membrane but also one for the loss of strength of the membrane.

Studies of the effect of cooling velocities on hemolysis have been made by cooling blood cells in different size capillary tubes to different temperatures. Figure 11.12 shows data by Gehenio *et al.* (1963) of hemolysis of blood in two different diameter capillaries, curve A for 1-mm i.d. and curve B for 0.5-mm i.d. Since these curves were obtained from data taken by sudden immersion in the cold bath, the lower temperatures represent higher cooling rates. The sample in the smaller capillary also freezes faster than that in the larger and therefore undergoes a higher cooling rate at the same immersion temperature. These data show the minimum of the hypothetical curve of Fig. 11.7 that resulted from the combination of dehydration rate versus time of chemical reaction within the cell.

Calibration measurements of the cooling rate for these data have been made by the same investigators (Luyet† *et al.* 1963) and the experimental

† Prof. B. J. Luyet (1897–1974) personally financed and privately published the journal *Biodynamica*. Prior to his death he donated all of the plates to the Biomedical Research Institute, Rockville, Maryland.

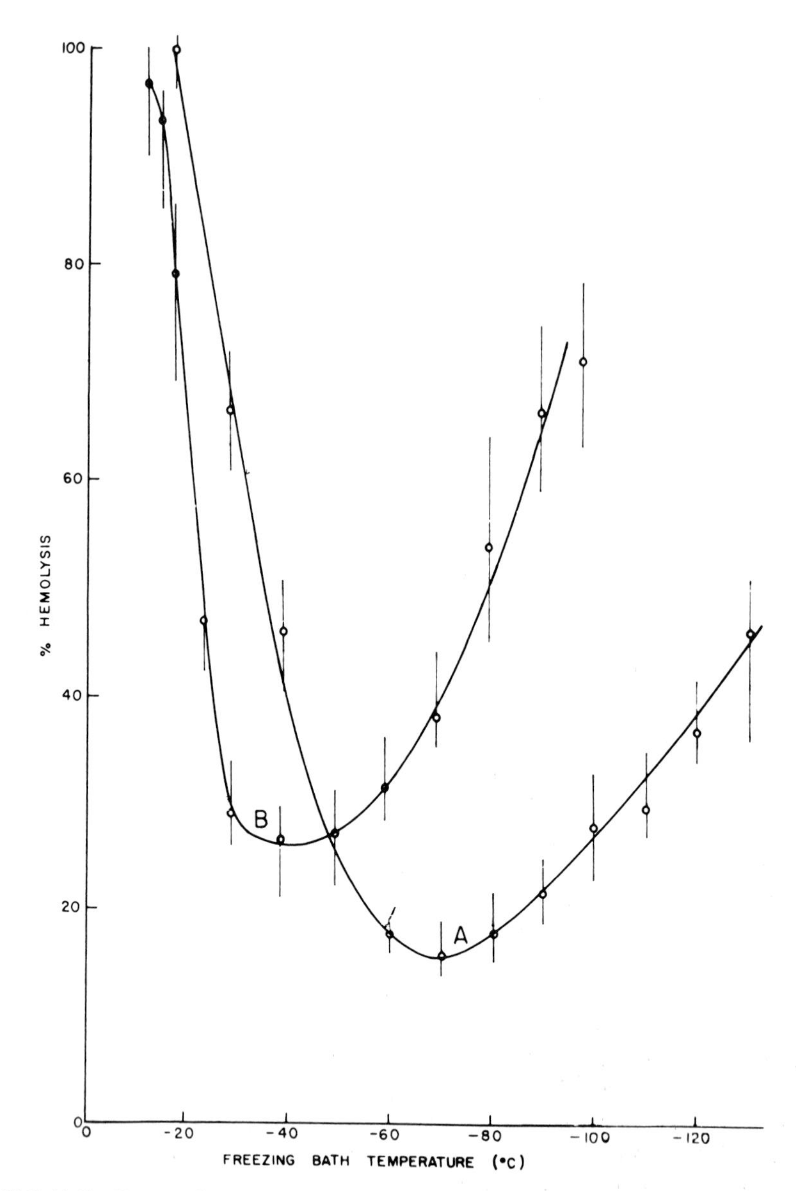

FIG. 11.12 Percent destruction (hemolysis) of red blood cells frozen in capillary tubes immersed in cooling baths at various low temperatures. Curve A, capillaries of 1-mm i.d.; curve B, capillaries of 0.5-mm i.d. [From Gehenio *et al.* (1963).]

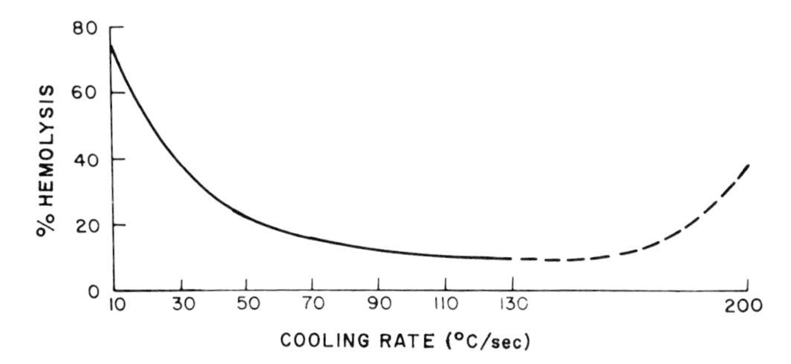

FIG. 11.13 Hemolysis in terms of cooling rate for data similar to that of Fig. 11.12. Dashed line is hemolysis in terms of approximate cooling rates calculated from the data. [From Luyet *et al.* (1963).]

results of Fig. 11.12 are expressed as cooling rates in Fig. 11.13. These numbers substantiate the prediction by computer analysis discussed earlier.

Interesting studies have been made on the use of protective additives. Two different types of behavior of additives have been observed, one that permeates the cell, e.g., glycerol (and dimethylsulfoxide), and an other that does not permeate, e.g., sucrose (and dextrose). Figure 11.14 illustrates these types of behavior. Curve 1 shows hemolysis versus temperature to which the specimens are cooled with no additives. Blood cells can be cooled to $-3°C$ with no hemolysis, but cooling to the region of $-10°C$ to $-20°C$ causes 100% hemolysis. The remainder of curve 1 is that of Fig. 11.12. The data of curve 2 were taken with 10% glycerol and that of curve 3 with 10% sucrose. It was observed that glycerol causes an increase in the number of

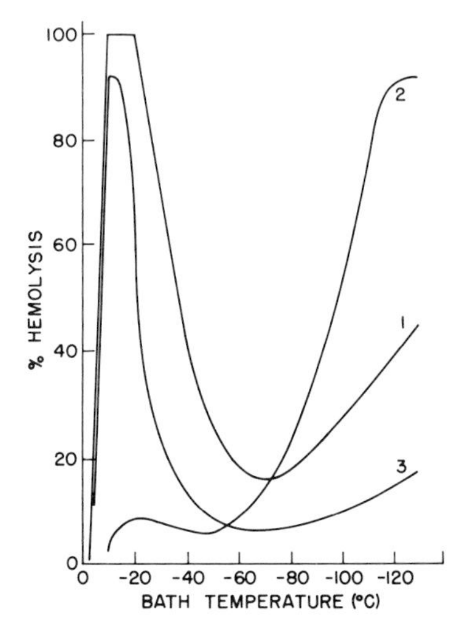

FIG. 11.14 Percent hemolysis of red blood cells versus cooling bath temperature (1) with no protective agent, (2) in a solution of 10% glycerol, (3) in a 10% sucrose solution. [Redrawn from Rapatz and Luyet (1965).]

ice nucleation sites and an increase in viscosity of the cell interior, while the effects of sucrose do not seem to be known. Clearly, there remain fertile fields for fundamental investigations of the role of solutes in nucleation theory and diffusion kinetics and, until such investigations are made, the effects of additives can only be hypothesized.

CRYOSURGERY IN OPHTHALMOLOGY

The use of cold probes in the body will generally kill tissue, and in certain cases it is a more convenient method than excision with a scalpel. Before proceeding to connect the theory of cell survival, discussed earlier, with surgical techniques, we will mention briefly two other uses of a cold probe that occur uniquely in ophthalmology (Bellows, 1971).

The cornea of the eye is very resistant to damage from freezing temperatures. Because of this resistance, infection of the cornea by herpes virus can be significantly reduced by touching the cornea for a few seconds with a probe filled with dry ice (−79°C). The contact is made for about 7–10 sec and defrosted with a stream of water. Contact is made several times with each infected part of the cornea and the cure rate is very high. In this procedure the cornea is unharmed by the freezing, while the virus is killed.

The second unique use of a cold probe in ophthalmology is in the removal of cataracts. In this situation it is desired to remove the lens behind the cornea. Surgically this is done by grasping the lens with tweezers, after a suitable incision is made, and pulling it free from the lens capsule. If a flat cold probe is placed in contact with the lens an ice column is formed that results in a larger area of contact with the lens than by tweezers. Furthermore, the ice column has a higher tensile strength than the unfrozen lens. These two effects result in both a stronger and a more positive bond between the lens and the extracting instrument, which facilitates removal.

The cell damage properties of freezing are utilized in two other ophthalmic procedures. A detached retina or a retinal tear can be mended by scar tissue resulting from cauterization. Much of this is now done by lasers, but in certain situations selective freezing of the spot is preferred. This is accomplished with a cold probe. In the condition of glaucoma the ciliary body produces an excess of fluid within the eye, thereby causing a greater than normal internal pressure. Selected application of a cold probe can kill parts of the ciliary epithelium and thereby reduce the rate of fluid formation. These latter two procedures are cases of cell destruction by freezing, which is discussed in the next section.

CRYOSURGERY

The transfer of heat can be calculated from Fick's laws of diffusion, Eqs. (3.4) and (3.5), where heat is the diffusing quantity rather than mass. The heat transported will be proportional to the product of the thermal conductivity, the cross-sectional area, and the temperature gradient. This problem can be solved for a variety of geometries with appropriate choice of coordinate systems, and many books contain methods of solution.

The situation is a bit more complicated for an advancing frozen interface in a solution, but it is still soluble. In this situation the thermal exchange at the freezing boundary must be considered. The general expression for the rate of boundary advance db/dt is

$$\Delta H(db/dt) = k_i(dT/ds)_i - k_w(dT/ds)_w \tag{11.34}$$

where k_i and k_w are the thermal conductivities of ice and water, respectively, ΔH is the latent heat, and dT/ds is the temperature gradient. The temperature gradient times the thermal conductivity equals the heat flow. Thus this formula states that the difference in heat flow into and out of the boundary equals the net heat production, the latent heat times the rate of boundary advance. When realistic conditions are considered, such as density change, heat capacities, etc., the differential equation is more complicated, although solutions exist for simple geometries (Walder, 1968, 1971). The mathematical situation becomes even more intractible for *in vivo* calculations in which other terms such as blood perfusion rate and metabolic heat rate are involved. Then a computer is necessary, and solutions of this more complex situation have been obtained by Cooper and Trezek (1971).

We will examine the principles involved by the following comparison with experiment. Meryman (1966) placed thermocouples 1-cm apart radially in a tube of starch solution (to reduce crystal formation). The tube was insulated at the ends and immersed in a bath at $-72°C$. Temperatures vs. time were then recorded. These data are shown in Fig. 11.15. From these curves it can be seen that there are different cooling rates at different distances. For example, the region 1 cm from the periphery cools from 0°C to $-30°C$ at a rate of about 10°/min, whereas the region 4 cm from the periphery cools at a rate of about 1°/min over the same temperature range.

Consider now a cold probe at $-100°C$ held at that temperature for 100 sec in tissue. The result may be diagrammed as in Fig. 11.16. The dashed line represents the boundary of the tissue to be destroyed and the perimeter of the ball of ice just encompasses this. The probe is allowed to warm and we assume that in 200 sec the ice ball has melted. There will be a temperature gradient through the ice ball, but for simplicity of discussion we will assume three temperature zones—A, B, and C.

The cells in region A have been cooled from 37°C to about $-80°C$ (average for region A) in 100 sec, or with an average rate of approximately 1.2°/sec. However, the curves of Fig. 11.15 show that such a linear average is not valid. Examination of the curve at 1 cm from the plate shows that the cooling rate between 0°C and $-30°C$ is about five times larger than rate between

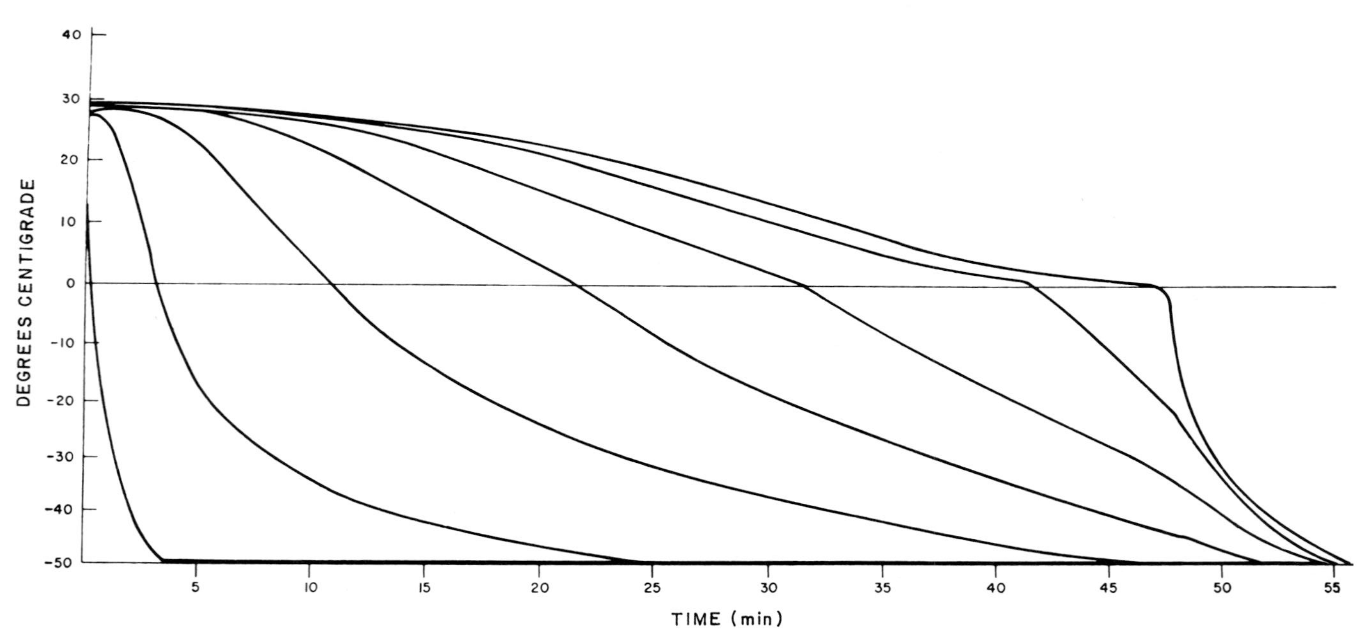

FIG. 11.15 Temperature curves recorded from thermocouples placed at 1-cm intervals from the periphery to the center of a cylinder filled with 7% starch solution. The cylinder was insulated at the ends and immersed in a bath at −72°C so that freezing took place from the periphery to the center. [From H. T. Meryman (1966).]

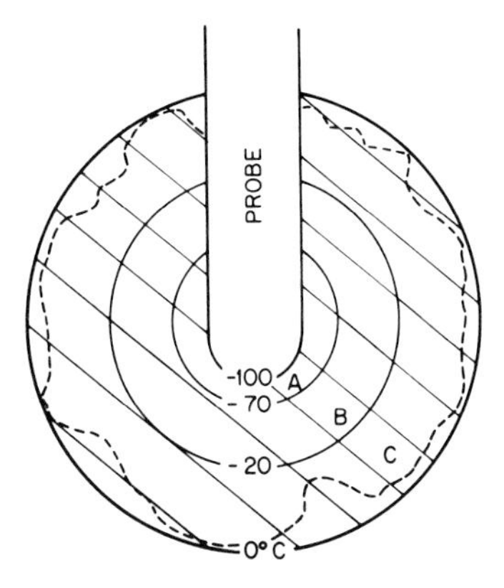

FIG. 11.16 Schematic representation of tissue subjected to cryosurgical freezing. The probe surface is assumed to be −100°C. The dotted line represents the boundary of the tissue to be destroyed. The solid line is the boundary of the sphere of ice. [From P. Mazur, *in* "Cryosurgery" (R. W. Rand, A. P. Rinfret and H. von Leden, eds.), 1968. Courtesy of Charles C. Thomas, Publisher, Springfield, Illinois.]

+30°C and −50°C. Since the region of interest for cell death is from 0°C down we should assume the higher cooling rate, or 5 × 1.2°C = 6°C/sec = 360°C/min. Although the melting requires 200 sec, Meryman (1966) has shown that, because of the high thermal diffusivity and low heat capacity of ice, it takes only 1/10 the time to warm the ice ball to the melting point as it does for complete melting. Thus the ice ball has warmed from −80°C to 0°C in 20 sec, or at a rate of 4°C/sec or 240°C/min. The middle curve of Fig. 11.17 for *S. cerevisiae* can be used for a rough estimate. It is seen that with these cooling and warming rates most cells in region A have been killed.

In region C the cells have been cooled to an average temperature of −10°C and again the critical question for survival is how fast they have been cooled from 0°C to −10°C. It should be noted, however, that cells are not killed totally until the temperature drops below −10°C regardless of the cooling rate. If we assume the temperature in the region 0°C to −10°C follows a curve of Fig. 11.15 between 1 and 2 cm then the cooling rate is between 5°C and 10°C/min. The warming rate will be 10°C in 20 sec or 30°C/min. The lower curve of Fig. 11.17 indicates that a considerable fraction of the cells would survive.

From these simple considerations Mazur (1968) concludes that the following procedure should be followed to lessen the survival of cells that are to be killed by freezing:

(1) Provide sufficient cooling so that the sphere of ice expands well beyond the estimated outer limits of the target area. To be more precise, the

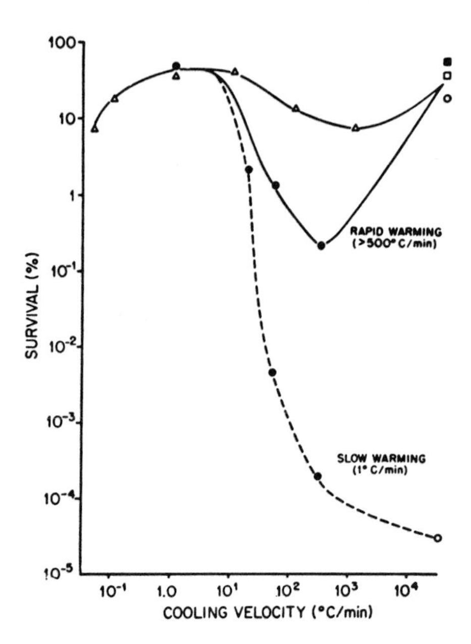

FIG. 11.17 Survival of *S. cerevisiae* as a function of cooling velocity and high, medium, and low rates of warming. [From Mazur (1968). See also Mazur and Schmidt (1968).]

ice sphere should be allowed to expand until all of the estimated target has been cooled to below −20°C, as measured by thermocouples.

(2) Allow the frozen sphere to warm slowly. It would be desirable to have at least 10 to 30 min elapse between the onset of warming and the disappearance of all ice.

The reasons for these recommendations have been developed in earlier sections of this chapter. We have seen in Table 11.1 that no cell can be supercooled below −20°. Most cells will be killed by intracellular freezing if cooled rapidly enough. Those that do survive will most likely be killed during the slow thawing because of exposure to a higher chemical concentration within the cell. The penalty for the larger ice ball is that some healthy tissue will be killed, so precision estimates of the ice ball size should be made. Several freezing cycles may be an advantageous procedure but this has not been sufficiently explored.

In conclusion it is appropriate to quote Meryman (1966) in his Foreword to the book "Cryobiology":

> Cryobiology is clearly at an early stage in its development, propelled headlong, by its applications, far beyond the limits of current

basic understanding. What cryobiology needs now is the participation at the basic research level of more than the present handful of investigators.

REFERENCES

Asahina, E. (1961). Intracellular freezing and frost resistance in egg-cells of the sea urchin, *Nature* (*London*) **191**, 1263.

Banin, A., and Anderson, D. M. (1975). A similar law may govern water freezing in minerals and living organisms, *Nature* (*London*) **255**, 261.

Bellows, J. G. (1971). Cryosurgery in opthalmology, *in* "Cryogenics in Surgery" (H. von Leden and W. G. Cahan, eds.). Medical Examination Publ., Flushing, New York.

Chambers, R., and Hale, H. P. (1932). The formation of ice in protoplasm, *Proc. Roy. Soc. London* **B110**, 336.

Cooper, T. E., and Trezek, G. J. (1971). Mathematical predictions of cryogenic lesions, *in* "Cryogenics in Surgery" (H. von Leden and W. G. Cahan, eds.). Medical Examination Publ., Flushing, New York.

Davies, J. T., and Rideal, E. K. (1961). "Interfacial Phenomena." Academic Press, New York.

Doebbler, G. F., Rowe, A. W., and Rinfret, A. P. (1966). Freezing of mammalian blood and its constituents, *in* "Cryobiology" (H. T. Meryman, ed.). Academic Press, New York.

Fletcher, N. H. (1962). "The Physics of Rain Clouds." Cambridge Univ. Press, London and New York.

Gehenio, P. M., Rapatz, G. L., and Luyet, B. J. (1963). Effects of freezing velocities in causing or preventing hemolysis, *Biodynamica* **9**, 77.

Goetz, A., and Goetz, S. S. (1938). Death by devitrification in yeast cells, *Biodynamica* **2**, 1.

Harris, R. J. C. (ed.) (1954). "Biological Applications of Freezing and Drying." Academic Press, New York.

Jackson, K. A., and Chalmers, B. (1958). Freezing of liquids in porous media with special reference to frost heave in soils, *J. Appl. Phys.* **29**, 1178.

Jacobs, M. H., Glassman, H. N., and Parpart, A. K. (1935). Osmotic properties of the erythrocyte, VII, The temperature coefficients of certain hemolytic processes, *J. Cell. Comp. Physiol.* **7**, 197.

Kingery, W. D. (1960). Regulation, surface diffusion and ice sintering, *J. Appl. Phys.* **31**, 833.

Kingery, W. D., and Berg, M. (1955). Study of the initial stages of sintering of solids by viscous flow, evaporation-condensation, and self diffusion, *J. Appl. Phys.* **26**, 1205.

Leibo, S. P., Farrant, J., Mazur, P., Hanna, M. G., Jr., and Smith, L. H. (1970). Effects of freezing on marrow stem cell suspensions: Interactions of cooling and warming rates in the presence of PVP, sucrose, or glycerol, *Cryobiology* **6**, 315.

Levitt, J. (1966). Winter hardiness in plants, *in* "Cryobiology" (H. T. Meryman, ed.). Academic Press, New York.

Lovelock, J. E. (1953). The haemolysis of human red blood cells by freezing and thawing, *Biochim. Byophys. Acta* **10**, 414.

Luyet, R. J., Rapatz, G. L., and Gehenio, P. M. (1963). On the mode of action of rapid cooling in the preservation of erythrocytes in frozen blood, *Biodynamica* **9**, 95.

Mazur, P. (1963a). Kinetics of water loss from cells at subzero temperatures and the likelihood of intracellular freezing, *J. Gen. Physiol.* **47**, 347.

Mazur, P. (1963b). Studies on rapidly frozen suspensions of yeast cells by differential thermal analysis and conductometry, *Biophys. J.* **3**, 323.

Mazur, P. (1966). Physical and chemical basis of injury in single-celled micro-organisms subjected to freezing and thawing, *in* "Cryobiology" (H. T. Meryman, ed.). Academic Press, New York.

Mazur, P. (1968). Physical–chemical factors underlying cell injury in cryosurgical freezing, *in* "Cryosurgery" (R. W. Rand, A. P. Rinfret, and H. von Leden, eds.). Springfield, Illinois.

Mazur, P. (1977). The role of intracellular freezing in the death of cells cooled at supraoptimal rates, *Cryobiology* **14**, 251.

Mazur, P., and Schmidt, J. J. (1968). Interactions of cooling velocity, temperature and warming velocity on the survival of frozen and thawed yeast, *Cryobiology* **5**, 1.

Meryman, H. T. (1966). Review of biological freezing, *in* "Cryobiology," (H. T. Meryman, ed.). Academic Press, New York.

Miller, R. H., and Mazur, P. (1976). Survival of frozen-thawed human red cells as a function of cooling and warming velocities, *Cryobiology* **13**, 404.

Rapatz, G., and Luyet, B. (1965). Effects of cooling rates on the preservation of erythrocytes in frozen blood containing various protective agents, *Byodynamica* **9**, 332.

Rasmussen, D. H., Macaulay, M. N., and Mackenzie, A. P. (1975). Supercooling and nucleation of ice in single cells, *Cryobiology* **12**, 325.

Salt, R. W. (1953). The influence of food on the cold hardiness of insects, *Can. Entomologist* **85**, 261.

Salt, R. W. (1963). Delayed inoculation freezing of insects, *Can. Entomolog.* **95**, 1190.

Smith, A. V., and Smiles, J. (1953). Microscopic studies of mammalian tissues during cooling to −79°C, *J. Roy. Microsc. Soc.* **73**, 134.

Stille, B. (1950). Untersuchungen über den Kältetod von Mikroorganismen, *Arch. Mikrobiol.* **14**, 554.

Walder, H. A. D. (1968). Experimental cryosurgery, *in* "Cryosurgery" (R. W. Rand, A. P. Rinfret, and H. von Leden, eds.). Thomas, Springfield, Illinois.

Walder, H. A. D. (1971). Experimental cryosurgery, *in* "Cryogenics in Surgery" (H. von Leden and W. G. Cahan, eds.). Medical Examination Publ., New York.

APPENDIX A

Chemical Thermodynamics

A.1 INTRODUCTION

There has been frequent use in this book of relations from the field of chemical thermodynamics in the development of physical concepts. In order to have all readers at the same level of knowledge and to avoid having to send them to other texts, the specific relations required in the chapters are all derived in this appendix.

A.2 FREE ENERGY

The first law of thermodynamics is a statement of the conservation of energy and is written as

$$dQ = dU + P\,dV \tag{A.1}$$

where Q is the total amount of heat, U is the internal energy, and PdV is the work done. The total amount of heat varies with the path chosen and can be divided in many ways between work and internal energy. However, if the transfer of heat at a given temperature T occurs reversibly then it is independent of path. This reversible change of heat per unit temperature dQ/T is defined as the change in entropy dS and if the process is irreversible it can be shown that $dS > dQ/T$. We may therefore substitute for dQ in Eq. (A.1) and write

$$T\,dS \geq dU + P\,dV \tag{A.2}$$

Thermodynamics can make no specific predictions about irreversible phenomena, so we consider here only reversible ones and write Eq. (A.2) as

$$dU + P\,dV - T\,dS = 0 \tag{A.3}$$

If we deal with condensed systems, which are essentially incompressible, we ignore the $P\,dV$ term and write

$$dU - T\,dS = 0 \tag{A.3$'$}$$

Note that this does not apply to condensed systems under high pressure. At constant temperature Eq. (A.3′) may be written

$$d(U - TS) \equiv dF = 0 \tag{A.4}$$

where $U - TS$ is defined as a function F, which is called the *Helmholz free energy*. If, however, work is done on or by the system we need to include it. We therefore define the *Gibbs free energy* with symbol G as

$$G = U + PV - TS$$

and

$$dG = dU + P\,dV + V\,dP - T\,dS - S\,dT \tag{A.5}$$

Using Eq. (A.3), we obtain for reversible processes

$$dG = V\,dP - S\,dT \tag{A.6}$$

If we have an isothermal system

$$dG = V\,dP \tag{A.6$'$}$$

Note that up to this point all of the relationships follow rigorously from the laws of thermodynamics. If now we use a particular molecular model such as the ideal gas law, which for 1 mole is

$$PV = RT, \tag{A.7}$$

and substitute Eq. (A.7) into Eq. (A.6′) and integrate we obtain the relation

$$\Delta G = \int_{G_1}^{G_2} dG = \int_{P_1}^{P_2} V\,dP = RT \int_{P_1}^{P_2} \frac{dP}{P} = RT \ln \frac{P_2}{P_1} \tag{A.8}$$

A.3 PARTIAL MOLAR QUANTITIES

The thermodynamic functions discussed above, such as heat content, free energy, etc., depend not only on the state, as defined by the temperature and pressure of the system, but also on the amount of material. Thus the entropy of n moles is n times the entropy of one mole, where n may be greater or less than unity. By contrast, the properties of temperature and pressure do not depend upon the mass of the system. Properties that depend upon mass are known as *extensive properties*; we represent them by X and let n_1, n_2, etc., be the numbers (or fractions) of moles making up the system. The complete differential dX is then given by

$$dX = (\partial X/\partial T)_{P,n_1,n_2,\ldots}\,dT + (\partial X/\partial P)_{T,n_1,n_2,\ldots}\,dP + (\partial X/\partial n_1)_{T,P,n_2,\ldots}\,dn_1 + (\partial X/\partial n_2)_{T,P,n_1,\ldots}\,dn_2 + \cdots \quad \text{(A.9)}$$

The quantities $\partial X/\partial n$ are called *partial molar quantities* and are written in an abbreviated form by writing a bar over the quantity. Thus, for example,

$$\overline{X}_1 = (\partial X/\partial n_1)_{T,P,n_2,\ldots}, \qquad \overline{X}_2 = (\partial X/\partial n_2)_{T,P,n_1,\ldots}, \cdots$$

Specifically, if X represents the volume of the system then

$$\overline{V}_1 = (\partial V/\partial n_1)_{T,P,n_2\ldots} \quad \text{(A.9')}$$

and $\overline{V}_1$ would be called the *partial molar volume* of constituent 1. In the same way, we can have the *partial molar free energy*, the *partial molar heat content*, etc. The physical significance of a partial molar quantity for a given substance is the increase of the property, such as volume, with the addition at constant T and P of 1 mole of the particular substance to such a large quantity that there is no appreciable change in the concentration.

If the above shorthand notation is used in Eq. (A.9) at constant T and P we obtain

$$dX = \overline{X}_1\,dn_1 + \overline{X}_2\,dn_2 + \cdots \quad \text{(A.10)}$$

which upon integration gives

$$X = \overline{X}_1 n_1 + \overline{X}_2 n_2 + \cdots \quad \text{(A.11)}$$

Eq. (A.11) shows that $\overline{X}$ may be regarded as the contribution per mole of each constituent to the total X.

If we differentiate Eq. (A.11) we obtain

$$\begin{aligned} dX &= (n_1\,d\overline{X}_1 + \overline{X}_1\,dn_1) + (n_2\,d\overline{X}_2 + \overline{X}_2\,dn) + \cdots \\ &= (n_1\,d\overline{X}_1 + n_2\,d\overline{X}_2 + \cdots) + (\overline{X}_1\,dn_1 + \overline{X}_2\,dn_2 + \cdots) \end{aligned}$$

and if we substitute Eq. (A.10) we obtain

$$n_1\, d\bar{X}_1 + n_2\, d\bar{X}_2 + \cdots = 0$$

which for a system of only two components yields

$$n_1\, d\bar{X}_1 = -\, n_2\, d\bar{X}_2 \tag{A.12}$$

This is called the *Gibbs–Duhem equation.*

A.4 CHEMICAL POTENTIAL

The Gibbs free energy is an extensive property represented by the symbol X above. Thus we can write the partial molar Gibbs free energy as

$$\bar{G}_1 = (\partial G/\partial n_1)_{T,P,n_2,\ldots} \tag{A.13}$$

for which another symbol is sometimes used:

$$\bar{G}_1 = \mu_1, \tag{A.13$'$}$$

where μ is called the *chemical potential.*

It is frequently useful to relate the partial molar volume to other properties. At constant T, Eq. (A.6) can be written as

$$(\partial G/\partial P)_{T,n_1,n_2,\ldots} = V \tag{A.14}$$

If we differentiate this with respect to n_1 we obtain

$$(\partial G/\partial P \partial n_1)_{T,n_2\ldots} = (\partial V/\partial n_1)_{P,T,n_2\ldots}$$

and if we differentiate Eq. (A.13′) with respect to P we obtain

$$(\partial G/\partial P \partial n_1)_{T,n_2,\ldots} = (\partial \mu_1/\partial P)_{T,n_1,n_2,\ldots}$$

By equating these two relations and noting Eq. (A.9′) we may write

$$\bar{V}_1 = (\partial \mu_1/\partial P)_{T,n_1,n_2} \tag{A.15}$$

In words, Eq. (A.15) means that the change in chemical potential of a given constituent with pressure at constant temperature and composition is the partial molar volume of that constituent. All of the standard relationships that are valid for integral thermodynamic properties are valid for their partial molar counterparts.

A.5 FUGACITY AND ACTIVITY

For 1 mole of a pure substance the free energy G and the chemical potential μ are identical, and thus Eq. (A.15) is

$$(\partial \mu/\partial P)_T = V \tag{A.15$'$}$$

where V is the molar volume. For an ideal gas

$$V = RT/P$$

so that integration of Eq. (A.15′) for an ideal gas gives

$$\mu = RT \ln P + \mu^0 \tag{A.16}$$

where the integration constant μ^0 depends only on the temperature and not on the nature of the gas. For a real gas, Eq. (A.16) does not hold, because of attractive or repulsive forces between the molecules. Instead of pressure a symbol P^* is used, called the *fugacity*, which approaches P as $P \to 0$ because the gas becomes more ideal. For most purposes it is more convenient to express the fugacity of a gas in terms of the ratio of the fugacity under any given condition compared with the value in the *standard state*: for gases this standard state is chosen as an ideal gas at unit pressure, i.e., when its fugacity is unity. Another term is used, called the *activity*, which is defined as the fugacity of a gas under any given condition compared with the value in a standard state. With this standard the activity of a gas is equal to its fugacity. Also we can define *activity coefficient* as the ratio of the activity to the pressure, i.e., P^*/P. This ratio, being unity for an ideal gas, is a measure of the departure from unity of an actual gas.

Suppose now that we have a mixture of ideal gases. By Dalton's law the total pressure P is equal to the sum of the partial pressures, i.e.,

$$P = P_1 + P_2 + P_3 + \cdots$$

and since $PV = nRT$ applies to n moles of any ideal gas it follows that

$$PV = (n_1 + n_2 + n_3 + \cdots)RT$$

We may also write that

$$P_1 = [n_1/(n_1 + n_2 + n_3 + \cdots)]P = x_1 P \quad \text{and}$$

$$P_2 = [n_2/(n_1 + n_2 + n_3 + \cdots)]P = x_2 P$$

where x is the mole fraction of any constituent gas, i.e., the number of moles of the constituent divided by the total number of moles of the gas. In the special case where $n_1 + n_2 + n_3 \cdots = 1$ mole, $n_1 = x_1$, etc., and

$$P_1 = x_1 RT/V, \qquad P_2 = x_2 RT/V, \quad \cdots$$

where V is the volume of 1 mole under pressure P. Thus we may write that the partial molar volume of any constituent such as n_1 is

$$\overline{V}_1 = (\partial V/\partial n_1)_{P,T,n_2,\ldots} = RT/P_1$$

From Eq. (A.15′), assuming the pressures of the other constituents remain unchanged, we may write

$$(\partial \mu_1/\partial P_1)_{T, n_1, n_2, \ldots} = \bar{V}_1 = RT/P_1 \tag{A.17}$$

Integrating Eq. (A.17) gives

$$\mu_1 = RT \ln P_1 + \mu_P{}^0 \tag{A.17′}$$

where $\mu_P{}^0$ now depends on the nature of the gas and its temperature, so that P_1 should be replaced by the fugacity $P_1{}^*$:

$$\mu_1 = RT \ln P_1{}^* + \mu_P{}^0 \tag{A.18}$$

Two other forms of Eq. (A.18) are useful. At constant T and composition, Eq. (A.17) can be expressed as

$$d\mu_1 = RT\, d \ln P_1 \tag{A.19}$$

If in a solution we define a concentration C of constituent 1 as $C_1 = n_1/V$ in moles per unit volume (unit volume = 1 l) P_1 can be replaced by RTC_1 in Eq. (A.17) and Eq. (A.19) would then be written as

$$d\mu_1 = RT\, d \ln C_1 \tag{A.20}$$

Upon integration, Eq. (A.20) becomes

$$\mu_1 = RT \ln C_1 + \mu_C{}^0 \tag{A.21}$$

Note that, by analogy, the concentration C must be corrected for the activity.

Another form is obtained by replacing P_1 by $x_1 P$ in Eq. (A.17), where x_1 is the mole fraction of constituent 1 and P the total pressure, so that at constant *temperature* and *pressure* Eq. (A.20) becomes

$$d\mu_1 = RT\, d \ln x_1 \tag{A.22}$$

and Eq. (A.21) is

$$\mu_1 = RT \ln x_1 + \mu_x{}^0 \tag{A.23}$$

where $\mu_x{}^0$ depends on temperature and pressure. For real gases, x_1 and C_1 should be replaced by their corresponding activities.

A.6 WORK IN ELECTROLYTES

It should be noted that the above expressions were derived with only PV energy in the Gibbs function. There are a variety of other energy forms, such as elastic energy, electrical or magnetic energy, etc. When these occur they are simply added to the Gibbs energy to express the total energy of the system. If the system causes charges to change their potential, such work also changes the free energy. Since the chemical system developed above is based on moles of substances, an electrical equivalent term is useful. One mole of singly charged ions can be expressed in coulombs of charge by multiplying Avogadro's number, 6.02×10^{23} molecules/mole, times the electronic charge, $e = 1.60 \times 10^{-19}$ C, to obtain 96,487 C/mole of single charges. This number is called the *Faraday constant* with the symbol F. Note that if an ion is doubly charged only half as many ions are required to make a Faraday and therefore the term *equivalent charge* is used. To remind one of this the Faraday is usually expressed as 96,487 C/g equivalent charge. Note also that many ionic compounds do not dissociate completely in solution and the number of ions present must be corrected for the degree of dissociation.

If work is done by the system to change the potential $\mathscr{V}$ of a mole of charges, the change in free energy of the system is (since potential $\mathscr{V}$ (in volts)† is work per coulomb)

$$\Delta G = -F\mathscr{V} \tag{A.24}$$

Since $\Delta G = \mu_{\text{state 1}} - \mu_{\text{state 2}}$,

$$\Delta G = RT \ln C_1 - RT \ln C_2 \tag{A.25}$$

where the C's are ion concentrations in two connecting solutions. Substituting Eq. (A.24) into Eq. (A.25) yields the relation

$$\mathscr{V} = (RT/F) \ln (C_2/C_1) \tag{A.26}$$

which is called the *Nernst equation*. Note that this relation applies only to solutions that can change concentration with neither heat nor volume change. Such solutions are called *ideal solutions*.

If the Boltzmann constant k is used in Eq. (A.25) instead of the gas constant R, it expresses the energy per degree per molecule for a unit charge:

$$\begin{aligned} \mathscr{V} \text{ (volts)} &= (kT/e) \ln (C_2/C_1) \\ &= \frac{(1.4 \times 10^{-23} \text{ J/deg})(311°\text{K})}{1.6 \times 10^{-19} \text{ C}} 2.3 \log(C_2/C_1), \end{aligned} \tag{A.27}$$

where body temperature of 38°C is used. Thus

$$\mathscr{V} = 61 \times 10^{-3} \log(C_2/C_1) \qquad \text{(volts)} \tag{A.28}$$

and, for example, if $C_2 = 10C_1$, $\mathscr{V} = 61 \times 10^{-3}$ volts of potential difference.

† In chemical literature E is often used in place of $\mathscr{V}$.

A.7 EQUILIBRIUM CONSTANTS

We see from Eqs. (A.13′), (A.16), and (A.21) that we may write the free energy of one mole of an ideal gas as

$$G = \mu^0 + RT \ln P$$

or, for a solution, as

$$G = \mu^0 + RT \ln C$$

where C is the concentration, with the recognition that a correction factor must be introduced for the activity coefficient. For quantities greater or smaller than one mole, one simply multiplies through by the corresponding factor.

If we have a chemical reaction in which some reactants A and B combine to make products C and D, where A, B, C, and D represent not only the identity of the species but also their concentrations in mole fractions, we write the reaction as

$$\mathrm{A} + \mathrm{B} \underset{K_2}{\overset{K_1}{\rightleftharpoons}} \mathrm{C} + \mathrm{D} \tag{A.29}$$

where K_1 is the rate constant for the forward reaction and K_2 is the rate constant for the decomposition, or back reaction. The free energy change of this reaction is

$$\begin{aligned}\Delta G &= G_{\text{products}} - G_{\text{reactants}} \\ &= G_{\mathrm{C}} + G_{\mathrm{D}} - G_{\mathrm{A}} - G_{\mathrm{B}} \\ &= \mu_{\mathrm{C}}{}^0 + \mu_{\mathrm{D}}{}^0 - \mu_{\mathrm{A}}{}^0 - \mu_{\mathrm{B}}{}^0 + RT \ln C + RT \ln D - RT \ln A - RT \ln \mathrm{B} \\ &= \Delta G^\circ + RT \ln (\mathrm{CD/AB}) \qquad \text{(A.30)}\end{aligned}$$

where $\Delta G^0 \equiv \mu_{\mathrm{C}}{}^0 + \mu_{\mathrm{D}}{}^0 - \mu_{\mathrm{A}}{}^0 - \mu_{\mathrm{B}}{}^0$. At equilibrium $\Delta G = 0$ for any small change and thus

$$\Delta G^\circ = -RT \ln (\mathrm{CD/AB}) \tag{A.31}$$

or

$$\mathrm{CD/AB} = \exp(-\Delta G^0/RT) \tag{A.32}$$

If we analyze the physical meaning of Eq. (A.29) we can say that whenever an A and a B come in contact they form a C and a D. The probability that they are in contact is the joint probability (product of probabilities) that they occupy adjacent sites in some coordinate grid in volume space. Their probabilities of occupancy of sites are defined as their concentrations and thus the joint probability of their being adjacent is simply the product of concentrations AB; the same argument applies to CD. We are interested in the rate at which the species move about because that determines the rate at which they may change positions to achieve an occupancy of adjacent sites. This involves a diffusion rate in a gas, or an atomic jump rate in a condensed system such as a liquid or a solid. All of this is contained in the rate constants K_1 and K_2, which have dimensions of reciprocal concentration times frequency. The rate at which A + B forms C + D is expressed as K_1AB. Similarly, a back reaction is taking place, expressed by K_2CD. Thus, if one asks how fast is A being lost from the system, its time rate of concentration change in differential form can be written as the difference between the loss rate and the formation rate of A:

$$d\mathrm{A}/dt = K_2\mathrm{CD} - K_1\mathrm{AB} \tag{A.33}$$

(The dimensionality of the K's mentioned above is evident in this equation.) And in equilibrium, in which the concentrations of A, B, C, and D remain constant, the rate of change of A is zero:

$$dA/dt = 0 = K_2\mathrm{CD} - K_1\mathrm{AB} \quad \text{and} \quad \mathrm{CD}/\mathrm{AB} = K_1/K_2 \equiv K \tag{A.34}$$

This is the same form as Eq. (A.32) and we may therefore write

$$K = \exp(-\Delta G^0/kT) \tag{A.34'}$$

Eq. (A.34) is readily generalized to any number of reactants and products and is known as the *law of mass action.*

A.8 OSMOTIC PRESSURE

Some membranes have the property of allowing one kind of molecule to pass through while blocking another; these are called *semipermeable membranes.* Although earlier ideas were that such membranes would pass small molecules but not large ones, experimental evidence has shown that this concept is incorrect.

Consider a cylinder with a semipermeable membrane barrier as in Fig. A.1. A pure solvent is on the left-hand side of the membrane and a solution of some other molecule dissolved in the solvent is on the right. There are two pistons on either side, one exerting a pressure P^0 and the other a pressure P. If the membrane is permeable to the solvent molecules but not to the

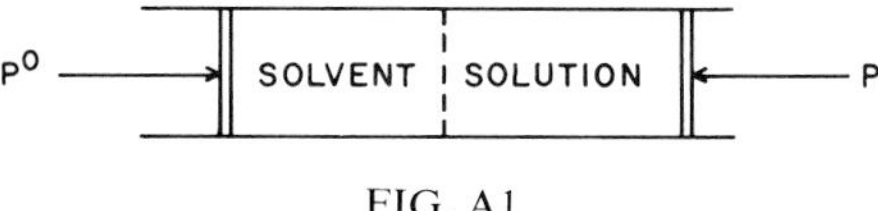

FIG. A1

solute molecules the pure solvent will tend to diffuse to the right with a greater pressure than the solvent in solution moving to the left. Thus, to maintain the original volumes the pressure P must be greater than the pressure P^0 and this pressure difference is called the *osmotic pressure*, π, where

$$\pi = P - P^0 \tag{A.35}$$

Without any loss of generality, P^0 can be taken as the ambient external pressure P_0 and we can write, instead of (A.35),

$$\pi = P - P_0 \tag{A.35$'$}$$

It has been shown that this osmotic pressure is the same as the kinetic pressure difference if the molecular mixture existed in the gaseous state. However, since we have developed the relations for chemical potential we will consider osmotic pressure from that viewpoint.

Define μ_0 as the chemical potential of the pure solvent and μ as the chemical potential of the solvent in solution, both at an external (ambient) pressure P_0. The chemical potential of the pure solvent at another pressure P is given by

$$\mu_0 = \mu + \int_{P_0}^{P} (\partial\mu/\partial P)_T \, dP \tag{A.36}$$

From Eq. (A.15), $\partial\mu/\partial P$ is equal to the partial molar volume $\overline{V}_1$, and therefore

$$\mu_0 = \mu + \int_{P_0}^{P} \overline{V}_1 \, dP \tag{A.37}$$

where $\overline{V}_1$ is the partial molar volume of the solvent in solution. Further, if we define p^0 and p as the partial vapor pressures of the solvent and the solvent in equilibrium with solution, respectively, then the chemical potentials of the vapors are

$$\mu_0(\text{vapor}) = \mu^0 + RT \ln p^0 \qquad \text{and} \qquad \mu(\text{vapor}) = \mu^0 + RT \ln p \tag{A.38}$$

This assumes that they behave ideally; if they don't, the vapor pressures should be replaced by the fugacities. Since the liquids and vapors are in equilibrium, μ_0(vapor) and μ(vapor) are also the chemical potentials of the pure solvent and the solvent in solution. Therefore Eq. (A.37) may be written, by substitution of Eq. (A.38), as

$$\Delta G = \mu_0 - \mu = RT \ln p^0 - RT \ln p = \int_{P_0}^{P} \overline{V}_1 \, dP$$

or

$$RT \ln(p_0/p) = \int_{P_0}^{P} \overline{V}_1 \, dP \tag{A.39}$$

For simplicity, assume $\overline{V}_1$ is independent of pressure, i.e., the solution is incompressible. Then Eq. (A.39) integrates to

$$RT \ln (p^0/p) = \overline{V}_1(P - P_0)$$

Since $P - P_0$ is the change in the applied pressure, in the present case it is the osmotic pressure of Eq. (A.35′). By substitution of Eq. (A.35′) we may write

$$\pi \overline{V}_1 = RT \ln (p^0/p) \tag{A.40}$$

We may express Eq. (A.40) in terms of solution concentration in the following way. We assume that Raoult's law, that partial pressures are proportional to the fractional concentrations, is obeyed. Then if x_1 is the fractional concentration of the solvent and x_2 that of the solute

$$p/p_0 = x_1 = 1 - x_2$$

Substituting this into Eq. (A.40) gives the relation

$$n\overline{V}_1 = -RT \ln (1 - x_2) \tag{A.41}$$

Note that the series expansion of a logarithm of the above form is

$$\log(1 + x) = x - \tfrac{1}{2}x^2 + \tfrac{1}{3}x^3 \cdots$$

For dilute solutions only the first term need be kept and

$$\pi \overline{V}_1 = RTx_2$$

Since

$$x_2 = n_2/(n_2 + n_1) \cong n_2/n_1,$$

we may write

$$\pi \overline{V}_1 = RT(n_2/n_1) \tag{A.42}$$

or

$$\pi n_1 \overline{V}_1 = RTn_2$$

The quantity $n_1 \overline{V}_1$ is the volume of the solvent that contains n_2 moles of solute. For dilute solutions this can be taken as approximately the volume of the total solution V, and we may write

$$\pi V = n_2 RT \tag{A.43}$$

or, defining the concentration of the solution as

$$C = n_2/V \quad \text{mole/liter}$$

we may write

$$\pi = CRT \tag{A.44}$$

It is interesting to note that the similarity of Eq. (A.44) to the ideal gas law led early investigators to believe that osmotic pressure arose from a molecular bombardment process. Although this may be analogous to a portion of the process, it should be recalled that Eq. (A.43) was obtained by approximations and therefore does not represent the entire mechanism.

A.9 DETERMINATION OF PARTIAL MOLAR VOLUME

A method of calculating the partial molar volume $\overline{V}_1$ of the solvent is as follows. If a solution contains n_1 moles of solvent of molecular weight M_1 and 1 mole of solute of molecular weight M_2 in V liters, the density of the solution is

$$\rho = (n_1 M_1 + M_2)/10^3 V \qquad \text{and} \qquad n_1 = (10^3 V\rho - M_2)/M_1$$

Upon differentiating,

$$dn_1 = 10^3(\rho\, dV + V\, d\rho)/M_1$$

and

$$dn_1/dV = (10^3/M_1)[\rho + (V\, d\rho/dV)] \tag{A.45}$$

Let C be the concentration of the solution in mols per liter; then $C = 1/V$ and

$$dC = -dV/V^2$$

Substituting for V and dV in Eq. (A.45) we obtain

$$dn_1/dV = (10^3/M_1)[\rho - (C\, d\rho/dC)]$$

and, upon inverting, we obtain $\overline{V}_1$:

$$\overline{V}_1 = \frac{dV}{dn_1} = \frac{M_1}{10^3} \frac{1}{(\rho - C\, d\rho/dC)} \tag{A.46}$$

It is thus seen that in order to evaluate $\overline{V}_1$ the density of the solution must be found as a function of concentration and the slope $d\rho/dC$ taken at the concentration of interest.

A.10 CLAPEYRON–CLAUSIUS EQUATION

We wish to obtain a relation between the thermodynamic variables when there is a change of phase, as in the vaporization of a liquid.

From the first law, Eq. (A.1), the heat absorbed by a system at constant pressure in going from state A to state B is

$$\begin{aligned} Q &= (U_{\mathrm{B}} - U_{\mathrm{A}}) + P(V_{\mathrm{B}} - V_{\mathrm{A}}) \\ &= (U_{\mathrm{B}} + PV_{\mathrm{B}}) - (U_{\mathrm{A}} + PV_{\mathrm{A}}) \end{aligned} \tag{A.47}$$

Since U, P, and V are properties of the state of the system and not on its previous history, the quantity $U + PV$ is a single-valued function of the variables. The symbol H is used to represent $U + PV$ and is called the *heat content* or *enthalpy* of the system. Eq. (A.47) can be rewritten as

$$Q = \Delta H$$

and the entropy change between two states can be rewritten from the de- nition

$$dS = dQ/T, \qquad S_{\mathrm{B}} - S_{\mathrm{A}} = \int_A^B dQ/T$$

as

$$\Delta S = \Delta H/T \tag{A.48}$$

We may obtain the other required relation from the definition of the Helmholz free energy F (Eq. (A.4)):

$$F = U - TS, \qquad dF = dU - T\,dS - S\,dT \tag{A.49}$$

Substituting the First Law, $dU = dQ - P\,dV$, and noting that for a reversible process $dQ = T\,dS$ we have $dF = -P\,dV - S\,dT$. From this we may write†

$$(\partial P/\partial T)_V = (\partial S/\partial V)_T \tag{A.50}$$

†This relation is obtained from a fundamental property of partial derivatives. Suppose z is a function of x and y. By definition

$$dz = (\partial z/\partial x)_y\,dx + (\partial z/\partial y)_x\,dy$$

Define $M = (\partial z/\partial x)_y$ and $N = (\partial z/\partial y)_x$

Then $(\partial M/\partial y)_x = \partial^2 z/\partial x \partial y$ and $(\partial N/\partial x)_y = \partial^2 z/\partial y \partial x$

and therefore

$$(\partial M/\partial y)_x = (\partial N/\partial x)_y$$

In Eq. (A.49), $M = -P$ and $N = -S$ and the relation of Eq. (A.50) can be written by analogy.

For our present purposes we wish to consider a system of a liquid and its vapor in equilibrium. Thus the pressure is the vapor pressure, which, while it is dependent on the temperature, is independent of the volume of either the liquid or vapor present. Therefore, in this instance, we may simplify the writing of the left-hand side of Eq. (A.50) to dP/dT instead of $(\partial P/\partial T)_V$. To obtain a simplification of the right-hand side of Eq. (A.50) consider a volume formed by a container with a movable piston. This volume has some liquid in equilibrium with its vapor and the system is in a constant temperature bath. If the piston is raised slightly, the pressure is temporarily reduced and some liquid will evaporate to restore the equilibrium vapor pressure. The energy for the molecules (heat of evaporation ΔH) is supplied by the constant temperature bath. It is clear that for each incremental unit of volume increase a constant increment of heat is required to restore the equilibrium vapor pressure. Therefore, since the heat supplied to the system is proportional to the volume change, by Eq. (A.48) the entropy change is also proportional to the volume change, at constant temperature. We may therefore simplify the right-hand side of Eq. (A.50) to $\Delta S/\Delta V$ instead of $(\partial S/\partial V)_T$. We now may write Eq. (A.50) as

$$dP/dT = \Delta S/\Delta V \tag{A.50'}$$

and substitute Eq. (A.48) to obtain

$$dP/dT = \Delta H/T\Delta V \tag{A.51}$$

in which for 1 mole of liquid ΔH is the molar heat of vaporization. This is known as the *Clapeyron–Clausius equation.*

If the transition is from the liquid to the vapor then Eq. (A.51) is written

$$dP/dT = \Delta H/[T(V_V - V_L)] \tag{A.52}$$

where V_V and V_L are the molar volumes of the vapor and liquid, respectively. At equilibrium the vapor essentially obeys the ideal gas law and the molar vapor volume can be written

$$V_V = RT/P$$

Substituting this into Eq. (A.52) and neglecting the liquid volume, since it is small compared with the volume of the gas, we write

$$dP/dT = \Delta HP/RT^2 \tag{A.53}$$

Equation (A.53) can be written in the form $d\,(\ln P)/dT = \Delta H/(RT^2)$ or

$$d\,(\ln P)/d\,(1/T) = -\Delta H/R \tag{A.53'}$$

The integrated form of Eq. (A.53′) is

$$\ln P = -\Delta H/(RT) + \text{constant} \tag{A.54}$$

which shows that a plot of ln P versus $1/T$ should give a straight line, the slope of which is $-\Delta H/R$.

A.11 DEPRESSION OF FREEZING POINT

If we integrate Eq. (A.53) with temperature limits T_1 and T_2 and corresponding vapor pressure limits of P_1 and P_2, assuming ΔH to be constant, we obtain

$$\ln \frac{P_1}{P_2} = -\frac{\Delta H}{R}\left(\frac{1}{T_1} - \frac{1}{T_2}\right) = \frac{\Delta H}{R}\left(\frac{T_1 - T_2}{T_1 T_2}\right) \tag{A.55}$$

It is useful at this point to note also that if we substitute activity a for pressure in Eq. (A.53) and integrate between T, the freezing point of a solution, and T_0, the freezing point of the pure solute, we obtain (recall that activity of the solvent in solution is the *ratio* of the fugacities)

$$\ln a = -\frac{\Delta H}{R}\left(\frac{1}{T} - \frac{1}{T_0}\right) = -\frac{\Delta H}{R}\frac{\Delta T_f}{T_0 T} \tag{A.56}$$

where $\Delta T_f = T_0 - T$ is the depression of the freezing point. For dilute solutions let $T_0 T = T_0^2$ and write

$$\ln a = -(\Delta H/R T_0^2)\Delta T_f \tag{A.57}$$

We therefore note that not only can the addition of solutes lower the freezing point of a solvent, but that the higher the concentration the lower the freezing point.

APPENDIX **B**

The Wave Equation

Consider a horizontal string stretched in the x direction. If a disturbance is made in the y direction by plucking it near the left-hand end at time zero the shape of the disturbance can be expressed as

$$y = F(x).$$

At a later time the disturbance has moved along the string in the x-direction and $F(x)$ is no longer the correct function to describe it because it is no longer in the same place. However, if some value θt is subtracted from x at time t we get exactly the same shape that occurred at time zero. Therefore the quantity $x - \theta t$ must be a constant for any point on the wavelike disturbance and the function F that describes the wave is

$$y = F(x - \theta t)$$

when the wave moves from left to right. Since the quantity $x - \theta t$ is a constant it follows that

$$\frac{d}{dt}(x - \theta t) = \frac{dx}{dt} - \theta = 0 \qquad \text{and} \qquad \theta = \frac{dx}{dt}$$

which identifies θ as the velocity at which the disturbance or wave travels along the string.

Now let some arbitrary quantity E be set equal to the function

$$E = x - \theta t$$

and use the chain rule of partial differentiation

$$\partial y/\partial x = (\partial y/\partial E)(\partial E/\partial x)$$

But, since $\partial E/\partial x = 1$

$$\partial y/\partial x = \partial y/\partial E$$

and, applying the chain rule a second time we obtain

$$\partial^2 y/\partial x^2 = \partial^2 y/\partial E^2$$

In the same manner

$$\partial y/\partial t = (\partial y/\partial E)(\partial E/\partial t)$$

and, since $\partial E/\partial t = -\theta$

$$\partial y/\partial t = -\theta(\partial y/\partial E)$$

Applying the chain rule a second time we obtain

$$\partial^2 y/\partial t^2 = \theta^2(\partial^2 y/\partial E^2), \qquad \text{or} \qquad \partial^2 y/\partial E^2 = (1/\theta^2)(\partial^2 y/\partial t^2)$$

Equating the two values of $\partial^2 y/\partial E^2$ results in

$$\partial^2 y/\partial x^2 = (1/\theta^2)(\partial^2 y/\partial t^2)$$

which is the wave equation, where θ is the wave velocity. This equation is valid for any wave propagation in a medium.

APPENDIX C

Binomial and Poisson Probability Distributions

C.1 BINOMIAL PROBABILITY DISTRIBUTION

Consider n objects $x_1, x_2, \ldots, x_n$ to be selected r at a time with choices called $a_1, a_2, \ldots, a_r$. For the first choice there are n possibilities. For the second choice there are $n - 1$ possibilities since the object chosen for a_1 may not be chosen again. Having chosen a_1 and a_2, there are now $n - 2$ choices for a_3 and so on until $n - r + 1$ choices for a_r. Hence for n things taken r at a time the number of choices ${}_nP_r$ can be written

$$ {}_nP_r = n(n-1)(n-2)\cdots(n-r+1) \tag{C.1} $$

Factorial notation is

$$ n! = n(n-1)\cdots 3\cdot 2\cdot 1 = n(n-1)! $$

Therefore one can write

$$ (n-1)! = n!/n $$

Note also that by definition $0! = 1$.

If Eq. (C.1) had the additional terms $(n - r)(n - r - 1)(n - r - 2)\cdots 1 = (n - r)!$ it could have been written as $n!$ This simplification can be achieved by multiplying and dividing Eq. (C.1) by $(n - r)!$ and writing

$$_nP_r = n!/(n - r)! \tag{C.2}$$

If, in previous selection of objects, there is an additional condition that the order of selection does not matter, i.e., the choices are indistinguishable, then the number $_nP_r$ is reduced by the number of possible permutations of choices $a_1, a_2, \ldots, a_r$ themselves, of which there are $r!$ permutations. For this situation $_nP_r$ is divided by $r!$ to give

$$_nP_r = n!/(n - r)!r! \tag{C.3}$$

This form is identifiable with the coefficients of the nth product of two terms such as $(x + y)^n$. Consider, for example, the case of $n = 3$:

$$\begin{aligned}(x + y)^3 &= (x + y)(x + y)(x + y)\\ &= xxx + xxy + xyx + xxy + xyy + yxy + yyx + yyy\\ &= x^3 + 3x^2y + 3xy^3 + y^3\end{aligned}$$

This example product of three factors shows that the coefficients of the last equation (with four terms) are the number of ways the x's and y's in each term can be arranged if the order selection does not matter, e.g., the x's are indistinguishable, or, as expressed above, the coefficients are the number of ways of choosing x from r of the factors and y from the rest. One can continue with this procedure to larger n and observe that this pattern holds. Not only does the $_nP_r$ relation determine the coefficients, but the factors themselves can be written $x^r y^{n-r}$. Therefore, any term of a binomial expansion can be expressed as

$$P_r = (x + y)^n = [n!/(n - r)!r!]x^r y^{n-r} \tag{C.4}$$

where the n subscript on the P has been deleted as unnecessary.

C.2 POISSON PROBABILITY DISTRIBUTION

This binomial expansion can be used to calculate the probability distribution in a system that has two possible outcomes, such as the tossing of a coin, firing at a target, emitting an ion from a biological system, etc. Although in tossing a coin the probability of a head is the same as that of a tail, there is no requirement that the two outcomes have equal probabilities. By definition, a probability cannot be greater than 1 or less than 0; therefore, in a two-outcome system, if x is the probability of one of the outcomes and

y the other then $y = 1 - x$. Define n as the pool or store of x and y. This could mean the number of trials in coin tossing, the number of ions in a cell that may or may not be released, etc. The total number of *successes* m (either x or y may be considered a success and the other a failure) will be the product of the number of trials n and the probability for success in each trial x:

$$m = nx \tag{C.5}$$

The quantity m is usually the measurable one in a biological experiment because it represents accumulated data. It is often difficult, however, to obtain both of the other parameters n and x; while n may be measurable, x may be the probability of a single ion being released or of a chromosome being struck by an irradiation particle. Therefore such a *two-parameter* probability distribution may not be appropriate for data analysis.

In situations where the probability of success is very low and many trials are made, i.e., x can be said to be nearly 0 while n is very large, the binomial distribution can be approximated by a *one-parameter* distribution. We will now examine this limiting condition of Eq. (C.4).

Substitute 0 for r in Eq. (C.4) and write

$$P_0 = y^n.$$

Since $x = m/n$ and $y = 1 - x$, $y = 1 - (m/n)$ and

$$P_0 = [1 - (m/n)]^n \tag{C.6}$$

This can be shown to approach an exponential for large n in the following way. Let $a = -m/n$ and therefore $n = -m/a$. Substitute this into Eq. (C.6) and write

$$P_0 = (1 + a)^{-m/a} = [(1 + a)^{1/a}]^{-m}$$

Since $a = -m/n$, as n approaches infinity a approaches zero and by definition of limits

$$\lim_{a \to 0} (1 + a)^{1/a} = e$$

and therefore

$$P_0 = y^n = e^{-m} \tag{C.7}$$

Consider the probability for a single success, i.e., $r = 1$, in Eq. (C.4):

$$P_1 = \frac{n!}{(n-1)!1!} xy^{n-1} = \frac{n(n-1)!}{(n-1)!} xy^{n-1}$$

$$= nxy^{n-1} = my^{n-1}$$

Since y^n has the limiting value of e^{-m} for large n so does y^{n-1} and we may write

$$P_1 = me^{-m}$$

For the probability of two successes

$$P_2 = \frac{n!}{(n-2)!2!} x^2 y^{n-2} = \frac{n(n-1)(n-2)!}{(n-2)!2!} x^2 y^{n-2} = \frac{n(n-1)}{2} x^2 y^{n-2}$$

$$\simeq \frac{n^2}{2} x^2 y^{n-2} = \frac{m^2}{2} m^{n-2} = \frac{m^2}{2} e^{-m}$$

For the probability of three successes, n still being large so that $n(n-1)(n-2) \simeq n^3$ and $y^{n-3} \simeq e^{-m}$,

$$P_3 = (m^3/6)e^{-m}$$

It is seen that for the large n approximation the general term of this series is

$$P_r = (m^r/r!)e^{-m} \tag{C.8}$$

This is a one-parameter approximation and it is called the *Poisson probability distribution.*

APPENDIX **D**

Differential Equations

Two types of elementary differential equations arise in this book; one is immediately integrable and the other can be made integrable. Their methods of solution are indicated here.

D.1 VARIABLES SEPARABLE

A very common equation is one in which a time rate of loss of a quantity is proportional to the quantity present. This occurs first in the solution of the Maxwell element in Chapter 2 in which the time rate of change of stress is proportional to the stress. Another familiar occurrence is in radioactive decay in which the decay rate is proportional to the quantity present. Both of these equations have the same form, but consider stress relaxation

$$(1/G)(d\sigma/dt) + (\sigma/\eta) = 0$$

which is Eq. (2.3) with an instantaneous displacement. Rewrite as

$$d\sigma/dt = -(G/\eta)\sigma \tag{D.1}$$

cross-multiply by dt, divide by σ, and integrate

$$\int (1/\sigma)\, d\sigma = -(G/\eta) \int dt$$

$$\ln \sigma = -(G/\eta)t + c \qquad \text{(D.2)}$$

$$\sigma = ce^{-Gt/\eta} \qquad \text{(D.3)}$$

The constant of integration c must be evaluated from the boundary conditions of the problem. In the present case the boundary conditions are the initial conditions; when $t = 0$ the stress σ is at an initial value of σ_0. Setting $t = 0$ and $\sigma = \sigma_0$ in Eq. (D.3) yields $c = \sigma_0$ and the solution is that of Eq. (2.4).

Frequently the constant G/η is written as $1/\tau$

$$\sigma = \sigma_0 e^{-t/\tau}$$

where τ is called the decay constant or relaxation time and is the time for an exponential decrease to reach $1/e$ of its original value (at $t = \tau$, $e^{-1} = 1/e$).

D.2 EXACT DIFFERENTIAL EQUATIONS

To obtain the solution of elementary differential equations an integrable form is required. If variables are not separable as above another method is to look for an exact form. Consider as an example the product xy. Its differential is

$$x\, dy + y\, dx$$

If this appeared on one side of an equation we would immediately recognize its integral to be xy and thereby write the solution. Situations of this sort are rare and do not occur in this book. However, a number of equations do occur in which a multiplier will make the equation exact. This is considered next.

D.3 INTEGRATING FACTORS

A very common differential equation is the linear one. An equation is linear when it is of first degree in the dependent variable and its derivative and may be written in the form

$$(dy/dx) + Py = Q \qquad \text{(D.4)}$$

where P and Q are functions of x alone, or constants.

We seek a function of x, called $R(x)$, by which we could multiply both sides of Eq. (D.4) with the result that the left-hand side is exact. The right-hand side, being originally a function of x, or a constant, remains integrable when multiplied by a function of x.

Multiplying the left-hand side of (D.4) by $R(x)$ yields

$$R(dy/dx) + RPy \tag{D.5}$$

Note that if we differentiate the product Ry we obtain

$$d(Ry)/dx = R(dy/dx) + y(dR/dx) \tag{D.6}$$

and therefore the integral of (D.5) is Ry if (D.6) can be made to correspond. It is seen that the first term on the right of (D.6) is the same as the first term of (D.5). The second term on the right of (D.6) will be the same as the second term of (D.5) if

$$dR/dx = RP \tag{D.7}$$

Since P is a function of x only, the variables of (D.7) are separable and the equation may be integrated

$$dR/R = P\,dx, \qquad \ln R = \int P\,dx$$

$$R = e^{\int P\,dx} \tag{D.8}$$

We have not introduced a constant of integration because we need only a single integrating factor. The function (D.8) is the multiplying factor which makes Eq. (D.4) integrable, and we may obtain y immediately by a little algebra. First multiply both sides of Eq. (D.4) by Eq. (D.8)

$$e^{\int P\,dx}(dy/dx) + e^{\int P\,dx}Py = Qe^{\int P\,dx}$$

The left-hand side is now exact and integrates to Ry or

$$\begin{aligned} e^{\int P\,dx}y &= \int Qe^{\int P\,dx}\,dx + c \\ y &= e^{-\int P\,dx}\int Qe^{\int P\,dx}\,dx + ce^{-\int P\,dx} \end{aligned} \tag{D.9}$$

and thus Eq. (D.9) is the formula for the solution of Eq. (D.4).

As an example let us apply this formula to solve Eq. (2.5)

$$\eta(dx/dt) + Gx = \sigma \tag{2.5}$$

This can be put in the form of Eq. (D.4) by the substitutions $x = y$, $t = x$, $G/\eta = P$, and $\sigma/\eta = Q$. First obtain the integral

$$\int P\,dx = (G/\eta)\int dt = (G/\eta)t$$

and write Eq. (D.9) with these substitutions

$$x = e^{-Gt/\eta}(\sigma/\eta)\int e^{Gt/\eta}\,dt + ce^{-Gt/\eta}$$

Performing the integration yields

$$x = e^{-Gt/\eta}(\sigma/G)e^{Gt/\eta} + ce^{-Gt/\eta} \tag{D.10}$$

$$x = (\sigma/G) + ce^{-Gt/\eta} \tag{D.11}$$

The boundary, or initial, condition is that when $t = 0$, $x = 0$. Substitution of these into Eq. (D.11) gives the value of the constant as $c = -\sigma/G$. Inserting this into Eq. (D.11) and factoring yields Eq. (2.6)

$$x = (\sigma/G)(1 - e^{-Gt/\eta}) \tag{2.6}$$

Index

C

D

E

F

G

O

P

W

X

A B C 8 D 9 E 0 F 1 G 2 H 3 I 4 J 5